Electrode Materials for Rechargeable Lithium Batteries

Electrode Materials for Rechargeable Lithium Batteries

Editor

Wenbo Liu

Basel • Beijing • Wuhan • Barcelona • Belgrade • Novi Sad • Cluj • Manchester

Editor
Wenbo Liu
Sichuan University
Chengdu, China

Editorial Office
MDPI
St. Alban-Anlage 66
4052 Basel, Switzerland

This is a reprint of articles from the Special Issue published online in the open access journal *Batteries* (ISSN 2313-0105) (available at: https://www.mdpi.com/journal/batteries/special_issues/electrode_materials_rechargeable_lithium_batteries).

For citation purposes, cite each article independently as indicated on the article page online and as indicated below:

Lastname, A.A.; Lastname, B.B. Article Title. *Journal Name* **Year**, *Volume Number*, Page Range.

ISBN 978-3-0365-8780-6 (Hbk)
ISBN 978-3-0365-8781-3 (PDF)
doi.org/10.3390/books978-3-0365-8781-3

© 2023 by the authors. Articles in this book are Open Access and distributed under the Creative Commons Attribution (CC BY) license. The book as a whole is distributed by MDPI under the terms and conditions of the Creative Commons Attribution-NonCommercial-NoDerivs (CC BY-NC-ND) license.

Contents

About the Editor vii

Preface ix

Renny Nazario-Naveda, Segundo Rojas-Flores, Luisa Juárez-Cortijo, Moises Gallozzo-Cardenas, Félix N. Díaz, Luis Angelats-Silva and Santiago M. Benites
Effect of x on the Electrochemical Performance of Two-Layered Cathode Materials $x\text{Li}_2\text{MnO}_3$–$(1-x)\text{LiNi}_{0.5}\text{Mn}_{0.5}\text{O}_2$
Reprinted from: *Batteries* 2022, *8*, 63, doi:10.3390/batteries8070063 1

Jian Gao, Mengxin Zhou, Xinyao Wang, Hong Wang, Zhen Yin, Xiaoyao Tan and Yuan Li
Preparing Co/N-Doped Carbon as Electrocatalyst toward Oxygen Reduction Reaction via the Ancient "Pharaoh's Snakes" Reaction
Reprinted from: *Batteries* 2022, *8*, 150, doi:10.3390/batteries8100150 17

Yangyang Wang, Guangya Zhang, Guangfeng Dong and Heng Zheng
Research Progress of Working Electrode in Electrochemical Extraction of Lithium from Brine
Reprinted from: *Batteries* 2022, *8*, 225, doi:10.3390/batteries8110225 29

Lixia Wang, Taibao Zhao, Ruiping Chen, Hua Fang, Yihao Yang, Yang Cao and Linsen Zhang
Molybdenum Nitride and Oxide Quantum Dot @ Nitrogen-Doped Graphene Nanocomposite Material for Rechargeable Lithium Ion Batteries
Reprinted from: *Batteries* 2023, *9*, 32, doi:10.3390/batteries9010032 39

Kai Zhang, Erwin Hüger, Yong Li, Harald Schmidt and Fuqian Yang
Review and Stress Analysis on the Lithiation Onset of Amorphous Silicon Films
Reprinted from: *Batteries* 2023, *9*, 105, doi:10.3390/batteries9020105 53

Fengjun Deng, Yuhang Zhang and Yingjian Yu
Conductive Metal–Organic Frameworks for Rechargeable Lithium Batteries
Reprinted from: *Batteries* 2023, *9*, 109, doi:10.3390/batteries9020109 79

Otavio J. B. J. Marques, Michael D. Walter, Elena V. Timofeeva and Carlo U. Segre
Effect of Initial Structure on Performance of High-Entropy Oxide Anodes for Li-Ion Batteries
Reprinted from: *Batteries* 2023, *9*, 115, doi:10.3390/batteries9020115 107

Hua Fang, Qingsong Liu, Xiaohua Feng, Ji Yan, Lixia Wang, Linghao He, et al.
Carbon-Coated Si Nanoparticles Anchored on Three-Dimensional Carbon Nanotube Matrix for High-Energy Stable Lithium-Ion Batteries
Reprinted from: *Batteries* 2023, *9*, 118, doi:10.3390/batteries9020118 119

Yinglu Hu, Li Liu, Jingwei Zhao, Dechao Zhang, Jiadong Shen, Fangkun Li, et al.
Lithiophilic Quinone Lithium Salt Formed by Tetrafluoro-1,4-Benzoquinone Guides Uniform Lithium Deposition to Stabilize the Interface of Anode and PVDF-Based Solid Electrolytes
Reprinted from: *Batteries* 2023, *9*, 322, doi:10.3390/batteries9060322 133

Wei Zou, Hua Fang, Tengbo Ma, Yanhui Zhao, Lixia Wang, Xiaodong Jia and Linsen Zhang
Three-Dimensional Nanoporous CNT@Mn_3O_4 Hybrid Anode: High Pseudocapacitive Contribution and Superior Lithium Storage
Reprinted from: *Batteries* 2023, *9*, 389, doi:10.3390/batteries9070389 145

About the Editor

Wenbo Liu

Prof. Dr. Wenbo Liu is the Associate Head of the Department of Materials Forming and Control Engineering, the School of Mechanical Engineering, Sichuan University. His current area of research is advanced metal functional materials, which mainly include the following aspects: (1) nanostructured metal materials from initial fabrication and analysis to potential applications (such as energy storage and conversion, environmental protection, catalysis, sensing, etc.) and (2) Mn-Cu-based damping alloys from material design to technological development and engineering applications. He has published over 100 journal papers in peer-reviewed journals, such as *Adv. Funct. Mater.*, *Chem. Eng. J.*, *EcoMat*, *J. Mater. Chem. A*, *Composites Part B-Eng.*, *J. Mater. Sci. Technol.*, *J. Power Sources*, etc.. His papers have a total ISI citation score of over 1300 and an H-index of 23 (web of science). He is the holder of 18 Chinese patents and has 8 pending patent applications. Liu has received many prizes and awards, including the Hong Kong Scholars Award and the Chengdu Natural Science Award, for his achievements in the fields of materials science and engineering. He is a Life Member of American Association for Science and Technology (AASCIT), as well as an Editorial Board Member OF 10 international journals, such as *NML*, *EEM*, *CCL*, *BE*, *FER*, etc.

Preface

With the rapid development of high-efficiency electrochemical energy storage devices, = lithium-ion batteries (LIBs) have been widely applied in various industrial and consumer applications, such as mobile phones, laptops, electric vehicles, smart grids, etc. However, traditional electrode materials hardly meet the expected demands of the energy and power densities of future energy storage systems due to their extremely limited specific capacities, short lifetimes, and poor safety. As a result, seeking alternative high-performance electrode materials will be a primary challenge for next-generation rechargeable lithium batteries (RLBs) in the future, such as advanced lithium-ion batteries, lithium metal batteries, lithium sulfur batteries, and lithium oxygen/air batteries.

This book's subject and scope covers the fabrication and synthesis of electrode materials, lithium dendrite growth and inhibition, polysulfide transformation, novel electrode structure design, electrode material failure, lithium storage mechanisms, electrochemical performance optimization, safety assessment and evaluation, advanced characterization techniques, and multi-scale computational modeling.

We believe that this book will offer the latest insights into the use of electrode materials to develop advanced RLBs. The book's audience may include, but is not limited to, researchers, engineers, industrialists, technicians, teachers, students, and industrial investors. Also, we sincerely hope that it will inspire and provide assistance to researchers and firms seeking to develop alternative high-performance electrode materials for use in next-generation rechargeable lithium batteries employed in various industrial and consumer fields.

Wenbo Liu
Editor

Article

Effect of x on the Electrochemical Performance of Two-Layered Cathode Materials $xLi_2MnO_3–(1-x)LiNi_{0.5}Mn_{0.5}O_2$

Renny Nazario-Naveda [1,*], Segundo Rojas-Flores [2], Luisa Juárez-Cortijo [3], Moises Gallozzo-Cardenas [4], Félix N. Díaz [2,5], Luis Angelats-Silva [6] and Santiago M. Benites [1]

1 Vicerrectorado de Investigación, Universidad Autónoma del Perú, Lima 15842, Peru; santiago.benites@autonoma.pe
2 Escuela de Ingeniería Mecánica Eléctrica, Universidad Señor de Sipán, Chiclayo 14000, Peru; segundo.rojas.89@gmail.com (S.R.-F.); diazfelixn@crece.uss.edu.pe (F.N.D.)
3 Grupo de Investigación en Ciencias Aplicadas y Nuevas Tecnologías, Universidad Privada del Norte, Trujillo 13007, Peru; luisa.juarez@upn.edu.pe
4 Facultad de Ciencias de la Salud, Universidad César Vallejo, Trujillo 13001, Peru; mmgallozzo@ucvvirtual.edu.pe
5 Dirección de Gestión Académica, Universidad Tecnológica del Perú, Lima 15419, Peru
6 Laboratorio de Investigación Multidisciplinario, Universidad Privada Antenor Orrego, Trujillo 13008, Peru; langelatss@upao.edu.pe
* Correspondence: scored731@gmail.com

Abstract: In our study, the cathodic material $xLi_2MnO_3–(1-x)LiNi_{0.5}Mn_{0.5}O_2$ was synthesized by means of the co-precipitation technique. The effect of x (proportion of components Li_2MnO_3 and $LiNi_{0.5}Mn_{0.5}O_2$) on the structural, morphological, and electrochemical performance of the material was evaluated. Materials were structurally characterized using X-ray diffraction (XRD), and the morphological analysis was performed using the scanning electron microscopy (SEM) technique, while charge–discharge curves and differential capacity and impedance spectroscopy (EIS) were used to study the electrochemical behavior. The results confirm the formation of the structures with two phases corresponding to the rhombohedral space group R3m and the monoclinic space group C2/m, which was associated to the components of the layered material. Very dense agglomerations of particles between 10 and 20 μm were also observed. In addition, the increase in the proportion of the $LiNi_{0.5}Mn_{0.5}O_2$ component affected the initial irreversible capacity and the Li_2MnO_3 layer's activation and cycling performance, suggesting an optimal chemical ratio of the material's component layers to ensure high energy density and long-term durability.

Keywords: lithium-ion battery; cathode material; layered composite; Li-rich Mn Ni based

1. Introduction

The world's dependence on fossil fuels is a subject that is still debated today. Although it is true that fossil fuels are necessary for the development of civilization as we know it, it is also important to take into account their impact on the environment [1]. The increase in the demand for energy and the concern for the care of the environment have increased the interest for the use of new energies [2–5]. The use of these energy systems is closely related to the development of increasingly efficient storage systems; hence, much research is focused on improving their performance. The characteristics of lithium batteries, such as their high specific capacity, energy density, and the possibility of using non-toxic reagents in their manufacture [6–9], have enabled them to be used in various technological devices, especially in the automotive area, such as electric vehicles and hybrid electric vehicles [10,11], where energy density plays an important role. In this aspect, lithium batteries surpass other energy storage systems.

Several cathode materials have been used in lithium battery systems [10,12]. Recently, lithium-rich layered oxide (LLO) materials have been considered as promising cathode materials because of their high capacities (200–300 mAhg^{-1}) with 3.6 V or larger operating voltages when they are charged to over 4.5 V at room temperature [13,14]. Many combinations for these composite materials have been proposed with a general expression such as xLi$_2$MnO$_3$–(1−x)LiMO$_2$ (M = Mn, Ni, Co, Fe, Cr, etc.) [15–19], where excess of Li can be incorporated into the layered structure through the integration of Li$_2$MnO$_3$ in LiMO$_2$, substituting transition metals M [20]. However, LLO materials often show a big decrease in cycling capacity and a low rate capability. These limitations arise because in order for the material to deliver large discharge capacity of more than 200 mAhg^{-1}, it is necessary that cathode be charged above 4.5 V versus Li/Li$^+$ [21,22]. To overcome these drawbacks, many authors have used some oxides, among other materials, as cover materials [15,18,19,23,24] or have proposed different synthesis techniques [25,26].

The compatibility of the structure is clearly seen, and Li$_2$MnO$_3$ can be considered as a particular case of LiMO$_2$. Li$_2$MnO$_3$ has a C2/m monoclinic layered structure and can be represented in the normal layered notation as Li[Li$_{1/3}$Mn$_{2/3}$]O$_2$, composed of alternating Li layers and Li/Mn layers and being structurally compatible with LiMO$_2$, but electrochemically inactive. In a xLi$_2$MnO$_3$–(1−x)LiMO$_2$ composite material, Li$_2$MnO$_3$ provides structural stabilization to LiMO$_2$ because of its electrochemical inactivity between 3 and 4 V vs. Li/Li$^+$. However, Li$_2$MnO$_3$ can transform into an active phase by charging at 4.5 V vs. Li/Li$^+$. The charge mechanism at a high voltage plateau has not yet been well-clarified [26–28]. Some authors presented a composition with the integration of Li$_2$MnO$_3$ domains in a LiMO$_2$ matrix [29–31]. They claim that the results show the existence of two different local structures for both layered materials. On the other hand, a homogeneous solid solution of Li$_2$MnO$_3$ and LiMO$_2$ was reported [26,32]. They stated that the XRD results show a superstructure in the transition metal layer with a Li ordering. Despite their claims being well-supported, there appears a problem when the symmetry is considered, and there is a debate whether it is a solid solution with R3m rhombohedral symmetry or a solid solution with C2/m monoclinic symmetry. Researchers are making efforts to clarify the issue of the structure of layered composite materials because it is essential to understand the relationship between a crystal's structure and electrochemical performance. Based on the latest studies on layered oxide materials, the aim of this research was to synthesize high energy density xLi$_2$MnO$_3$–(1−x)LiMn$_{0.5}$Ni$_{0.5}$O$_2$ cathode materials by means of the co-precipitation method and study the effect of the ratio of transition metals on morphological characteristics and electrochemical performance in terms of current density, voltage, capacity, and ionic impedance.

2. Experimental

2.1. Synthesis of Precursor

In this first step 0.2 M of nickel (II) sulphate hexahydrate [NiSO$_4$·6H$_2$O] (Sigma-Aldrich, St. Louis, MI, U.S., 99%) and 0.2 M of manganese (II) sulphate monohydrate [MnSO$_4$·H$_2$O] (Sigma-Aldrich 98%) were separately dissolved in distilled water and then were mixed together for 10 min with constant stirring. This solution is denoted as Solution A. Separately, a solution of 1 M of sodium bicarbonate [NaHCO$_3$] (Sigma-Aldrich 99%) in distilled water was prepared, and this solution is denoted as Solution B. Solution A was slowly added drop by drop into Solution B by constant magnetic stirring with a temperature of 60 °C. The mixed solution was kept at a pH of 8.5 in order to create the precipitate of Ni$_{(1-x)/2}$Mn$_{(1+x)/2}$CO$_3$; this was obtained by adding an ammonium hydroxide solution [NH$_4$OH] (Sigma Aldrich 28–30%) drop by drop. The mixture was left for 12 h at the same temperature with constant magnetic stirring in order to secure the complete precipitation of the Ni and Mn sulphate solutions on the carbonate solution. After that, the mixture was filtered and washed three times with distilled water. The wet powder was dried for 12 h at 100 °C and then grounded with a mortar and pestle to get the nickel manganese carbonate, the final precursor powder. The quantities of NiSO$_4$·6H$_2$O and MnSO$_4$·H$_2$O varied stoichiometrically, depending on the x value of xLi$_2$MnO$_3$–(1−x)LiNi$_{0.5}$Mn$_{0.5}$O$_2$.

2.2. Synthesis of xLi_2MnO_3–$(1-x)LiNi_{0.5}Mn_{0.5}O_2$ Powder

Obtained nickel manganese carbonate powder was mixed with a stoichiometric amount of lithium carbonate [Li_2CO_3] (Strem Chemicals, Newburyport, MA, U.S., 99.999%) and ground using a mortar and pestle. After intense grinding, the mixture was annealed at 950 °C for 12 h, for which a 6" W × 6" D × 6" H Kerr 666 furnace was used. After 12 h, the calcined powder was quenched at room temperature to keep the oxidation states of the transition metals at 4^+ and 2^+ for Mn and Ni, respectively. The calcined powder was ground again to obtain the xLi_2MnO_3–$(1-x)LiNi_{0.5}Mn_{0.5}O_2$ final powder.

2.3. Electrochemical Coin Cell Fabrication

Coin-type cells (CR2032) were assembled in order to evaluate the electrochemical performance. The cathode material was prepared using 80 wt% of xLi_2MnO_3–$(1-x)LiNi_{0.5}Mn_{0.5}O_2$ powder, 10 wt% of carbon black, and 10 wt% of polyvinylidene fluoride. A slurry of the mixed powders was made by using N-Methyl-2-Pyrrolidinone, and it was then coated on aluminum foil (Alfa Aesar, Haverhill, MA, U.S., 99.999%) with a thickness of 0.025. The coated aluminum foil was put in a furnace at 100 °C for 12 h and cut in small circles. For the anode, Li foil with a thickness of 0.75 mm (Sigma Aldrich 99.9%) was used, and the separator was a polypropylene membrane (Celgard 2500, Charlotte, NC, U.S.) with a thickness of 25 µm. A standard electrolyte with 1 mol/kg $LiPF_6$ dissolved in ethylene carbonate (EC) and dimethyl carbonate (DMC) in a 1:2 wt ratio was used. The coin cells were assembled inside of an argon-filled MBraun, Stratham, NH, U.S., glove box.

2.4. Characterization Techniques

In this work, a powder structural analysis was carried out using Siemens D5000 X-ray diffractometer, U.S. (XRD) with Cu Kα radiation (1.5405 Å) in the 2θ angle range from 15° to 75° with a step of 0.02°. A scanning electron microscope (SEM) was used to study the morphology of materials, and the analysis was conducted with a JEOL 7600 FESEM, U.S., system interfaced to a Thermo-Electron System Microanalysis Systems. The samples were adhered to the top of the sample holder with double-sided conductive tape; for better resolution, a Au coating was applied. The electrochemical measurements were evaluated with a Gamry Instruments, Warminster, PA, U.S., G/PC14 potentiostat system. The charge–discharge measurements were performed at room temperature at a voltage range of 2.0–4.8 V using selected current density.

3. Results and Discussion

The X-ray diffraction patterns of the xLi_2MnO_3–$(1-x)LiNi_{0.5}Mn_{0.5}O_2$ (x = 0.3, 0.5 and 0.7) synthesized materials are presented in Figure 1a. The typical diffraction pattern of this composed material is shown, where most peaks agree well with the R3m space group related to the rhombohedral type structure for $LiMO_2$ (JCPDS# 09-0063) [31]. The low intensity peaks between 20° and 30° show characteristic peaks of the monoclinic phase, with the space group C2/m associated with the layered composite material Li_2MnO_3 (JCPDS# 84-1634) [32,33]. The XRD patterns confirm the superlattice phase corresponding to Li_2MnO_3. The patterns became more obvious when the fraction of the Li_2MnO_3 layered material was incremented in xLi_2MnO_3–$(1-x)LiMO_2$. The presence of the small peaks (110), (020), (−111), and (021) is related to the existence of the superstructure caused by the reordering of Li and Mn in the main structure, where lithium ions entered into the transition metal layers positions as expected [32]. Furthermore, the presence of the (006) and (012) peaks confirm the successful formation of a crystalline layered structure with no spinal structure, using co-precipitation synthesis method for all values of x [34,35]. It is clear then that the two structural components of this layered composite prove the coexistence of two compatible phases.

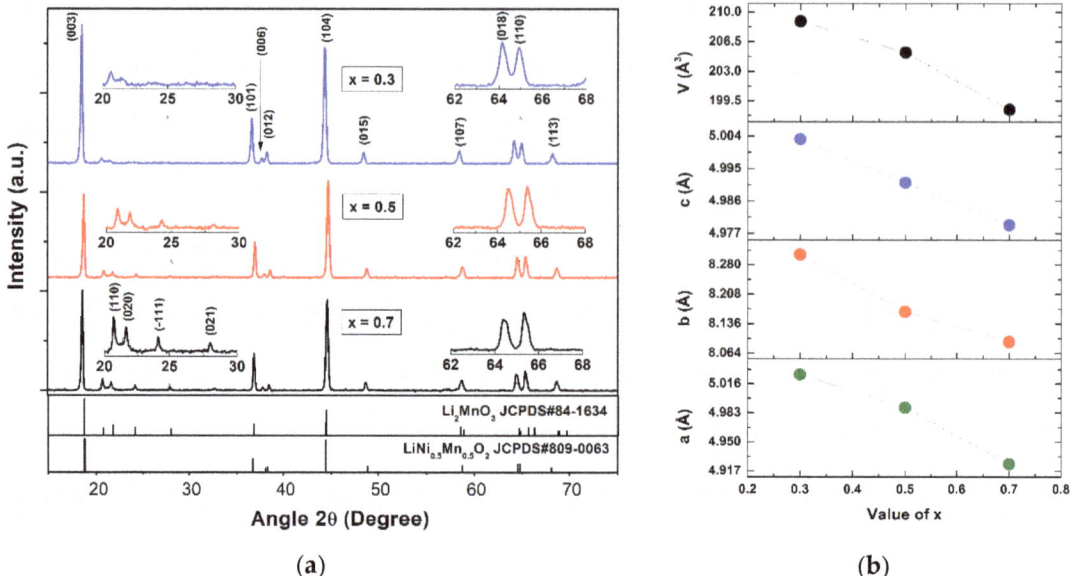

Figure 1. (a) X-ray diffraction patterns and (b) monoclinic structure lattice parameters of $x\text{Li}_2\text{MnO}_3$–$(1-x)\text{LiNi}_{0.5}\text{Mn}_{0.5}\text{O}_2$.

Figure 1b shows the unit cell parameters calculated for all the synthesized materials, based on a monoclinic structure belonging to the C2/m spaces group. The lattice parameters a, b, and c decreased with the x value, specifically with the reduction of Ni in the stoichiometric equation, which corresponded to the increase of the Li content. The stoichiometric ratio of the components' materials was determined when the formula was expressed as $\text{Li}[\text{Li}_y\text{Mn}_{1-y-z}\text{Ni}_z]\text{O}_2$, where $y = x/(2 + x)$ and $z = (1 - x)/(2.5)$. This is shown in Table 1. These decreases in the lattice parameters may be related to both the distribution of lithium ions in the metal layers changing the characteristic order with Mn and the substitution of Mn by Ni [30–33]. Figure 2 shows the crystal structures of layered Li_2MnO_3 and LiMO_2, where the compatibility of the structure is clearly seen. Li_2MnO_3 can be considered a particular case of LiMO_2, as it is described when it is expressed in layered notation. Additionally, the average size of the crystallite obtained using the Scherrer equation from the diffraction peaks (003) indicated that by means of the co-precipitation method used, nanocrystals at 36.0 nm, 37.8 nm, and 43.9 nm can be obtained for the x values of 0.3, 0.5, and 0.7 respectively.

Table 1. Stoichiometry of Li-rich layered oxides in the most common formulations.

Value of x	$x\text{Li}_2\text{MnO}_3$–$(1-x)\text{LiMn}_{0.5}\text{Ni}_{0.5}\text{O}_2$	$\text{Li}[\text{Li}_y\text{Mn}_{1-y-z}\text{Ni}_z]\text{O}_2$
0.3	$0.3\text{Li}_2\text{MnO}_3$–$0.7\text{LiNi}_{0.5}\text{Mn}_{0.5}\text{O}_2$	$\text{Li}[\text{Li}_{0.13}\text{Mn}_{0.59}\text{Ni}_{0.28}]\text{O}_2$
0.5	$0.5\text{Li}_2\text{MnO}_3$–$0.5\text{LiNi}_{0.5}\text{Mn}_{0.5}\text{O}_2$	$\text{Li}[\text{Li}_{0.20}\text{Mn}_{0.60}\text{Ni}_{0.20}]\text{O}_2$
0.7	$0.7\text{Li}_2\text{MnO}_3$–$0.3\text{LiNi}_{0.5}\text{Mn}_{0.5}\text{O}_2$	$\text{Li}[\text{Li}_{0.26}\text{Mn}_{0.62}\text{Ni}_{0.12}]\text{O}_2$

Scanning electron microscopy (SEM) was used to verify the morphology and topology of the synthesized material in order to find some relation between its features and the after-studied electrochemical behavior. As previously explained, the precursor material nickel manganese carbonate $\text{Ni}_{(1-x)/2}\text{Mn}_{(1+x)/2}\text{CO}_3$ had to be previous synthesized. Figure 3 shows the SEM micrographs of the as-synthesized powders of (a) $\text{Ni}_{0.35}\text{Mn}_{0.65}\text{CO}_3$, (b) $\text{Ni}_{0.25}\text{Mn}_{0.75}\text{CO}_3$, and (c) $\text{Ni}_{0.15}\text{Mn}_{0.85}\text{CO}_3$. With a zoom of x5000, it was possible to observe a nearly sphere-like morphology with a uniform particle distribution, similar to that found by other authors [33,36,37]. Figure 3 shows the precursors' present forms of nearly spherical solids; however, these reduced their size with the increase of Li and Mn (increase in the value of x). In Figure 3a, the agglomerations are larger and less spherical. In addition, small pores and a rougher surface can be observed, likely due to the release of CO_2 in the heat treatment process and the subsequent formation of oxides [38]. Figure 3c shows agglomerations similar to those mentioned above but smaller, with inhomogeneous growth and the presence of some impurities in some parts, associated with the greater formation of MnO_2 [33]. For x = 0.5 (Figure 3b), the agglomerations presented more compact spheres with smooth particles, no pores, and reduced impurities. Figure 4 shows the EDS spectra of precursor materials, confirming the purity of the sample and the absence of additional elements. The intensity represents the proportions of the materials used to make the precursor materials. Ni decreased with x increasing, according to the stoichiometric equation. However, the presence of oxygen was high for x = 0.3, which would explain the presence of the CO_2 formation that decreases as x increases, as shown in the SEM images.

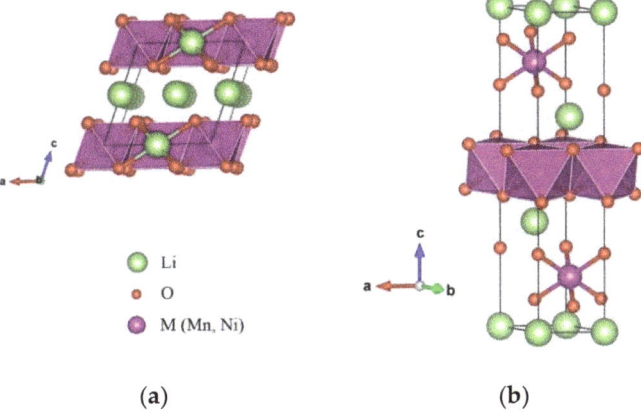

(a)　　　　　　　　　　　　　(b)

Figure 2. Crystal structure of (**a**) monoclinic Li_2MnO_3 structure with space group C2/m and (**b**) rhombohedral $LiMO_2$ structure. M = Ni, Mn with space group R3m using VESTA software.

(a)

(b)

Figure 3. *Cont.*

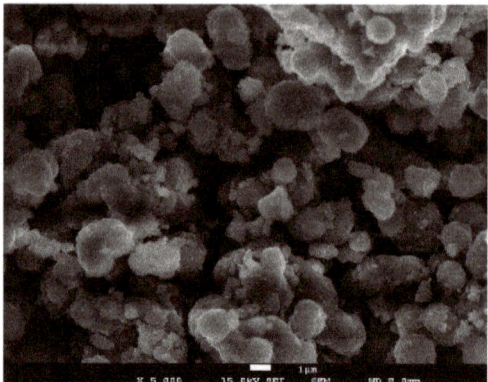

Figure 3. SEM micrographs of (**a**) Ni$_{0.35}$Mn$_{0.65}$CO$_3$, (**b**) Ni$_{0.25}$Mn$_{0.75}$CO$_3$, and (**c**) Ni$_{0.15}$Mn$_{0.85}$CO$_3$ powders, the results of the first step of the synthesis process.

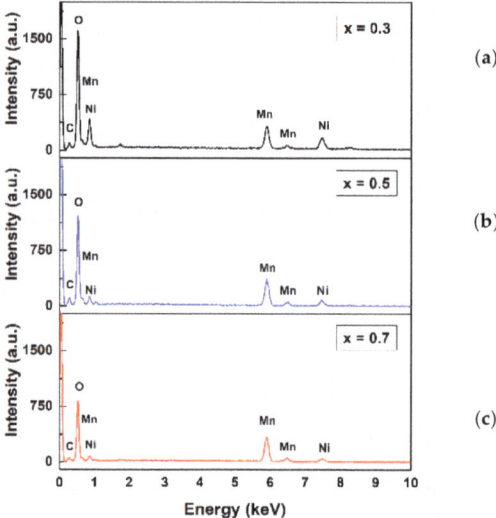

Figure 4. EDS spectra of (**a**) Ni$_{0.35}$Mn$_{0.65}$CO$_3$, (**b**) Ni$_{0.25}$Mn$_{0.75}$CO$_3$, and (**c**) Ni$_{0.15}$Mn$_{0.85}$CO$_3$ powders, the results of the first step of the synthesis process.

Micrographs of the synthesized 0.3Li$_2$MnO$_3$–0.7LiNi$_{0.5}$Mn$_{0.5}$O$_2$, 0.5Li$_2$MnO$_3$–0.5LiNi$_{0.5}$Mn$_{0.5}$O$_2$, and 0.7Li$_2$MnO$_3$–0.3LiNi$_{0.5}$Mn$_{0.5}$O$_2$ final powders after calcination of precursors with Li$_2$CO$_3$ are showed in Figure 5. The primary particles are micrometric in size and concentrated in irregularly polyhedral-shaped agglomerations. Relatively smaller primary particle agglomerates allow better charge–discharge capacity due to their higher trap density, which shortens lithium diffusion [39]. The pristine x = 0.5 material composite presented a lower density of agglomerations compared with the other synthesized material. The precursor spherical shape could be the reason why a better formation of the final product was possible, which may help improve the electrochemical performance. The size of these agglomerations increased with the increase of x. For x = 0.7 particles that exceeded a micrometer in length, the characteristics of the precursor seemed to influence the growth mechanism for the formation of the final material [37,40]. The micrographs also show free space between agglomerations, giving a high material porosity, which is good for the ions and electrons' mobility [40–42].

Figure 5. SEM micrographs of (**a**) 0.3Li$_2$MnO$_3$–0.7LiNi$_{0.5}$Mn$_{0.5}$O$_2$, (**b**) 0.5Li$_2$MnO$_3$–0.5LiNi$_{0.5}$Mn$_{0.5}$O$_2$, and (**c**) 0.7Li$_2$MnO$_3$–0.3LiNi$_{0.5}$Mn$_{0.5}$O$_2$ cathode material powders.

The charge–discharge profiles and differential capacities for the 1st, 5th, 10th, and 40th cycles at a current density of 15 mA/g for (a) 0.3Li$_2$MnO$_3$–0.7LiNi$_{0.5}$Mn$_{0.5}$O$_2$, (b) 0.5Li$_2$MnO$_3$–0.5LiNi$_{0.5}$Mn$_{0.5}$O$_2$, and (c) 0.7Li$_2$MnO$_3$–0.3LiNi$_{0.5}$Mn$_{0.5}$O$_2$ cathode material are shown in Figure 6. For the x = 0.3 electrode material (Figure 6(a1)), a clear two-step reaction appeared at the first charge process. There was a plateau voltage between 4–4.5 V corresponding to the Li extraction from LiMO$_2$ component, reaching a capacity of 120 mAh/g. The Li extraction from the Li$_2$MnO$_3$ component started from ~4.5 V as was predicted, and the capacity contributed to the total capacity is 103 mAh/g [43]. A possible explanation of low capacity is the current density used. High current densities deliver lower capacities, as some literatures present [20,39,42], and a lower current density can make Li extraction more effective. One other possibility is the formation of some spinel phase among the structure of the electrode material; this option is enhanced by the change to some clearly spinel phase at 40th cycles seen in the discharge process at ~3 V [44,45]. The decrease of capacities at high cycle numbers could be explained by the change from layered to spinel phase transformation, due to the migration of transition metal ions to Li sites without much disarrangement of the main layered structure [44]. A differential capacity plot (Figure 6(a2)) showed two sharp oxidation peaks at 4.1 V and 4.6 V, corresponding to the first charge two-step reactions of the LiMO$_2$ and Li$_2$MnO$_3$ components, respectively. The second peak represents the irreversible loss of capacity when Li was extracted in the form of Li$_2$O [46]. The first peak, which was shifted to ~3.8 V in the next cycles, corresponds to the Ni oxidation from Ni^{2+} to Ni^{4+}. The manganese-corresponding oxidation peaks were not even visible in the 40th cycle. In the reduction zone, the predominant peak is around 3.8 V; this peak is associated with Ni reduction from Ni^{4+} to Ni^{2+}. At 40 cycles, a new peak arose at ~2.93 V, supporting the idea of some spinel phase formation. This peak is a characteristic peak of Mn reduction from Mn^{4+} to Mn^{3+} in the spinel structure [47].

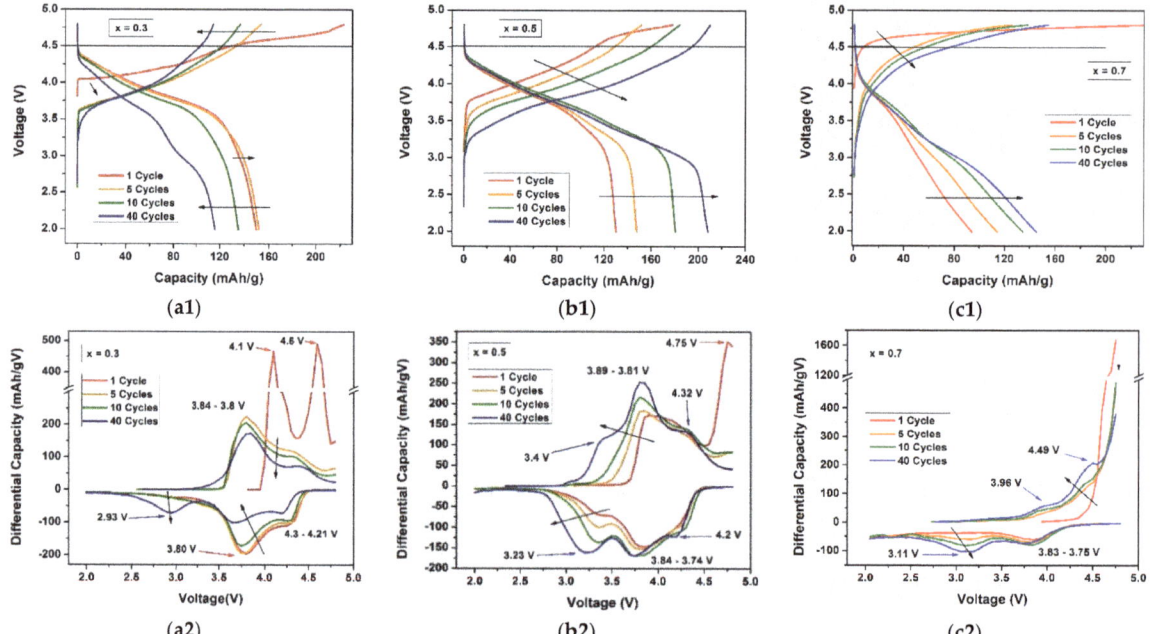

Figure 6. Charge–discharge profiles (**a1**–**c1**) and differential capacity plots (**a2**–**c2**) for the 1st, 5th, 10th, and 40th cycles for (**a**) 0.3Li$_2$MnO$_3$–0.7LiNi$_{0.5}$Mn$_{0.5}$O$_2$, (**b**) 0.5Li$_2$MnO$_3$–0.5LiNi$_{0.5}$Mn$_{0.5}$O$_2$, and (**c**) 0.7Li$_2$MnO$_3$–0.3LiNi$_{0.5}$Mn$_{0.5}$O$_2$ cathode materials.

For the x = 0.5 electrode material (Figure 6(b1)), a two-step reaction also appeared for the first charge process. In the first step (2–4.5 V) related to the Li extraction from LiMO$_2$, a capacity of 123 mAh/g was achieved near the theoretical value for the charge capacity for this component. In the second step (4.5 V), a capacity of 55 mAh/g was delivered due to the activation of Li$_2$MnO$_3$ component. This value was significantly small compared with the theoretical value, revealing that not all Li$_2$MnO$_3$ material could be activated at this current density. Li$_2$MnO$_3$ shows to be electrochemically inactive before 4.5 V vs. Li/Li$^+$, because manganese has a 4$^+$ oxidation state, making the Li$^+$ intercalation difficult [35,47]. Electrochemical activity at a higher voltage is associated with the loss of oxygen by removing Li$_2$O, allowing a change in the phase to MnO$_2$ [48,49]. The first discharge capacity was 130 mAh/g, even lower when compared to the theoretical value, but still in the range of values reported in the literature [35,39]. Despite that, the capacity progressively increased through the cycles, achieving a regular capacity of around 210 mAh/g for the 40th cycle, showing the progressive activation of Li$_2$MnO$_3$ component. This was near the value of others obtained experimentally in some of the literature [35,37,39]. The differential capacity plot (Figure 6(b2)) showed a peak associated with the initial activation of Li$_2$MnO$_3$, which appeared around 4.75 V. Both the peaks of oxidation and reduction related to Ni and Mn reactions were present. The Ni oxidation from Ni^{2+} to Ni^{4+} was associated with the peak placed at ~3.81 V in the first cycle, which stabilized at

~3.89 V, indicating an increasing of activated Ni step-by-step during cycling. Similarly, Mn oxidation increased from Mn^{3+} to Mn^{4+}, as represented by the raised peak at 3.4 V. The reductions occurred at the potentials of ~3.84 for the Ni reduction and at the raised peak at ~3.23 V for the Mn reduction. A greater progressive activation of Mn was observed, more than that of Ni, confirming the idea of the step-by-step Li_2MnO_3 activation through cycling [40,47,50,51]. For the x = 0.7 electrode material (Figure 6(c1)), only one reaction was observed for the first charge process, related to the Li extraction from $LiMO_2$. It was not possible to observe the activation of the Li_2MnO_3 layer in the first cycle. For the following cycles, an increase in the discharge capacity was observed, until the capacity achieved 140 mAh/g in cycle 40. Li_2MnO_3 proved to be electrochemically inactive, possibly due to high impedance making the Li^+ intercalation difficult [49]. A differential capacity plot (Figure 6(c2)) showed the behavior of the first loading process, where it was not possible to observe peaks related to the complete activation of the material. After 10 cycles, peaks of oxidation and reduction related to Ni and Mn reactions appeared but shifted to the high voltage in the charge process and to the left in the discharge process, indicating the formation of spinel structures [45,52]. In all cases, the differential capacity values were very low, indicating an inefficient charge transfer process between the material and the electrolyte, possibly due to the size of the conglomerates that formed the material [32,39].

Figure 7 shows the charge–discharge profiles and differential capacity plots of the (a) $0.3Li_2MnO_3$–$0.7LiNi_{0.5}Mn_{0.5}O_2$, (b) $0.5Li_2MnO_3$–$0.5LiNi_{0.5}Mn_{0.5}O_2$, and (c) $0.7Li_2MnO_3$–$0.3LiNi_{0.5}Mn_{0.5}O_2$ cathode materials at different current densities of 15, 30, and 45 mA/g. The 20th cycle was taken for this analysis. For the x = 0.3 cathode material (Figure 7(a1)), the capacity was reduced from ~150 mAh/g (15 mA/g) to ~120 mAh/g (30 mA/g) and ~100 mAh/g (45 mA/g). An inflection point is shown between 3.5 and 4 V in the discharge process, where the capacity dropped drastically. A cathode material with x = 0.5 showed similar behavior (Figure 7(b1)) but reached higher capacities, i.e., ~210 mAh/g (15 mA/g), ~190 mAh/g (30 mA/g), and ~170 mAh/g (45 mA/g). For the x = 0.7 cathode materials, only one activation process was seen (Figure 7(c1)). The capacities achieved were ~140 mAh/g (15 mA/g), ~64 mAh/g (30 mA/g), and ~57 mAh/g (45 mA/g). The differential capacities of the cathode materials showed all peaks decreasing their intensities with the increase of the current density; additionally, a degradation of the voltage occurred when the current density increased [53,54]. These reductions of electrochemical performance could be explained by considering that, at high current densities, the amount of lithium insertion/extraction is affected due to the structure's surface not allowing to Li ions to leave their sites, resulting in low capacities [46]. For example, for x = 0.5 (Figure 7(b2)), the Mn oxidation from M^{3+} to Mn^{4+} at ~3.4 V was only seen at 15 mA/g and not at 45 mA/g. As we observed, for the lithium extraction process of xLi_2MnO_3–$(1-x)LiNi_{0.5}Mn_{0.5}O_2$, the charging process was carried out in two different steps. The first one was when lithium was extracted from the $LiMO_2$ component, typically at potentials < 4.5 V, accompanied by the oxidation of the M cations. A total of $(1-x)Li$ were removed from the $LiMO_2$ irreversible component to form MnO_2. The second step was the extraction of lithium from the Li_2MnO_3 component at potentials > 4.5 V. In this last step, oxygen was also extracted from the structure, giving a net loss of Li_2O. Next, xLi was extracted from the Li_2MnO_3 component to form MnO_2 [39,48,49]. In both reactions, the total lithium extracted rose to $(1+x)Li$. In the discharge process, lithium was reinserted into the structure. $(1-x)Li$ was reinserted into the MO_2 component, recovering the $LiMO_2$ initial component, but on the other hand, only xLi was reinserted into the MnO_2 component, resulting in $LiMnO_2$ with rock salt [48,49]. This was caused by its inability to reinsert oxygen into the structure during the discharge process, which means there must always be an irreversible capacity loss of xLi in the first cycle [40].

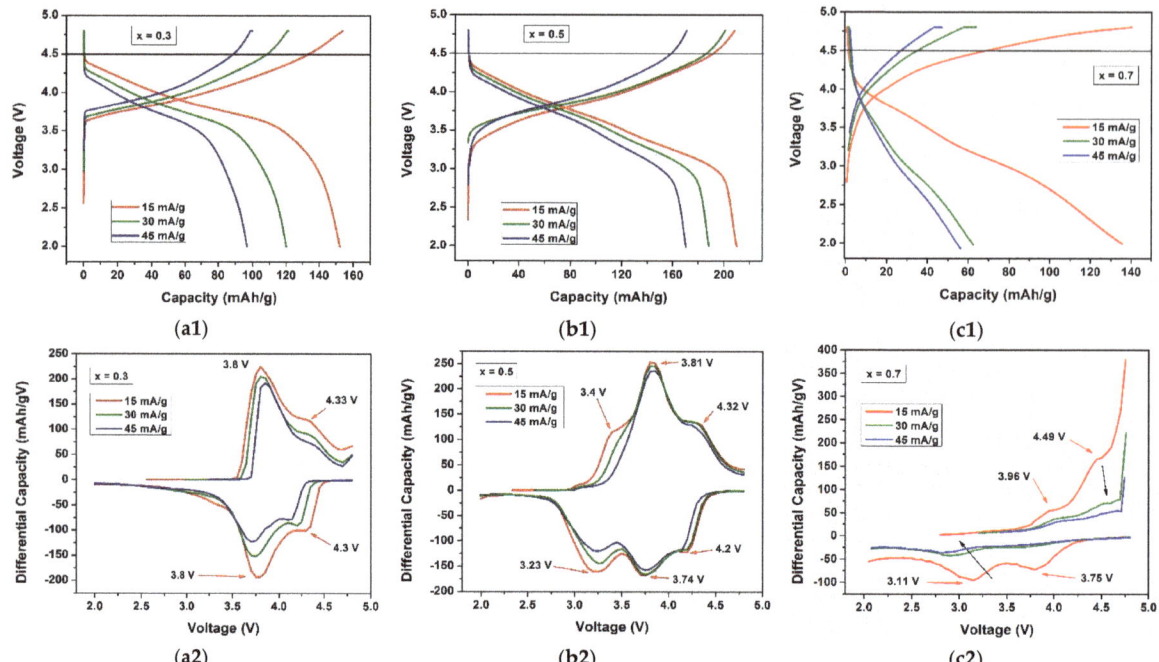

Figure 7. Charge–discharge profiles (**a1–c1**) and differential capacity plots (**a2–c2**) for 15, 30, and 45 mA/g current density for (**a**) $0.3Li_2MnO_3$–$0.7LiNi_{0.5}Mn_{0.5}O_2$, (**b**) $0.5Li_2MnO_3$–$0.5LiNi_{0.5}Mn_{0.5}O_2$, and (**c**) $0.7Li_2MnO_3$–$0.3LiNi_{0.5}Mn_{0.5}O_2$ cathode materials.

The variations of the charge–discharge capacity and coulombic efficiency through cycling for the (a) $0.3Li_2MnO_3$–$0.7LiNi_{0.5}Mn_{0.5}O_2$, (b) $0.5Li_2MnO_3$–$0.5LiNi_{0.5}Mn_{0.5}O_2$, and (c) $0.7Li_2MnO_3$–$0.3LiNi_{0.5}Mn_{0.5}O_2$ cathode materials are shown in Figure 8. For the x = 0.3 (Figure 8a) first cycle, the irreversible process was measured at a coulombic efficiency of around 68%. The discharge capacity was almost constant for the first 10 cycles, with a slight decrease in the last cycles, using a current density of 15 mA/g, although certain stability was also observed at higher current densities. For x = 0.5 (Figure 8b), the first-cycle charge–discharge capacities difference produced a higher coulombic efficiency of 73%. In this material, the discharge capacity using 15 mA/g gradually increased through cycling until the capacity was almost 220 mAh/g in the 19th cycle. This could be caused by an incomplete activation of the cathode material, or it may be due to the size of the particle and the long diffusion distance of Li+, which made it more difficult to activate them. A decrease in the next cycle to 210 mAh/g was then observed, remaining almost constant at this value. This is shown in Figure 6(b1), where at first charge the Li_2MnO_3 layer was not activated at all but activated slowly and continuously in the following cycles. This behavior is completely different compared with x = 0.3 (Figure 8a), which showed decreasing capacity, because the plateau seen on Figure 6(a1) was longer with a more complete activation of Li_2MnO_3, representing an irreversible loss [51]. For x = 0.7 (Figure 8c), the first-cycle charge–discharge capacities difference produced a low coulombic efficiency, around 20% The cycling was almost constant at any current density but had low capacity. At the current densities of 30 and 45 mA/g, there was no significant difference, perhaps because of the high content of Li_2MnO_3 associated with cycling performance deterioration and its disposition to change structure to spinel form [40]. Additionally, it seemed that a low reversible capacity was possible due to a longer diffusion distance, which led to a slower rate of diffusion than was reflected in lower capacities, as seen in this study [55].

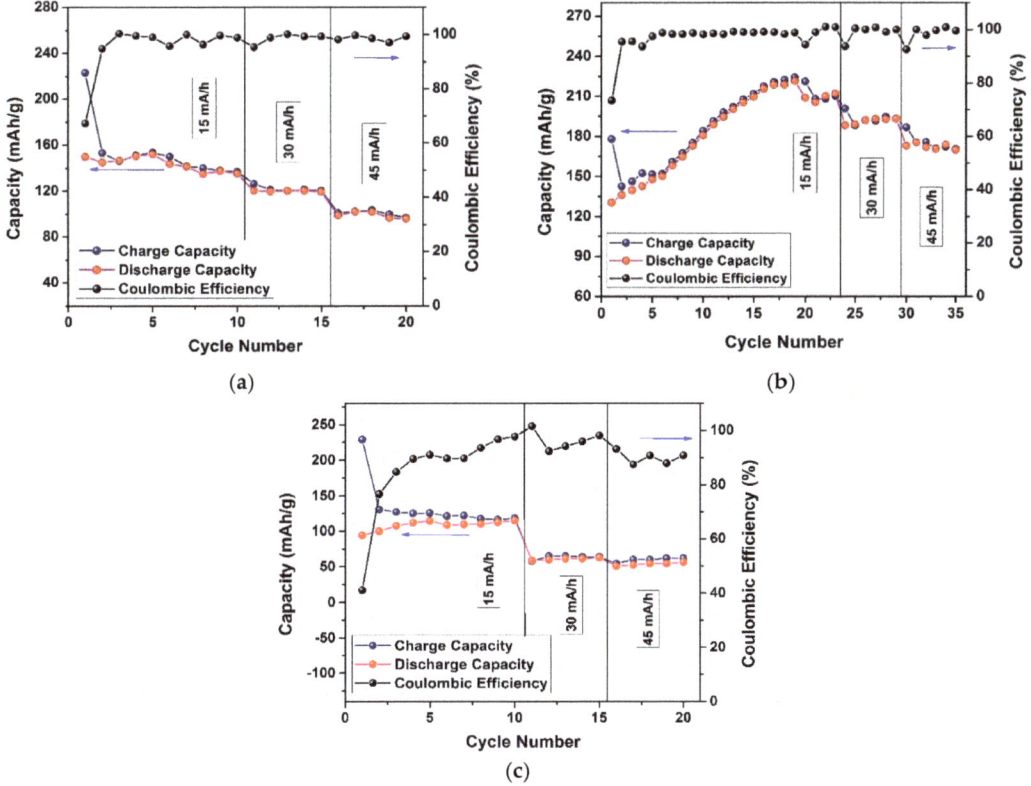

Figure 8. Charge–discharge capacity and coulombic efficiency through cycling for (**a**) $0.3Li_2MnO_3$–$0.7LiNi_{0.5}Mn_{0.5}O_2$, (**b**) $0.5Li_2MnO_3$–$0.5LiNi_{0.5}Mn_{0.5}O_2$, and (**c**) $0.7Li_2MnO_3$–$0.3LiNi_{0.5}Mn_{0.5}O_2$ cathode materials.

Electrochemical impedance spectroscopy (EIS) measurements were carried out on the electrochemical cells prepared with the synthesized materials, and the Nyquist plot results are shown in Figure 9. EIS allows us to understand the various situations that occur at the interface of the electrode and the electrolyte through a non-destructive technique that determines the impedance of the electrochemical cell based on small AC signals, with constant DC voltages in a wide range of frequencies [56]. The impedance spectra were obtained within the frequency range of 10 mHz to 100 kHz. Each EIS spectrum presented two overlapped semicircles due to the combination of a capacitor and a resistor element connected in parallel and an inclined line in the low-frequency region. This characteristic plot suggested the contribution of two different resistive elements to the total impedance of the cell. The high-frequency semicircle represents the impedance of the surface electrolyte interface (SEI) film (R_{sf}), the medium frequency semicircle denotes the charge transfer impedance across the interface (R_{ct}), and the small interrupt in the high frequency corresponds to the electrolyte impedance (R_e) [29,40]. All parameters were represented by plot fitting, with the electrochemical equivalent circuit (EEC) shown on Figure 10. Where CPE_1 represents the capacitance of SEI film, which grows thicker by electrolyte–electrode reaction during charge–discharge and deteriorates the performance, CPE_2 corresponds to charge–transfer capacitance, and Z_w is the Warburg impedance related to the lithium ions' diffusion. The impedance parameters' (R_{sf}, R_{ct} and R_e) values after fitting are summarized in Table 1.

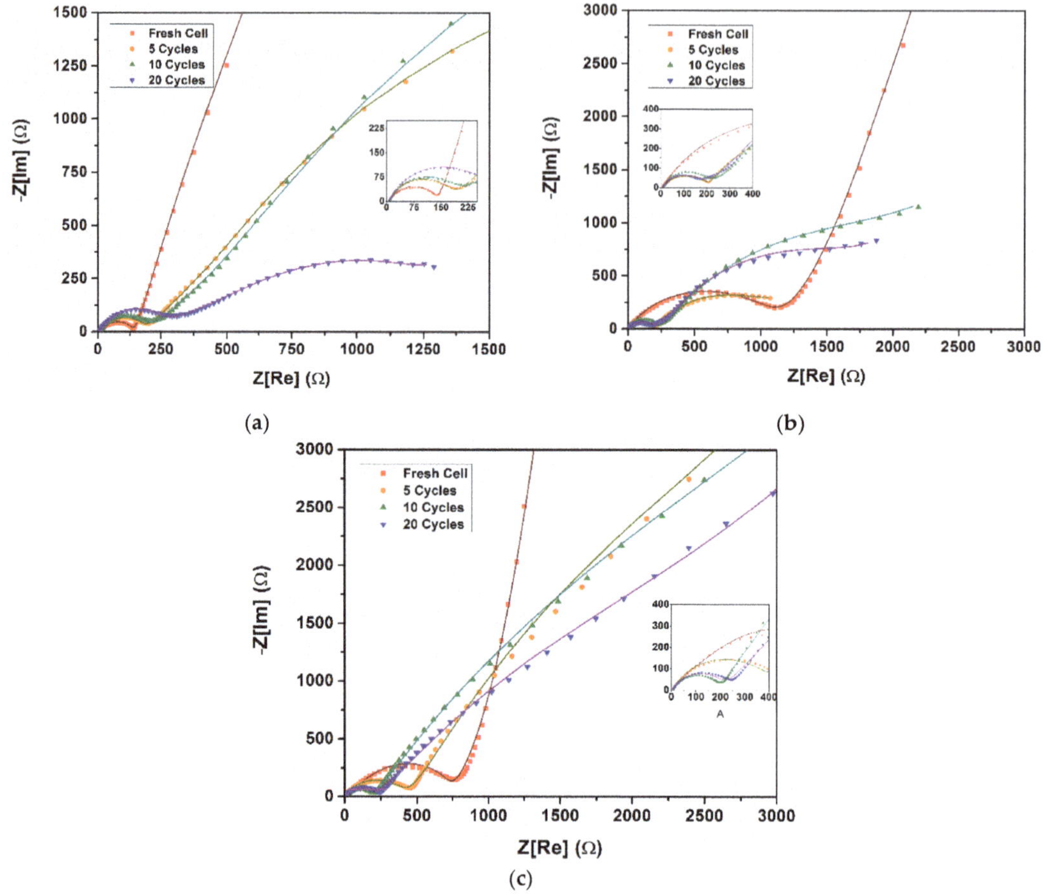

Figure 9. Impedance spectrum of the (a) $0.3Li_2MnO_3$–$0.7LiNi_{0.5}Mn_{0.5}O_2$, (b) $0.5Li_2MnO_3$–$0.5LiNi_{0.5}Mn_{0.5}O_2$, and (c) $0.7Li_2MnO_3$–$0.3LiNi_{0.5}Mn_{0.5}O_2$ cathode materials.

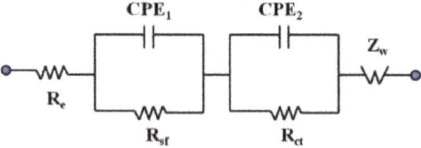

Figure 10. Electrochemical electric circuit (EEC) used to fit the EIS data.

For electrochemical cells with the x = 0.3 layered material (Figure 9a), an increase in the area of semicircle was observed as the number of cycles increased. An increase in the R_{sf} was expected because of the growth of the SEI layer at the interface of the electrolyte and electrode surface; nevertheless, a high resistance value was found at the first cycle, as shown in Table 2. This can explain the late activation of the $LiMO_2$-associated layer around 4 V. For subsequent cycles, the impedance was drastically reduced to 900 Ω to increase when reaching 20 cycles. In the other hand, the R_{ct} impedance increased with cycles; this behavior could explain the decrease in discharge capacity observed in Figure 6a and poor cycling stability [57]. Electrochemical cells with x = 0.5 (Figure 9b) showed a large reduction in semicircle diameter of the high-frequency region with the cycle increasing, related to a reduction of the charge–transfer resistance, which could be beneficial for improving the discharge capacity [25]. A large difference in diameter between the first cycle and the following was clearly seen. For the fifth, tenth, and twentieth, diameter was highly similar, revealing that the SEI film formed mostly during the first charge–discharge process and remained stable during the following charge–discharge cycles. Additionally, since the R_{ct} value decreased during large cycles, with this material it is possible to reach high capacity values, as well as the composite layer structure becoming more stable, which improves cyclability during the charge–discharge process, as shown in Figure 8b. Finally, for the cells with x = 0.7, although the diameters of the semicircles in the high frequency region had a behavior similar to that analyzed above, they did not present the second semicircle in the low frequency region. Instead, almost straight "tails" were observed (Figure 9c). Since the first cycle was a very inclined one and exceeded 45°, this was associated with the large size of the particles [40], which could explain the low capacity reached in the electrochemical cells using this material.

Table 2. Impedance parameters obtained after fitting the EIS experimental data based on the EEC model shown in Figure 9.

Impedance Parameter	x\Cycle	1st	5th	10th	20th
R_{sf} (Ω)	0.3	14,760	900	900	1067
	0.5	505.4	81.03	65.37	2087
	0.7	435.1	2688	5000	1500
R_{ct} (Ω)	0.3	130	167.9	197.5	221.1
	0.5	0.9519	1758	965.9	66.63
	0.7	265.7	400	180	200
R_e (Ω)	0.3	7.901	7.217	7.539	3.273
	0.5	9.380	4.45	4.807	6.000
	0.7	4.408	7.501	1.532	5.699

4. Conclusions

In this study, a xLi_2MnO_3–$(1-x)LiNi_{0.5}Mn_{0.5}O_2$ layered composite material for x = 0.3, 0.5, and 0.7 was synthesized by a carbonate-assisted co-precipitation method. The X-ray diffraction patterns confirmed the formation of the two-phase layered crystal structure, the rhombohedral $LiMO_2$ structure with R3m space group, and the monoclinic Li_2MnO_3 structure with C2/m space group. We also observed a decrease in the size of the lattice parameters with the decrease in the amount of Ni. The CR2032 coin cells were successfully assembled in an Ar-filled glove box in order to test the electrochemical behavior of the synthesized electrode materials. The electrochemical results show the two-step Li extraction, one at potential <4.5 V related to the Li extraction from $LiMO_2$, and the other at potential >4.5, which corresponds to the activation of Li_2MnO_3 by means of the extraction of Li in the form of LiO_2. As x increased, the charge and discharge reactions assigned to the component similar to Li_2MnO_3 increased, while those corresponding to $LiMO_2$ decreased, producing an activation that allowed high capacities to be reached. However, there was a limit related to the size of the particle agglomerates and some spinel phase formation within the structure of the first material, which was reflected in a slow diffusion rate, inefficient Li_2MnO_3 activation, and low capacities. Better electrochemical performance was found for the x = 0.5 electrode material with a first-charge capacity of 173 mAh/g and a higher discharge capacity of 210 mAh/g.

Author Contributions: Conceptualization, R.N.-N.; methodology, R.N.-N., S.R.-F. and L.J.-C.; software, M.G.-C. and F.N.D.; validation, R.N.-N., L.A.-S. and S.M.B.; formal analysis, R.N.-N.; investigation, R.N.-N. and L.J.-C.; data curation, M.G.-C. and F.N.D.; writing—original draft preparation, R.N.-N.; writing—review and editing, S.R.-F. and M.G.-C.; project administration, R.N.-N. and S.M.B. All authors have read and agreed to the published version of the manuscript.

Funding: This research received no external funding.

Data Availability Statement: Not applicable.

Acknowledgments: The authors thank Maharaj S. Tomar for his guidance and support in this research.

Conflicts of Interest: The authors declare no conflict of interest.

References

1. Ghosh, S.; Bhattacharjee, U.; Bhowmik, S.; Martha, D.; Surendra, K. A Review on High-Capacity and High-Voltage Cathodes for Next-Generation Lithium-ion Batteries. *J. Energy Power Technol.* **2022**, *4*, 1–59. [CrossRef]
2. Rojas-Flores, S.; Pérez-Delgado, O.; Nazario-Naveda, R.; Rojales-Alfaro, H.; Benites, S.M.; La Cruz-Noriega, D.; Otiniano, N.M. Potential Use of Papaya Waste as a Fuel for Bioelectricity Generation. *Processes* **2021**, *9*, 1799. [CrossRef]
3. Segundo, R.F.; La Cruz-Noriega, D.; Milly Otiniano, N.; Benites, S.M.; Esparza, M.; Nazario-Naveda, R. Use of Onion Waste as Fuel for the Generation of Bioelectricity. *Molecules* **2022**, *27*, 625. [CrossRef] [PubMed]
4. Nazario-Naveda, R.; Benites, S.M. Sugar Industry Waste for Bioelectricity Generation. *Environ. Res. Eng. Manag.* **2021**, *77*, 15–22. [CrossRef]
5. Segundo, R.F.; Renny, N.N.; Moises, G.C.; Daniel, D.N.; Natalia, D.D.; Karen, V.R. Generation of Bioelectricity from Organic Fruit Waste. *Environ. Res. Eng. Manag.* **2021**, *77*, 6–14. [CrossRef]
6. Hayner, C.M.; Zhao, X.; Kung, H.H. Materials for rechargeable lithium-ion batteries. *Annu. Rev. Chem. Biomol. Eng.* **2012**, *3*, 445–471. [CrossRef]
7. Konishi, H.; Hirano, T.; Takamatsu, D.; Gunji, A.; Feng, X.; Furutsuki, S.; Takahashi, S.; Terada, S. Potential hysteresis between charge and discharge reactions in $Li_{1.2}Ni_{0.13}Mn_{0.54}Co_{0.13}O_2$ for lithium ion batteries. *Solid State Ion.* **2017**, *300*, 120–127. [CrossRef]
8. Konishi, H.; Hirano, T.; Takamatsu, D.; Gunji, A.; Feng, X.; Furutsuki, S.; Okumura, T.; Terada, S.; Tamura, K. Mechanisms responsible for two possible electrochemical reactions in $Li_{1.2}Ni_{0.13}Mn_{0.54}Co_{0.13}O_2$ used for lithium ion batteries. *J. Solid State Chem.* **2018**, *258*, 225–231. [CrossRef]
9. Luo, K.; Roberts, M.R.; Guerrini, N.; Tapia-Ruiz, N.; Hao, R.; Massel, F.; Pickup, D.M.; Ramos, S.; Liu, Y.S.; Guo, J.; et al. Anion redox chemistry in the cobalt free 3d transition metal oxide intercalation electrode $Li[Li_{0.2}Ni_{0.2}Mn_{0.6}]O_2$. *J. Am. Chem. Soc.* **2016**, *138*, 11211–11218. [CrossRef]
10. Nisa, S.S.; Rahmawati, M.; Yudha, C.S.; Nilasary, H.; Nursukatmo, H.; Oktaviano, H.S.; Muzayanha, S.U.; Purwanto, A. A Fast Approach to Obtain Layered Transition-Metal Cathode Material for Rechargeable Batteries. *Batteries* **2022**, *8*, 4. [CrossRef]
11. Klink, J.; Hebenbrock, A.; Grabow, J.; Orazov, N.; Nylén, U.; Benger, R.; Beck, H.-P. Comparison of Model-Based and Sensor-Based Detection of Thermal Runaway in Li-Ion Battery Modules for Automotive Application. *Batteries* **2022**, *8*, 34. [CrossRef]
12. Xu, B.; Qian, D.; Wang, Z.; Meng, Y.S. Recent progress in cathode materials research for advanced lithium ion batteries. *Mater. Sci. Eng. R Rep.* **2012**, *73*, 51–65. [CrossRef]
13. Yu, H.; Zhou, H. High-energy cathode materials (Li_2MnO_3–$LiMO_2$) for lithium-ion batteries. *J. Phys. Chem. Lett.* **2013**, *4*, 1268–1280. [CrossRef] [PubMed]
14. Konishi, H.; Hirano, T.; Takamatsu, D.; Gunji, A.; Feng, X.; Furutsuki, S.; Okumura, T.; Terada, S. Suppression of potential hysteresis between charge and discharge reactions in lithium-rich layer-structured cathode material by increasing nickel/manganese ratio. *Solid State Ion.* **2017**, *308*, 84–89. [CrossRef]
15. Hou, X.; Wang, Y.; Song, J.; Gu, H.; Guo, R.; Liu, W.; Mao, Y.; Xie, J. Electrochemical behavior of Mn-based Li-rich cathode material $Li_{1.15}Ni_{0.17}Co_{0.11}Mn_{0.57}O_2$ fluorinated by NH_4F. *Solid State Ion.* **2018**, *325*, 1–6. [CrossRef]
16. Li, Y.C.; Xiang, W.; Wu, Z.G.; Xu, C.L.; Xu, Y.D.; Xiao, Y.; Yang, Z.G.; Wu, C.J.; Lv, G.P.; Guo, X.D. Construction of homogeneously Al^{3+} doped Ni rich Ni-Co-Mn cathode with high stable cycling performance and storage stability via scalable continuous precipitation. *Electrochim. Acta* **2018**, *291*, 84–94. [CrossRef]
17. Wang, C.C.; Lin, Y.C.; Chiu, K.F. Alleviation of voltage fade of lithium-rich layered oxide cathodes of Li-ion battery by incorporation of Cr. *J. Alloy. Compd.* **2017**, *721*, 600–608. [CrossRef]
18. Zou, T.; Qi, W.; Liu, X.; Wu, X.; Fan, D.; Guo, S.; Wang, L. Improvement of the electrochemical performance of $Li_{1.2}Ni_{0.13}Co_{0.13}Mn_{0.54}O_2$ cathode material by Al_2O_3 surface coating. *J. Electroanal. Chem.* **2020**, *859*, 113845. [CrossRef]
19. Zhong, J.; Yang, Z.; Yu, Y.; Liu, Y.; Li, J.; Kang, F. Surface substitution of polyanion to improve structure stability and electrochemical properties of lithium-rich layered cathode oxides. *Appl. Surf. Sci.* **2020**, *512*, 145741. [CrossRef]
20. Kim, D.; Sandi, G.; Croy, J.R.; Gallagher, K.G.; Kang, S.H.; Lee, E.; Slater, M.D.; Johnson, C.S.; Thackeray, M.M. Composite layered-layered-spinel cathode structures for lithium-ion batteries. *J. Electrochem. Soc.* **2012**, *160*, A31. [CrossRef]
21. Xiao, L.; Xiao, J.; Yu, X.; Yan, P.; Zheng, J.; Engelhard, M.; Bhattacharya, P.; Wang, C.; Yang, X.Q.; Zhang, J.G. Effects of structural defects on the electrochemical activation of Li_2MnO_3. *Nano Energy* **2015**, *16*, 143–151. [CrossRef]
22. Zhang, K.; Li, B.; Zuo, Y.; Song, J.; Shang, H.; Ning, F.; Xia, D. Voltage decay in layered Li-rich Mn-based cathode materials. *Electrochem. Energy Rev.* **2019**, *2*, 606–623. [CrossRef]
23. Baggetto, L.; Dudney, N.J.; Veith, G.M. Surface chemistry of metal oxide coated lithium manganese nickel oxide thin film cathodes studied by XPS. *Electrochim. Acta* **2013**, *90*, 135–147. [CrossRef]
24. Zhou, D.; Liu, R.; He, Y.B.; Li, F.; Liu, M.; Li, B.; Yang, Q.H.; Cai, Q.; Kang, F. SiO_2 hollow nanosphere-based composite solid electrolyte for lithium metal batteries to suppress lithium dendrite growth and enhance cycle life. *Adv. Energy Mater.* **2016**, *6*, 1502214. [CrossRef]

25. Shojan, J.; Chitturi, V.R.; Torres, L.; Singh, G.; Katiyar, R.S. Lithium-ion battery performance of layered $0.3Li_2MnO_3$–$0.7LiNi_{0.5}Mn_{0.5}O_2$ composite cathode prepared by co-precipitation and sol–gel methods. *Mater. Lett.* **2013**, *104*, 57–60. [CrossRef]
26. Xiang, Y.; Jiang, Y.; Liu, S.; Wu, J.; Liu, Z.; Zhu, L.; Xiong, L.; He, Z.; Wu, X. Improved electrochemical performance of $0.5Li_2MnO_3$-$0.5LiNi_{0.5}Mn_{0.5}O_2$ cathode materials for lithium ion batteries synthesized by ionic-liquid-assisted hydrothermal method. *Front. Chem.* **2020**, *8*, 729. [CrossRef]
27. Bettge, M.; Li, Y.; Sankaran, B.; Rago, N.D.; Spila, T.; Haasch, R.T.; Petrov, I.; Abraham, D.P. Improving high-capacity $Li_{1.2}Ni_{0.15}Mn_{0.55}Co_{0.1}O_2$-based lithium-ion cells by modifiying the positive electrode with alumina. *J. Power Sources* **2013**, *233*, 346–357. [CrossRef]
28. Klein, A.; Axmann, P.; Yada, C.; Wohlfahrt-Mehrens, M. Improving the cycling stability of Li_2MnO_3 by surface treatment. *J. Power Sources* **2015**, *288*, 302–307. [CrossRef]
29. Lin, J.; Mu, D.; Jin, Y.; Wu, B.; Ma, Y.; Wu, F. Li-rich layered composite $Li[Li_{0.2}Ni_{0.2}Mn_{0.6}]O_2$ synthesized by a novel approach as cathode material for lithium ion battery. *J. Power Sources* **2013**, *230*, 76–80. [CrossRef]
30. Wu, Q.; Zhang, X.; Sun, S.; Wan, N.; Pan, D.; Bai, Y.; Zhu, H.; Hu, Y.S.; Dai, S. Improved electrochemical performance of spinel $LiMn_{1.5}Ni_{0.5}O_4$ through MgF_2 nano-coating. *Nanoscale* **2015**, *7*, 15609–15617. [CrossRef]
31. Song, L.; Tang, Z.; Chen, Y.; Xiao, Z.; Li, L.; Zheng, H.; Li, B.; Liu, Z. Structural analysis of layered Li_2MnO_3–$LiMO_2$ (M= $Ni_{1/3}Mn_{1/3}Co_{1/3}$, $Ni_{1/2}Mn_{1/2}$) cathode materials by Rietveld refinement and first-principles calculations. *Ceram. Int.* **2016**, *42*, 8537–8544. [CrossRef]
32. Redel, K.; Kulka, A.; Plewa, A.; Molenda, J. High-performance Li-rich layered transition metal oxide cathode materials for Li-ion batteries. *J. Electrochem. Soc.* **2019**, *166*, A5333. [CrossRef]
33. Gabrielli, G.; Marinaro, M.; Mancini, M.; Axmann, P.; Wohlfahrt-Mehrens, M. A new approach for compensating the irreversible capacity loss of high-energy Si/C | $LiNi_{0.5}Mn_{1.5}O_4$ lithium-ion batteries. *J. Power Sources* **2017**, *351*, 35–44. [CrossRef]
34. Hua, W.; Chen, M.; Schwarz, B.; Knapp, M.; Bruns, M.; Barthel, J.; Yang, X.; Sigel, F.; Azmi, R.; Senyshyn, A.; et al. Lithium/Oxygen Incorporation and Microstructural Evolution during Synthesis of Li-Rich Layered $Li[Li_{0.2}Ni_{0.2}Mn_{0.6}]O_2$ Oxides. *Adv. Energy Mater.* **2019**, *9*, 1803094. [CrossRef]
35. Konishi, H.; Hirano, T.; Takamatsu, D.; Okumura, T. Electrochemical reaction mechanism of two components in xLi_2MnO_3–(1−x)$LiNi_{0.5}Mn_{0.5}O_2$ and effect of x on the electrochemical performance in lithium ion battery. *J. Electroanal. Chem.* **2020**, *873*, 114402. [CrossRef]
36. Jiang, Y.; Yang, Z.; Luo, W.; Hu, X.; Huang, Y. Hollow $0.3Li_2MnO_3$-$0.7LiNi_{0.5}Mn_{0.5}O_2$ microspheres as a high-performance cathode material for lithium–ion batteries. *Phys. Chem. Chem. Phys.* **2013**, *15*, 2954–2960. [CrossRef] [PubMed]
37. Xiang, Y.; Li, J.; Liao, Q.; Wu, X. Morphology and particle growth of Mn-based carbonate precursor in the presence of ethylene glycol for high-capacity Li-rich cathode materials. *Ionics* **2019**, *25*, 81–87. [CrossRef]
38. Guo, J.; Deng, Z.; Yan, S.; Lang, Y.; Gong, J.; Wang, L.; Liang, G. Preparation and electrochemical performance of $LiNi0.5Mn1.5O4$ spinels with different particle sizes and surface orientations as cathode materials for lithium-ion battery. *J. Mater. Sci.* **2020**, *55*, 13157–13176. [CrossRef]
39. Singh, G.; West, W.C.; Soler, J.; Katiyar, R.S. In situ Raman spectroscopy of layered solid solution Li_2MnO_3–$LiMO_2$ (M = Ni, Mn, Co). *J. Power Sources* **2012**, *218*, 34–38. [CrossRef]
40. Liu, C.; Wu, M.; Zong, Y.; Zhang, L.; Yang, Y.; Yang, G. Synthesis and structural properties of xLi_2MnO_3-(1-x)$LiNi_{0.5}Mn_{0.5}O_2$ single crystals towards enhancing reversibility for lithium-ion battery/pouch cells. *J. Alloy. Compd.* **2019**, *770*, 490–499. [CrossRef]
41. Xiang, Y.; Sun, Z.; Li, J.; Wu, X.; Liu, Z.; Xiong, L.; Yin, Z. Improved electrochemical performance of $Li_{1.2}Ni_{0.2}Mn_{0.6}O_2$ cathode material for lithium ion batteries synthesized by the polyvinyl alcohol assisted sol-gel method. *Ceram. Int.* **2017**, *43*, 2320–2324. [CrossRef]
42. Yu, C.; Li, G.; Guan, X.; Zheng, J.; Li, L. Composites $Li_{1+x}Mn_{0.5+0.5\,x}Ni_{0.5-0.5x}O_2$ ($0.1 \leq x \leq 0.4$): Optimized preparation to yield an excellent cycling performance as cathode for lithium-ion batteries. *Electrochim. Acta* **2012**, *61*, 216–224. [CrossRef]
43. Wang, J.; He, X.; Paillard, E.; Laszczynski, N.; Li, J.; Passerini, S. Lithium-and Manganese-Rich Oxide Cathode Materials for High-Energy Lithium Ion Batteries. *Adv. Energy Mater.* **2016**, *6*, 1600906. [CrossRef]
44. Gu, M.; Belharouak, I.; Zheng, J.; Wu, H.; Xiao, J.; Genc, A.; Amine, K.; Thevuthasan, S.; Baer, D.R.; Zhang, J.G.; et al. Formation of the spinel phase in the layered composite cathode used in Li-ion batteries. *ACS Nano* **2013**, *7*, 760–767. [CrossRef]
45. Hy, S.; Liu, H.; Zhang, M.; Qian, D.; Hwang, B.J.; Meng, Y.S. Performance and design considerations for lithium excess layered oxide positive electrode materials for lithium ion batteries. *Energy Environ. Sci.* **2016**, *9*, 1931–1954. [CrossRef]
46. Hy, S.; Felix, F.; Rick, J.; Su, W.N.; Hwang, B.J. Direct In situ observation of Li_2O evolution on Li-Rich high-capacity cathode material, $Li[Ni_xLi_{(1-2x)/3}Mn_{(2-x)/3}]O_2$ ($0 \leq x \leq 0.5$). *J. Am. Chem. Soc.* **2014**, *136*, 999–1007. [CrossRef]
47. Buzlukov, A.; Mouesca, J.M.; Buannic, L.; Hediger, S.; Simonin, L.; Canevet, E.; Colin, J.F.; Gutel, T.; Bardet, M. Li-Rich Mn/Ni Layered Oxide as Electrode Material for Lithium Batteries: A 7Li MAS NMR Study Revealing Segregation into (Nanoscale) Domains with Highly Different Electrochemical Behaviors. *J. Phys. Chem. C* **2016**, *120*, 19049–19063. [CrossRef]
48. Oishi, M.; Yogi, C.; Watanabe, I.; Ohta, T.; Orikasa, Y.; Uchimoto, Y.; Ogumi, Z. Direct observation of reversible charge compensation by oxygen ion in Li-rich manganese layered oxide positive electrode material, $Li_{1.16}Ni_{0.15}Co_{0.19}Mn_{0.50}O_2$. *J. Power Sources* **2015**, *276*, 89–94. [CrossRef]

49. Zhao, S.; Yan, K.; Zhang, J.; Sun, B.; Wang, G. Reaction Mechanisms of Layered Lithium-Rich Cathode Materials for High-Energy Lithium-Ion Batteries. *Angew. Chem. Int. Ed.* **2021**, *60*, 2208–2220. [CrossRef]
50. Yang, P.; Li, H.; Wei, X.; Zhang, S.; Xing, Y. Structure tuned $Li_{1.2}Mn_{0.6}Ni_{0.2}O_2$ with low cation mixing and Ni segregation as high-performance cathode materials for Li-ion batteries. *Electrochim. Acta* **2018**, *271*, 276–283. [CrossRef]
51. Hy, S.; Su, W.N.; Chen, J.M.; Hwang, B.J. Soft X-ray absorption spectroscopic and Raman studies on Li1.2Ni0.2Mn0.6O2 for lithium-ion batteries. *J. Phys. Chem. C* **2012**, *116*, 25242–25247. [CrossRef]
52. Peng, H.; Zhao, S.X.; Huang, C.; Yu, L.Q.; Fang, Z.Q.; Wei, G.D. In Situ construction of spinel coating on the surface of a lithium-rich manganese-based single crystal for inhibiting voltage fade. *ACS Appl. Mater. Interfaces* **2020**, *12*, 11579–11588. [CrossRef] [PubMed]
53. Konishi, H.; Gunji, A.; Feng, X.; Furutsuki, S. Effect of transition metal composition on electrochemical performance of nickel-manganese-based lithium-rich layer-structured cathode materials in lithium-ion batteries. *J. Solid State Chem.* **2017**, *249*, 80–86. [CrossRef]
54. Konishi, H.; Terada, S.; Okumura, T. Effect of Lithium/Transition-Metal Ratio on the Electrochemical Properties of Lithium-Rich Cathode Materials with Different Nickel/Manganese Ratios for Lithium-Ion Batteries. *ChemistrySelect* **2019**, *4*, 9444–9450. [CrossRef]
55. Yang, F.; Zhang, Q.; Hu, X.; Peng, T. Synthesis of layered $xLi_2MnO_3 \cdot (1-x)LiMnO_2$ nanoplates and its electrochemical performance as Li-rich cathode materials for Li-ion battery. *Electrochim. Acta* **2015**, *165*, 182–190. [CrossRef]
56. Meddings, N.; Heinrich, M.; Overney, F.; Lee, J.S.; Ruiz, V.; Napolitano, E.; Seitz, S.; Hinds, G.; Raccichini, R.; Gaberšček, M.; et al. Application of electrochemical impedance spectroscopy to commercial Li-ion cells: A review. *J. Power Sources* **2020**, *480*, 228742. [CrossRef]
57. Nie, Y.; Xiao, W.; Miao, C.; Xu, M.; Wang, C. Effect of calcining oxygen pressure gradient on properties of $LiNi_{0.8}Co_{0.15}Al_{0.05}O_2$ cathode materials for lithium ion batteries. *Electrochim. Acta* **2020**, *334*, 135654. [CrossRef]

Article

Preparing Co/N-Doped Carbon as Electrocatalyst toward Oxygen Reduction Reaction via the Ancient "Pharaoh's Snakes" Reaction

Jian Gao [1,2,*], Mengxin Zhou [1], Xinyao Wang [1], Hong Wang [1], Zhen Yin [3], Xiaoyao Tan [1,*] and Yuan Li [1,*]

[1] Tianjin Key Laboratory of Green Chemical Engineering Process Engineering, Tiangong University, Tianjin 300387, China
[2] Haian Nanjing University High Tech Institute, 428, Zhennan Road, Haian 226600, China
[3] College of Chemical Engineering and Materials Science, Tianjin University of Science and Technology, 29 13th Avenue, TEDA, Tianjin 300457, China
* Correspondence: gaojian@tiangong.edu.cn (J.G.); tanxiaoyao@tiangong.edu.cn (X.T.); liyuan@tiangong.edu.cn (Y.L.)

Abstract: The oxygen reduction reaction (ORR) is of great importance for clean energy storage and conversion techniques such as fuel cells and metal–air batteries (MABs). However, the ORR is kinetically sluggish, and expensive noble metal catalysts are required. The high price and limited preservation of noble metal catalysts has largely hindered the wide application of clean power sources such as fuel cells and MABs. Therefore, it is important to prepare non-expensive metal catalysts (NPMC) to cut the price of the fuel cells and MABs for wide application. Here, we report the preparation of a Co_3O_4 carried on the N-doped carbon (Co/N-C) as the ORR NPMC with a facile Pharaoh's Snakes reaction. The gas generated during the reaction is able to fabricate the porous structure of the resultant carbon doped with heteroatoms such as Co and N. The catalyst provides a high electrocatalytic activity towards ORR via the 4-e pathway with an onset and half-wave potential of 0.98 and 0.79 V (vs. RHE), respectively, in an electrolyte of 0.1 M KOH. The onset and half-wave potentials are close to those of the commercial Pt/C. This work demonstrates the promising potential of an ancient technology for preparing NPMCs toward the ORR.

Keywords: oxygen reduction reaction; non-precious metal catalyst; Pharaoh's snakes; electrocatalysts; Li-air battery

1. Introduction

The rapid depletion of fossil fuels and severe ecological deterioration have made it urgent to develop sustainable and clean energy sources to address the issues regarding energy and environmental [1]. Due to their environmental friendliness and high energy density, fuel cells and metal–air batteries (MABs, such as the Zn-Air and Li-Air batteries) are considered to be excellent candidates for various applications such as stationary power generation, mobile power source, and transportation in the future [2,3]. In fuel cells and MABs, the chemical energy in the fuels is transformed directly into electric power free of the Carnot recycle. Therefore, the fuel cells and MABs possess efficiencies higher than those of tradition technologies that generate electric power based on combustion.

Despite these encouraging advantages, fuel cells and MABs still face several challenges. One of the challenges is their high price because of their dependence on noble metal catalysts. In both fuel cells and MABs, the O_2 or the O_2 in the air is reduced via the oxygen reduction reaction (ORR) on the cathode during discharge [4–6]. According to the ORR mechanism model proposed by Wroblowa et al. in 1976 [7], the ORR takes place by means of a complicated process, resulting in the sluggishness of the ORR. The sluggish ORR necessitates the use of the corresponding catalysts [8]. The state-of-the-art catalysts for the

ORR are mostly platinum-based materials. However, the high cost of platinum has thus far prohibited the wide use of the catalyst and, consequently, the corresponding energy conversion and storage devices. Apart from the high price, the limited preserve of Pt is also a significnat challenge that has also severely impeded the wide application of Pt catalysts and new energy systems such as fuel cells and MABs. One approach to solving this problem is to reduce the usage of Pt catalysts by improving the Pt use efficiency in the catalyst layers. The use efficiency can be improved by alloying Pt with non-precious metals such as Fe, Co, Ni, etc., or by supporting the catalyst on carrier materials. However, the reduction in Pt loading has been totally offset by increasing Pt prices, rendering efforts devoted to reducing Pt loading in the last two decades almost totally ineffective [9]. Therefore, Pt loading reduction seems not to be a long-term solution for reducing the cost of catalysts for fuel cells or MABs. Another promising approach is to develop non-precious-metal electrocatalysts (NPMCs) from non-Pt metals (non-PGM) [9,10]. Since no precious metals are needed in the catalyst, this approach is an ideal strategy in the long run, making possible the wide application of clean and efficient storage and conversion devices such as the fue cell and MABs [9,11].

In recent years, transition metals carried on heteroatom-doped carbon (M/N-C) materials have been widely investigated as NPMCs for ORR [9]. In this field, great efforts have been devoted to developing methods for preparing the NPMCs [9,12]. Generally, materials rich in porous structures are more likely to provide high catalytic activity. A tremendous amount of work has been conducted on the preparation of porous catalysts for ORR. Some examples of this are as follows: aerogels were heated to prepare porous catalysts for ORR with an onset potential of 0.92 V vs. RHE (V_{RHE}) [13]. Metal–organic frameworks (MOFs) were investigate as self-sacrificing precursors or templates for the preparation of nano-porous carbons [14]. Hard templates and NH_3 were used to prepare porous ORR catalysts [15], and an onset potential of ~1 V_{RHE} was obtained. While trying to prepare efficient porous catalysts, some scientists have been inspired by ancient knowledge, and encouraging results have been achieved [12,16,17]. The "Pharaoh's snakes" reaction is a famous reaction for generating foam materials. In this reaction, great amounts of gases are generated, resulting in the fabrication of a porous, foam-like product. Therefore, this "Pharaoh's snakes" reaction holds great promise for providing an ORR catalyst with high activity. With sugar, melamine and iron nitrate as precursors, Fe/N-C has been prepared using the "Pharaoh's snakes" reaction that exhibited high performance [12]. However, the Fe ions dissolved from Fe/N-C were detrimental to battery durability. The Fe ions catalyze radical formation from H_2O_2 via Fenton reactions [8]. These radicals degrade battery components such as organic ionomers, membranes and the Fe/N-C catalyst itself [18]. Therefore, it is necessary to replace Fe with metals that do not catalyze Fenton reactions [19], and Co is such a promising alternative choice [20]. However, the preparation of Co/N-C catalysts using the "Pharaoh's snakes" reaction has rarely been performed.

In this study, porous Co/N-C was prepared using a slightly adjusted "Pharaoh's snakes" method, and was investigated as a catalyst for ORR in alkaline electrolyte. The master catalyst sample shows an onset and a half-wave potential of 0.98 and 0.79 V_{RHE}, respectively. These results demonstrate an interesting application of an ancient reaction to solve a modern problem.

2. Experimental Section
2.1. Preparation of Electrocatalysts

As displayed in Figure 1, sugar powder (3 g), NH_4HCO_3 (0.8 g), melamine (2.4 g), $Co(NO_3)_2 \cdot 6H_2O$ (0.1 g) were mixed with purified water to prepare a solution, which was then dried, ground in a mortar, and burned to obtain a snake-like foam ash. The ash was then pyrolyzed in a tubular furnace with a nitrogen stream of 60 mL min^{-1} at 900 °C for 1 h with a rate of temperature increase of 5 °C min^{-1}. A black carbon material (Co/N-C-900) was obtained when the furnace was cooled to room temperature. The temperature–time curve is presented in Figure S1 in the Supplementary Materials (ESI). The Co/N-C was also prepared with pyrolyzing temperatures of 800 (Co/N-C-800) and 1000 °C (Co/N-C-1000) to investigate the influence of the pyrolyzing temperature on catalytic activity.

Figure 1. A schematic illustration of the preparation of the Co/N-C using the "Pharaoh's snakes" reaction.

2.2. Characterization

The catalyst morphologies were examined using a scanning electron microscope (FESEM, JEOL JSM-7500F) and a field emission transmission electron microscope (TEM, JEOL JEM-2100F). X-ray diffraction (XRD) patterns were collected with Cu Kα radiation (λ = 1.5406 Å) (Panalytical, Empyrean). A confocal micro-Raman spectrometer (Jobin Yvon Laboratory RAM HR1800) was used to determine the Raman spectra through a 10× microscope objective with backscattered geometry. An Ar+ laser was used as an excitation source, emitting at a wavelength of 633 nm.

The specific surface area and pore size were measured using nitrogen adsorption/desorption isotherms (Micromeritics ASAP 2020 V3.00 H) by means of the Brunauer–Emmett–Teller (BET) and Barrett–Joyner–Halenda (BJH) methods, respectively. To investigate the surface composition of the catalyst, X-ray photoelectron spectroscopy (XPS) characterization was performed on a VG ESCALAB 220i-XL instrument with a monochromatic AlKα X-ray source.

2.3. Electrochemical Measurements

The catalytic activity was investigated by means of electrochemical methods conducted on an electrochemical work station (CHI 760E, Chenhua, Shanghai) in a three-electrode system. To prepare the working electrode: the catalyst (2.5 mg) was dispersed in ethanol (450 μL) and Nafion solution (50 μL) and treated ultrasonically for 30 min to obtain a homogeneous catalyst ink. Subsequently, the catalyst ink (10 μL) was added onto the GC electrode, yielding a working electrode with a catalyst loading of 0.25 mg cm^{-2}. For comparison, a working electrode with Pt/C (20 Wt.% Pt) was also prepared and tested. A platinum wire and Hg/HgO were used as the counter and reference electrodes, respectively. The potentials were calibrated to a reversible hydrogen electrode (V_{RHE}). Cyclic voltammetry (CV) and linear sweep voltammetry (LSV) techniques were applied for the catalytic performance investigation in the electrolyte of 0.1 M KOH saturated with N_2 or O_2. The CVs obtained in the N_2 saturated electrolyte at various scanning rate were used to evaluate the electrochemical active surface area (ECSA). The rotating ring disk electrode (RRDE) tests were conducted at a rate of 10 mV s^{-1} with a ring potential of 1.5 V_{RHE}. The catalyst stability was studied with a chronoamperometric current–time (i-t) curve at −0.2 V (vs. Hg/HgO) in the O_2-saturated electrolyte with a rotation speed of 400 rpm.

Electrochemical impedance spectroscopy (EIS) was measured in the frequency range of 0.01 to 1,000,000 Hz at 0.7 V_{RHE} with an amplitude of 0.005 V.

3. Results and Discussion

Figure 1 schematically shows the preparation process. The NH_4HCO_3 decomposes at ~60 °C, generating a great mount of gases such as NH_3 and CO_2. The gases result in the fabrication of pores in solid precursor. As the temperature increases, the sugar will melt to form a viscous sol, wrapping up the nitrate. As the temperature increases, the cobalt nitrate wrapped in the sol will decompose, and more porous structures will be fabricated by the gases generated by the decomposition of the nitrate, resulting in a foam-like porous carbon being obtained. At the same time, N from the melamine can be doped into the carbon. As a result, Co-carrying N-doped carbon is obtained.

The SEM image (Figure 2a) shows that the Co/N-C-900 is randomly scattered as irregular particles with various sizes. Figure 2b shows the SEM image at higher magnification. Channels connecting the particles can be observed. Figure 2c is the SEM image at an even higher magnification, and pores can be observed in some particles. To demonstrate the pores more clearly, an enlarged image of the circled area in Figure 2c is presented in Figure 2d.

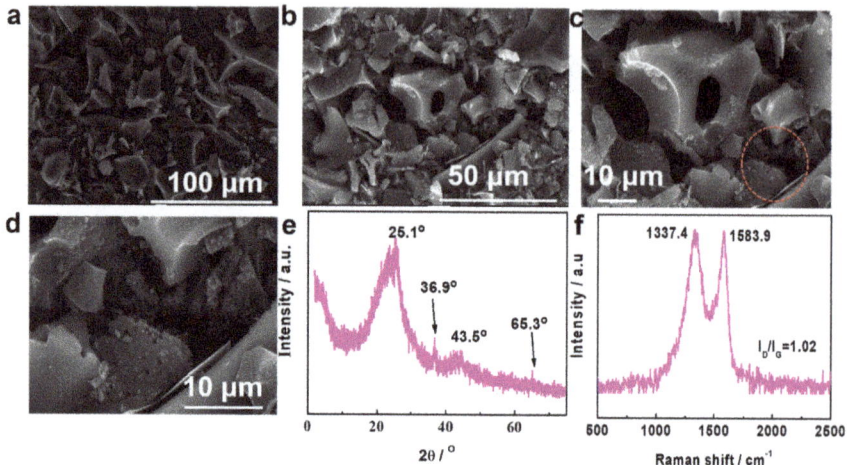

Figure 2. (a–d) SEM images at various magnifications for Co/N-C-900 (d presents enlarged image of the part in the red circle of c), (e) XRD pattern, (f) Raman spectra.

The crystallography of the Co/N-C-900 was studied using XRD (Figure 2e). The peaks at around 25.1 and 43.5° are related to the graphite (002, ICDD#41-1487) and (100) planes, respectively [21]. The two peaks indicate the formation of the graphite phase. The broad XRD peaks imply that crystallization is relatively low [22], and this is supported by the HRTEM images (Figure 3b). The peaks at 36.9 and 65.3° (ICDD # 00-042-1467) can be attributed to Co_3O_4 [23], implying that Co has been doped into the carbon matrix. Figure 2f shows the Raman spectra for Co/N-C-900. Two peaks appear at 1337.4 and 1583.9 cm^{-1} for the D and G bands, respectively. The peak for the D band is related to the disordered structure, implying that the graphite was successful dopped with heteroatoms [24]. The G band peak is related to the sp^2 hybridized carbon atoms [25]. The I_D/I_G for Co/N-C-900 is 1.02. This value is close to or higher than those of many other catalysts [26]. The high I_D/I_G suggests the richness of defective domains. This demonstrates that the Pharaoh's Snakes reaction is effective for fabricating the defective domains of the carbon material, and the defective domains are beneficial to catalytic performance [27].

The microstructure of the Co/N-C-900 was investigated further using TEM (Figure 3). Figure 3a confirms the structure of the channel-connected particles of the Co/N-C-900, which is in good agreement with the morphology, as shown in the SEM images (Figure 2). The channels and pores of the catalyst facilitate the transport of the reacting species to the active sites, which is helpful for improving catalytic peformance. As shown in the HRTEM image (Figure 3b), a crystal lattice of 0.21 nm can be observed for graphite [28]. However, the crystal lattice is not continuous, implying that the carbon is amorphous. This is supported by the XRD pattern (Figure 2e), which has broad peaks. A crystal lattice spacing of 0.46 nm can also be observed. This lattice spacing is attributed to the Co_3O_4 (111) plane [29], implying the existence of the Co_3O_4. The elemental distribution of Co/N-C-900 was analyzed using energy dispersive spectroscopy (EDS) with scanning transmission electron microscopy (STEM). As shown in the STEM images in Figure 3c–f, the nitrogen (green) is distributed uniformly on the carbon (red), suggesting the efficient doping of the N into the carbon matrix. Figure 2f shows that there are Co species carried on the carbon. However, the Co distribution is not uniform. During the preparation, sugar powder, NH_4HCO_3 and $Co(NO_3)_2 \cdot 6H_2O$ can dissolved in the water. Therefore, the $Co(NO_3)_2 \cdot 6H_2O$ can be distributed uniformly on the carbon source of the sugar. However, the melamine cannot be dissolved in water. As a result, the $Co(NO_3)_2 \cdot 6H_2O$ cannot be distributed uniformly on the carbon and nitrogen source of the melamine. Therefore, the Co is not uniformly distributed on the carbon from the melamine. A longer grounding time for the solid precursor can improve the distribution, as shown in Figure S2.

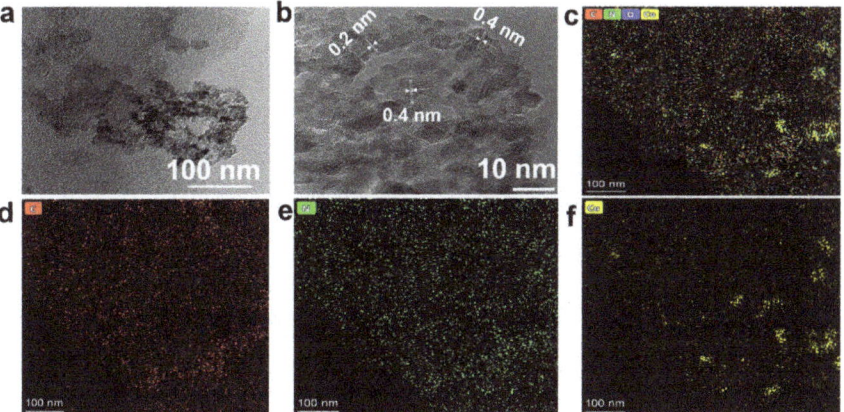

Figure 3. (**a**) TEM image of Co/N-C-900, (**b**) HRTEM image of Co/N-C-900, (**c**) EDS merged mapping image of C, N and Co, (**d**–**f**) EDS mapping images for C, N and Co, respectively.

The surface area as well as the pore width of the Co/N-C-900 were studied using the N_2 adsorption/desorption isotherm. A type IV isotherm is displayed in Figure 4a, suggesting a mesoporous structure, and this is backed by the distribution curve of the pore size (Figure 4b), demonstrating that the pore size is mainly 3.8 nm. The Brunauer–Emmett–Teller (BET) surface area of the Co/N-C-900 is 457.2 $m^2\,g^{-1}$. The carbonized sugar surface area was increased as a result of the gas generated via the decomposition of the nitrate and the bicarbonate in the mixture. X-ray photoelectron spectroscopy (XPS, Figure 4c–f) was applied to study the surface composition of Co/N-C-900. XPS detected the carbon, nitrogen and cobalt content of 81.75, 5.58 and 1.36 at %, respectively, on the surface of Co/N-C-900. The C1s high-resolution spectrum (Figure 4d) can be deconvoluted into carboxyl (289.8 eV) [30], C-N=C (286.0 eV), C-N (284.9 eV), and C=C (284.3 eV) [31–33], implying that the N has been successfully doped into the carbon matrix. The N is more electrically negative than carbon, resulting in a re-distribution of the charge of the neighboring carbon, which changes the O_2 chemisorption mode, weakening the O-O bonding and facilitating the ORR process [34]. Nitrogen doping also introduces spin density asymmetry, making it possible for the N-doped carbon to show electroncatalytic activities toward the ORR [35].

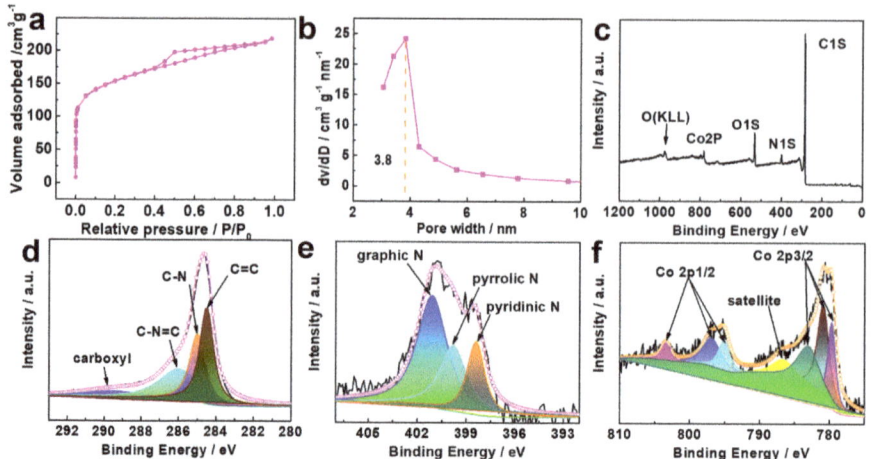

Figure 4. (a) Nitrogen sorption/desorption isotherm of the Co/N-C-900, (b) the corresponding pore width distribution curve, (c) XPS spectra survey of Co/N-C-900, (d) C1s XPS spectra, (e) N1s XPS spectra, (f) Co2p XPS spectra.

Figure 4e shows the high-resolution N1s spectra of Co/N-C-900, which can be deconvoluted into three peaks of pyridinic N (398.8 eV) [36], pyrrolic N (399.8 eV), and graphitic N(401.2 eV) [37]. Both the pyridinic N (20.1 at %) and the pyrrolic N (27.8 at %) have been considered to contribute ORR active sites [38,39]. Graphitic nitrogen is believed to be able to facilitate the electron transfer to the antibonding orbitals of oxygen from the carbon electronic bands [40,41]. The high-resolution Co spectrum (Figure 4f) can be resolved into Co $2p_{3/2}$ and Co $2p_{1/2}$ of $Co^{2+/3+}$. The peaks at 780.94, 795.04 [42], 796.9 [43], and 805.5 eV [44] can be attributed to Co^{2+}. The peaks at 779.74 [45], 780.94, and 786.73 eV can be attributed to $Co2p_{3/2}$ and the related satellite peaks [46]. The peak at 781.8 eV can be attributed to Co-N [47]. According to previous investigations [1], the M-N_x (M: metal) and N-C moieties are both supposed to be active sites for ORR.

The catalytic performance was characterized using electrochemical methods. The CV tests were conducted first to evaluate the electrochemical activity. In the N_2-saturated electrolyte, a featureless CV curve (Figure 5a) without peaks can be observed. However, a reduction peak can be observed in the O_2-saturated electrolyte. These results suggest a facile ORR process on the Co/N-C-900. Subsequently, the catalytic activity on Co/N-C-900 toward ORR was characterized using rotating disk electrode (RDE) technology, and the LSVs are displayed in Figure 5b. The LSVs show an onset potential of 0.98 V_{RHE}. This value is close to or higher than many reported values, such as (~0.74 V_{RHE}) [48], as well as with respect to those reported for biomass such as typha orientalis (~0.87 V_{RHE}) [49], eichhornia crassipes (~0.98 V_{RHE}) [50], hair (~0.96 V_{RHE}) [51], polyaniline with phytic acid (~0.94 V_{RHE}) [26], and coconut shells (~0.87 V_{RHE}) [52]. The half-wave potential of the Co/N-C-900 is 0.79 V_{RHE}. The catalytic performance of commercial Pt/C was also investigated, and the LSV curve is presented in Figure 5b for comparison. The Co/N-C-900 demonstrated a catalytic activity close to that of the commercial Pt/C (20. Wt%), the onset and half-wave potentials of which were 0.98 and 0.81 V_{RHE}, respectively. The limit current density of the Co/N-C-900 is 5.73 mA cm^{-2}, which is almost same as that of the commercial Pt/C (20. Wt%), of 5.74 mA cm^{-2}. Therefore, the "Pharaoh's snakes" reaction holds great promise as a method for the preparation of Co, N co-doped NPMC toward the ORR.

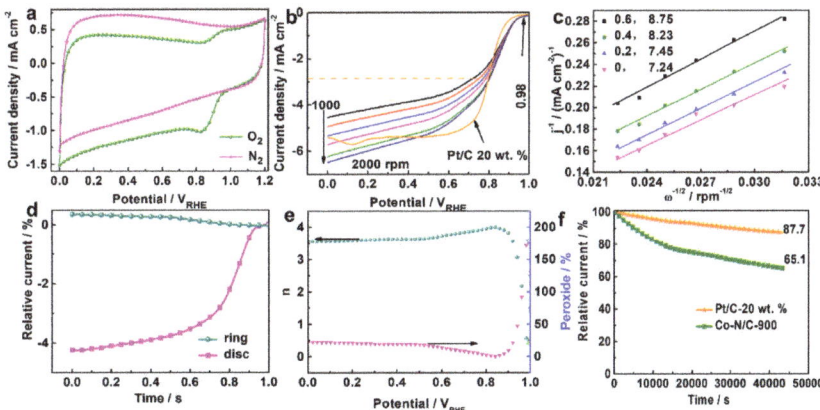

Figure 5. (a) CV on Co/N-C-900 in N$_2$, O$_2$-saturated electrolyte of 0.1M KOH solution, (b) LSV for Co/N-C-900 and Pt/C at 1600 rpm, (c) K-L lines. (d) LSV of the RRDE measurements (1600 rpm) of Co/N-C-900 (e) n and H$_2$O$_2$% at different potentials, (f) current–time curve 0.755 V$_{RHE}$ of Co/N-C-900 for ORR.

An increase in the current density, as shown in Figure 5b, can be observed with increasing rotational rate, resulting in higher oxygen flux to the electrode. The K-L lines (Koutecky–Levich lines, Figure 5c) were plotted as the inverse current density (j^{-1}) vs. the inverse of the rotation speed square root ($\omega^{1/2}$) at various electrode potentials. Based on the K-L lines, the apparent number of electrons transferred (n) was investigated with the help of the K-L line slope at various potentials. The n was then calculated using Equations (1) and (2), in which j_k and j are the kinetic and measured current density, respectively, B is the reciprocal K-L line slope (found in the range of 7.24~8.75), as shown in Figure 5c, F is the Faraday constant (96,485 C mol^{-1}), C$_0$ is the bulk O$_2$ concentration (1.1 × 10^{-3} mol L^{-3}), D$_0$ is the O$_2$ diffusion coefficient in 0.1 M KOH (1.9 × 10^{-5} cm^2 s^{-1}), and γ is the electrolyte kinetic viscosity (0.01 cm^2 s^{-1}). The constant adopted when rpm is used to express the rotational rate was taken as 0.2 [53].

$$\frac{1}{j} = \frac{1}{j_k} + \frac{1}{B\omega^{1/2}} \tag{1}$$

$$B = 0.2nFC_O D_O^{2/3} \gamma^{-1/6} \tag{2}$$

The n varies slightly from 3.24 to 3.88, falling within the range 3.2~4.1 for the four-electron ORR pathway [54,55]. Therefore, the ORR on the Co/N-C-900 is largely a four-electron process. The linearity of the K-L line and the n value indicate the rapid kinetics with a predominant four-electron pathway in the investigated potential range. The four-electron ORR pathway is preferred in fuel cells and MABs for its high electron efficiency and low H$_2$O$_2$ generation.

RRDE technology as applied to further evaluate the pathways for the ORR on the Co/N-C-900. As shown in Figure 5d, a low ring current density (~0.25 mA cm^{-2}) for peroxide oxidation and high disk current density (~4.2 mA cm^{-2}) for O$_2$ reduction can be observed, implying a small amount of H$_2$O$_2$ generation and high catalytic performance toward ORR [56]. The n and HO$_2^-$% were calculated according to Equations (3) and (4), respectively [57].

$$n = 4 \frac{i_d}{i_d + i_{r/N}} \tag{3}$$

$$HO_2^- \% = 200 \frac{\frac{i_r}{N}}{i_d + i_{r/N}} \tag{4}$$

where i_d and i_r are the ring and disk current, respectively. N is the current collection efficiency of the Pt ring, and was determined to be 0.39. The n and HO_2^-% at various potentials are shown in Figure 5e. The average n is higher than 3.5 in the potential range of 0~0.9 V_{RHE}, indicating the mainly four-electron pathway ORR pathway, which is in good agreement with the results from the RDE. The HO_2^-% is lower than 25% in the range of 0 to 0.9 V_{RHE}. The HO_2^- generation is close to the values reported for other materials, such as 24.6% [12] and ~20% [56]. Therefore, Co/N-C-900 is a promising ORR NPMC for fuel cells and MABs.

The ORR stability of Co/N-C-900 was investigated via chronoamperometry at 0.755 V_{RHE} in an O_2-saturated electrolyte of 0.1 M KOH solution. The current–time curve for Co/N-C-900 in Figure 5g remains at around 90% and at around 65% of the initial current value after stability tests of 3600 s and 432,000 s, respectively. The stability is still lower than that of the commercial Pt/C, as shown in Figure 5f, which shows a stability of 87.7% after the 42,000 s stability test. The Tafel plots (Figure S3 in the ESI) for Co/N-C-900 and Pt/C were investigated in the corresponding potential range between the onset and half-wave potentials. The Tafel slope of Co/N-C-900 was 136.9 mV dec^{-1}, which is higher than that obtained for Pt/C. The higher Tafel slope implies a faster overpotential increase with current density [58]. The stability and Tafel mean that the Co/N-C-900 possesses lower catalytic activity than the commercial Pt/C. However, the lower price of Co/N-C-900 still makes it a promising candidate as an ORR catalyst.

The effect of the pyrolyzing temperature on the catalytic activity was investigated. Figure 6a shows the LSV for the ORR occurring on the catalysts obtained using pyrolyzing temperatures of 800, 900 and 1000 °C. The catalyst prepared with a temperature of 900 °C showed the highest catalytic performance. The results imply that 900 °C is the optimal pyrolyzing temperature. Usually, a high pyrolyzing temperature is favorable for the carbonization of the catalyst [59] and helpful for improving the conductivity and electrocatalytic performance. However, high temperatures can cause a decrease in the number of heteroatoms [59], resulting in a reduction in catalytic performance. Therefore, there is usually an optimal pyrolyzing temperature for catalysts prepared via the pyrolization process, similar to what has been reported in other work [60].

CV tests were carried out at different scanning rates to measure the ECSA. The CV curves for the Co/N-C-800, 900 and 1000 are displayed in Figure S4 in the ESI. The lines in Figure 6b were determined by plotting the CV curve for the closed area vs. the scanning rate. The ECSA can be evaluated using line slopes proportional to the corresponding ECSAs [61]. The slope for Co/N-C-900 was higher than those of Co/N-C-800 and Co/N-C-1000. Therefore, the Co/N-C-900 possessed the highest ECSA. To further investigate the mechanism for the differences in catalytic performance among the Co/N-C-800, 900 and 1000, EIS measurements were conducted for the three samples, and the EIS plots are presented in Figure 6c. An enlarged figure is presented in Figure 6d showing the EIS plots for high frequencies more clearly. The EIS plots in Figure 6c,d indicate that the conductivity of Co/N-C-900 was higher than those of Co/N-C-900 and Co/N-C-1000. The high values of ECSA and conductivity may also be reasons explaining the high catalytic activity of Co/N-C-900.

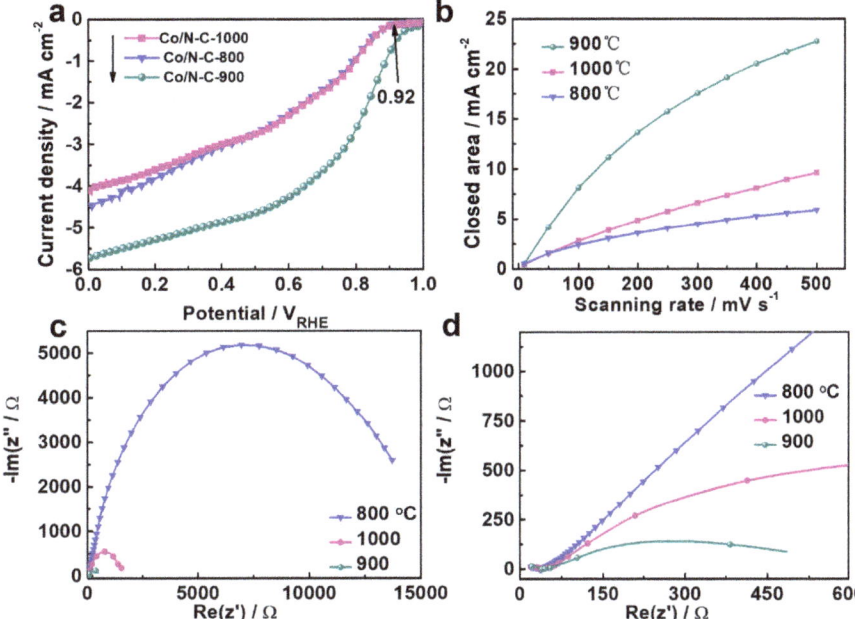

Figure 6. (**a**) LSV for the Co/N-C prepared at various temperatures, (**b**) CV closed area vs. scanning rate, (**c**) EIS, (**d**) enlarged EIS of the high-frequency part.

4. Conclusions

The ancient chemical magic "Pharaoh's snakes" reaction was used to prepare Co_3O_4 carrying N-doped porous carbon (Co/N-C) as a non-precious metal catalyst for ORR, an important electrode reaction in fuel cells and metal–air batteries. The "Pharaoh's snakes" reaction possesses the advantages of being low cost, facile, and fast in the preparation of Co/N-C catalyst for ORR. The obtained catalyst showed considerable activity towards ORR, with an onset potential of 0.98 V_{RHE} and a half-wave potential of 0.79 V_{RHE} in 0.1 M KOH solution as electrolyte. This investigation provides a promising alternative method for the preparation of Co, N co-doped non-precious-metal catalyst for ORR, an important electrode reaction for fuel cells and metal–air batteries such as Zn–air and Li–air batteries.

Supplementary Materials: The following supporting information can be downloaded at: https://www.mdpi.com/article/10.3390/batteries8100150/s1, Figure S1:Temperature-time curve; Figure S2: SEM mapping, (a) SEM image, (b) C, (c) N, (d) Co; Figure S3: Tafel plots of Co/N-C-900 and Pt/C 20 wt. %; Figure S4: CV curves at various scanning rates for Co/N-C-800(a), Co/N-C -900(b) and Co/N-C -1000 (c).

Author Contributions: Conceptualization, J.G. and Z.Y.; methodology, H.W.; validation, X.T.; formal analysis, Y.L.; investigation, M.Z., X.W. All authors have read and agreed to the published version of the manuscript.

Funding: This research received no external funding.

Data Availability Statement: Not applicable.

Acknowledgments: This work was supported by National Natural Science Foundation of China (No. 21776219, 21872104) and Tianjin Key Laboratory of Green Chemical Engineering Process Engineering, Tiangong University, GCEPE20190106. Jian Gao appreciate the support of from China Postdoctoral Science Foundation (No. 2018M631746) and Basic Research Program of Jiangsu (SBK20201213).

Conflicts of Interest: The authors declare no conflict of interest.

References

1. Wang, D.; Pan, X.; Yang, P.; Li, R.; Xu, H.; Li, Y.; Meng, F.; Zhang, J.; An, M. Transition metal and nitrogen Co-doped carbon-based electrocatalysts for the oxygen reduction reaction: From active site insights to the rational design of precursors and structures. *ChemSusChem* **2021**, *14*, 33–55. [CrossRef] [PubMed]
2. Li, G.; Wang, X.L.; Fu, J.; Li, J.D.; Park, M.G.; Zhang, Y.N.; Lui, G.; Chen, Z.W. Pomegranate-inspired design of highly active and durable bifunctional electrocatalysts for rechargeable metal-air batteries. *Angew. Chem. Int. Ed.* **2016**, *55*, 4977–4982. [CrossRef] [PubMed]
3. Liu, Z.; Zhao, Z.; Peng, B.; Duan, X.; Huang, Y. Beyond extended surfaces: Understanding the oxygen reduction reaction on nanocatalysts. *J. Am. Chem. Soc.* **2020**, *142*, 17812–17827. [CrossRef] [PubMed]
4. Li, J.; Hou, L.; Yu, M.; Li, Q.; Zhang, T.; Sun, H. Review and recent advances of oxygen transfer in Li-air batteries. *ChemElectroChem* **2021**, *8*, 3588–3603. [CrossRef]
5. Liu, B.; Sun, Y.; Liu, L.; Xu, S.; Yan, X. Advances in manganese-mased oxides cathodic electrocatalysts for Li–Air batteries. *Adv. Funct. Mater.* **2018**, *28*, 1704973. [CrossRef]
6. Xiao, Y.; Wang, J.; Wang, Y.; Zhang, W. A new promising catalytic activity on blue phosphorene nitrogen-doped nanosheets for the ORR as cathode in nonaqueous Li–air batteries. *Appl. Surf. Sci.* **2019**, *488*, 620–628. [CrossRef]
7. Wroblowa, H.S.; Razumney, G. Electroreduction of oxygen: A new mechanistic criterion. *J. Electroanal. Chem. Interfacial Electrochem.* **1976**, *69*, 195–201. [CrossRef]
8. Luo, E.; Chu, Y.; Liu, J.; Shi, Z.; Zhu, S.; Gong, L.; Ge, J.; Choi, C.H.; Liu, C.; Xing, W. Pyrolyzed M–N x catalysts for oxygen reduction reaction: Progress and prospects. *Energy Environ. Sci.* **2021**, *14*, 2158–2185. [CrossRef]
9. Chen, Z.W.; Higgins, D.; Yu, A.P.; Zhang, L.; Zhang, J.J. A review on non-precious metal electrocatalysts for PEM fuel cells. *Energy Environ. Sci.* **2011**, *4*, 3167–3192. [CrossRef]
10. Sun, Z.; Yuan, M.; Lin, L.; Yang, H.; Nan, C.; Sun, G.; Li, H.; Yang, X. Perovskite $La_{0.5}Sr_{0.5}CoO_{3-\delta}$ grown on Ti_3C_2T x MXene nanosheets as bifunctional efficient hybrid catalysts for Li–Oxygen batteries. *ACS Appl. Energy Mater.* **2019**, *2*, 4144–4150. [CrossRef]
11. Clark, S.; Latz, A.; Horstmann, B. A review of model-based design tools for metal-air batteries. *Batteries* **2018**, *4*, 5. [CrossRef]
12. Ren, G.Y.; Gao, L.L.; Teng, C.; Li, Y.A.; Yang, H.Q.; Shui, J.L.; Lu, X.Y.; Zhu, Y.; Dai, L.M. Ancient chemistry "Pharaoh's Snakes" for efficient Fe-/N-doped carbon electrocatalysts. *ACS Appl. Mater. Interfaces* **2018**, *10*, 10778–10785. [CrossRef] [PubMed]
13. Zion, N.; Cullen, D.A.; Zelenay, P.; Elbaz, L. Heat-treated aerogel as a catalyst for the oxygen reduction reaction. *Angew. Chem. Int. Ed.* **2020**, *132*, 2504–2510. [CrossRef]
14. Yang, L.; Zeng, X.; Wang, W.; Cao, D. Recent progress in MOF-derived, heteroatom-doped porous carbons as highly efficient electrocatalysts for oxygen reduction reaction in fuel cells. *Adv. Funct. Mater.* **2018**, *28*, 1704537. [CrossRef]
15. Liang, H.-W.; Zhuang, X.; Brüller, S.; Feng, X.; Müllen, K. Hierarchically porous carbons with optimized nitrogen doping as highly active electrocatalysts for oxygen reduction. *Nat. Commun.* **2014**, *5*, 4973. [CrossRef]
16. Ingo, G.M.; Guida, G.; Angelini, E.; Di Carlo, G.; Mezzi, A.; Padeletti, G. Ancient mercury-based plating methods: Combined use of surface analytical techniques for the study of manufacturing process and degradation phenomena. *Acc. Chem. Res.* **2013**, *46*, 2365–2375. [CrossRef]
17. Buelens, L.C.; Galvita, V.V.; Poelman, H.; Detavernier, C.; Marin, G.B. Super-dry reforming of methane intensifies CO_2 utilization via Le Chatelier's principle. *Science* **2016**, *354*, 449–452. [CrossRef]
18. Xie, X.H.; He, C.; Li, B.Y.; He, Y.H.; Cullen, D.A.; Wegener, E.C.; Kropf, A.J.; Martinez, U.; Cheng, Y.W.; Engelhard, M.H.; et al. Performance enhancement and degradation mechanism identification of a single-atom Co-N-C catalyst for proton exchange membrane fuel cells. *Nat. Catal.* **2020**, *3*, 1044–1054. [CrossRef]
19. Wang, X.X.; Prabhakaran, V.; He, Y.; Shao, Y.; Wu, G. Iron-free cathode catalysts for proton-exchange-membrane fuel cells: Cobalt catalysts and the peroxide mitigation approach. *Adv. Mater.* **2019**, *31*, 1805126. [CrossRef]
20. Wang, J.; Gao, R.; Zhou, D.; Chen, Z.; Wu, Z.; Schumacher, G.; Hu, Z.; Liu, X. Boosting the electrocatalytic activity of Co_3O_4 nanosheets for a Li-O_2 battery through modulating inner oxygen vacancy and exterior Co^{3+}/Co^{2+} ratio. *ACS Catal.* **2017**, *7*, 6533–6541. [CrossRef]
21. Gao, J.; Wang, Y.; Wu, H.; Liu, X.; Wang, L.; Yu, Q.; Li, A.; Wang, H.; Song, C.; Gao, Z.; et al. Construction of a sp^3/sp^2 carbon interface in 3D N-doped nanocarbons for the oxygen reduction reaction. *Angew. Chem. Int. Ed.* **2019**, *58*, 15089–15097. [CrossRef]
22. Chen, J.; Wang, X.; Cui, X.; Yang, G.; Zheng, W. Amorphous carbon enriched with pyridinic nitrogen as an efficient metal-free electrocatalyst for oxygen reduction reaction. *Chem. Commun.* **2014**, *50*, 557–559. [CrossRef] [PubMed]
23. Gao, J.; Ma, N.; Zheng, Y.; Zhang, J.; Gui, J.; Guo, C.; An, H.; Tan, X.; Yin, Z.; Ma, D. Cobalt/nitrogen-doped porous carbon nanosheets derived from polymerizable ionic liquids as bifunctional electrocatalyst for oxygen evolution and oxygen reduction reaction. *ChemCatChem* **2017**, *9*, 1601–1609. [CrossRef]
24. Wei, D.C.; Liu, Y.Q.; Wang, Y.; Zhang, H.L.; Huang, L.P.; Yu, G. Synthesis of N-doped graphene by chemical vapor deposition and its electrical properties. *Nano Lett.* **2009**, *9*, 1752–1758. [CrossRef] [PubMed]
25. Wang, J.; Gao, D.; Wang, G.; Miao, S.; Wu, H.; Li, J.; Bao, X. Cobalt nanoparticles encapsulated in nitrogen-doped carbon as a bifunctional catalyst for water electrolysis. *J. Mater. Chem. A* **2014**, *2*, 20067–20074. [CrossRef]
26. Zhang, J.; Zhao, Z.; Xia, Z.; Dai, L. A metal-free bifunctional electrocatalyst for oxygen reduction and oxygen evolution reactions. *Nat. Nanotechnol.* **2015**, *10*, 444–452. [CrossRef] [PubMed]

27. Yan, X.; Jia, Y.; Odedairo, T.; Zhao, X.; Jin, Z.; Zhu, Z.; Yao, X. Activated carbon becomes active for oxygen reduction and hydrogen evolution reactions. *Chem. Commun.* **2016**, *52*, 8156–8159. [CrossRef] [PubMed]
28. Li, M.; Hu, C.; Yu, C.; Wang, S.; Zhang, P.; Qiu, J. Organic amine-grafted carbon quantum dots with tailored surface and enhanced photoluminescence properties. *Carbon* **2015**, *91*, 291–297. [CrossRef]
29. Hemamalini, S.; Manimekalai, R. Synthesis, structural, magnetic, textural, optical investigation and photocatalytic performance of undoped and doped cobaltite nanoparticles. *J. Coord. Chem* **2020**, *73*, 3431–3451. [CrossRef]
30. Osbeck, S.; Bradley, R.; Liu, C.; Idriss, H.; Ward, S. Effect of an ultraviolet/ozone treatment on the surface texture and functional groups on polyacrylonitrile carbon fibres. *Carbon* **2011**, *49*, 4322–4330. [CrossRef]
31. Chao, H.; Chang, Y.; Mingyu, L.; Xiuna, W.; Qiang, D.; Gang, W.; Jieshan, Q. Nitrogen-doped carbon dots decorated on graphene: A novel all-carbon hybrid electrocatalyst for enhanced oxygen reduction reaction. *Chem. Commun.* **2015**, *51*, 3419–3422.
32. Tahir, M.; Mahmood, N.; Zhang, X.X.; Mahmood, T.; Butt, F.K.; Aslam, I.; Tanveer, M.; Idrees, F.; Khalid, S.; Shakir, I.; et al. Bifunctional catalysts of Co_3O_4@GCN tubular nanostructured (TNS) hybrids for oxygen and hydrogen evolution reactions. *Nano Res.* **2015**, *8*, 3725–3736. [CrossRef]
33. Zhou, G.-W.; Wang, J.; Gao, P.; Yang, X.; He, Y.-S.; Liao, X.-Z.; Yang, J.; Ma, Z.-F. Facile spray drying route for the three-dimensional graphene-encapsulated Fe_2O_3 nanoparticles for lithium ion battery anodes. *Ind. Eng. Chem. Res.* **2013**, *52*, 1197–1204. [CrossRef]
34. Gong, K.; Du, F.; Xia, Z.; Durstock, M.; Dai, L. Nitrogen-doped carbon nanotube arrays with high electrocatalytic activity for oxygen reduction. *Science* **2009**, *323*, 760–764. [CrossRef] [PubMed]
35. Zhang, L.; Xia, Z. Mechanisms of Oxygen Reduction Reaction on Nitrogen-Doped Graphene for Fuel Cells. *J. Phys. Chem. C* **2011**, *115*, 11170–11176. [CrossRef]
36. Kurak, K.A.; Anderson, A.B. Nitrogen-treated graphite and oxygen electroreduction on pyridinic edge sites. *J. Phys. Chem. C* **2009**, *113*, 6730–6734. [CrossRef]
37. Xiang, Z.; Xue, Y.; Cao, D.; Huang, L.; Chen, J.F.; Dai, L. Highly efficient electrocatalysts for oxygen reduction based on 2D covalent organic polymers complexed with non-precious metals. *Angew. Chem. Int. Ed.* **2014**, *53*, 2433–2437. [CrossRef] [PubMed]
38. Cui, X.; Yang, S.; Yan, X.; Leng, J.; Shuang, S.; Ajayan, P.M.; Zhang, Z. Pyridinic-nitrogen-Dominated graphene aerogels with Fe–N–C coordination for highly efficient oxygen reduction reaction. *Adv. Funct. Mater.* **2016**, *26*, 5708–5717. [CrossRef]
39. Kang, G.-S.; Jang, J.-H.; Son, S.-Y.; Lee, Y.-K.; Lee, D.C.; Yoo, S.J.; Lee, S.; Joh, H.-I. Pyrrolic N wrapping strategy to maximize the number of single-atomic Fe-Nx sites for oxygen reduction reaction. *J. Power Sources* **2022**, *520*, 230904. [CrossRef]
40. Zhu, J.; He, C.; Li, Y.; Kang, S.; Shen, P.K. One-step synthesis of boron and nitrogen-dual-self-doped graphene sheets as non-metal catalysts for oxygen reduction reaction. *J. Mater. Chem. A* **2013**, *1*, 14700–14705. [CrossRef]
41. Wang, P.; Wang, Z.; Jia, L.; Xiao, Z. Origin of the catalytic activity of graphite nitride for the electrochemical reduction of oxygen: Geometric factors vs. electronic factors. *Phys. Chem. Chem. Phys.* **2009**, *11*, 2730–2740. [CrossRef] [PubMed]
42. Ma, T.; Zhang, M.; Liu, H.; Wang, Y. Three-dimensional sulfur-doped graphene supported cobalt-molybdenum bimetallic sulfides nanocrystal with highly interfacial storage capability for supercapacitor electrodes. *Electrochim. Acta* **2019**, *322*, 134762. [CrossRef]
43. Hong, Q.-L.; Zhai, Q.-G.; Liang, X.-L.; Yang, Y.; Li, F.-M.; Jiang, Y.-C.; Hu, M.-C.; Li, S.-N.; Chen, Y. Holey cobalt oxyhydroxide nanosheets for the oxygen evolution reaction. *J. Mater. Chem. A* **2021**, *9*, 3297–3302. [CrossRef]
44. Ren, X.; Ge, R.; Zhang, Y.; Liu, D.; Wu, D.; Sun, X.; Du, B.; Wei, Q. Cobalt–borate nanowire array as a high-performance catalyst for oxygen evolution reaction in near-neutral media. *J. Mater. Chem. A* **2017**, *5*, 7291–7294. [CrossRef]
45. Béjar, J.; Álvarez-Contreras, L.; Ledesma-García, J.; Arjona, N.; Arriaga, L.G. An advanced three-dimensionally ordered macroporous $NiCo_2O_4$ spinel as a bifunctional electrocatalyst for rechargeable Zn–air batteries. *J. Mater. Chem. A* **2020**, *8*, 8554–8565. [CrossRef]
46. Liang, Z.; Dong, X. Co_2P nanosheet cocatalyst-modified $Cd_{0.5}Zn_{0.5}S$ nanoparticles as 2D-0D heterojunction photocatalysts toward high photocatalytic activity. *J. Photochem. Photobiol. A Chem.* **2021**, *407*, 113081. [CrossRef]
47. Niu, K.; Yang, B.; Cui, J.; Jin, J.; Fu, X.; Zhao, Q.; Zhang, J. Graphene-based non-noble-metal Co/N/C catalyst for oxygen reduction reaction in alkaline solution. *J. Power Sources* **2013**, *243*, 65–71. [CrossRef]
48. Zhang, J.; Zhang, G.; Jin, S.; Zhou, Y.; Ji, Q.; Lan, H.; Liu, H.; Qu, J. Graphitic N in nitrogen-doped carbon promotes hydrogen peroxide synthesis from electrocatalytic oxygen reduction. *Carbon* **2020**, *163*, 154–161. [CrossRef]
49. Chen, P.; Wang, L.-K.; Wang, G.; Gao, M.-R.; Ge, J.; Yuan, W.-J.; Shen, Y.-H.; Xie, A.-J.; Yu, S.-H. Nitrogen-doped nanoporous carbon nanosheets derived from plant biomass: An efficient catalyst for oxygen reduction reaction. *Energy Environ. Sci.* **2014**, *7*, 4095–4103. [CrossRef]
50. Liu, X.; Zhou, Y.; Zhou, W.; Li, L.; Huang, S.; Chen, S. Biomass-derived nitrogen self-doped porous carbon as effective catalysts for oxygen reduction reaction. *Nanoscale* **2015**, *7*, 6136–6142. [CrossRef]
51. Chaudhari, K.N.; Song, M.Y.; Yu, J.S. Transforming hair into heteroatom-doped carbon with high surface area. *Small* **2014**, *10*, 2625–2636. [CrossRef] [PubMed]
52. Borghei, M.; Laocharoen, N.; Kibena-Põldsepp, E.; Johansson, L.-S.; Campbell, J.; Kauppinen, E.; Tammeveski, K.; Rojas, O.J. Porous N, P-doped carbon from coconut shells with high electrocatalytic activity for oxygen reduction: Alternative to Pt-C for alkaline fuel cells. *Appl. Catal. B Environ.* **2017**, *204*, 394–402. [CrossRef]
53. Wang, S.; Yu, D.; Dai, L. Polyelectrolyte functionalized carbon nanotubes as efficient metal-free electrocatalysts for oxygen reduction. *J. Am. Chem. Soc.* **2011**, *133*, 5182–5185. [CrossRef]

54. Wang, Y.; Jiang, X. Facile preparation of porous carbon nanosheets without nemplate and their excellent electrocatalytic property. *ACS Appl. Mater. Interfaces* **2013**, *5*, 11597–11602. [CrossRef] [PubMed]
55. Yang, W.; Fellinger, T.-P.; Antonietti, M. Efficient metal-free oxygen reduction in alkaline medium on high-surface-area mesoporous nitrogen-doped carbons made from ionic liquids and nucleobases. *J. Am. Chem. Soc.* **2011**, *133*, 206–209. [CrossRef]
56. Chen, P.; Zang, J.; Zhou, S.; Jia, S.; Tian, P.; Cai, H.; Gao, H.; Wang, Y. N-doped 3D porous carbon catalyst derived from biowaste Triarrhena sacchariflora panicle for oxygen reduction reaction. *Carbon* **2019**, *146*, 70–77. [CrossRef]
57. Wu, C.-H.; Wang, K.-C.; Chang, S.-T.; Chang, Y.-C.; Chen, H.-Y.; Yamanaka, I.; Chiang, T.-C.; Huang, H.-C.; Wang, C.-H. High performance of metal-organic framework-derived catalyst supported by tellurium nanowire for oxygen reduction reaction. *Renew. Energ.* **2020**, *158*, 324–331. [CrossRef]
58. Kim, S.; Myles, T.D.; Kunz, H.R.; Kwak, D.; Wang, Y.; Maric, R. The effect of binder content on the performance of a high temperature polymer electrolyte membrane fuel cell produced with reactive spray deposition technology. *Electrochim. Acta* **2015**, *177*, 190–200. [CrossRef]
59. Lin, L.; Zhu, Q.; Xu, A.-W. Noble-metal-free Fe-N/C catalyst for highly efficient oxygen reduction reaction under both alkaline and acidic conditions. *J. Am. Chem. Soc.* **2014**, *136*, 11027–11033. [CrossRef]
60. Li, S.; Zhang, L.; Liu, H.; Pan, M.; Zan, L.; Zhang, J. Heat-treated cobalt–tripyridyl triazine (Co–TPTZ) electrocatalysts for oxygen reduction reaction in acidic medium. *Electrochim. Acta* **2010**, *55*, 4403–4411. [CrossRef]
61. McCrory, C.C.L.; Jung, S.; Peters, J.C.; Jaramillo, T.F. Benchmarking heterogeneous electrocatalysts for the oxygen evolution reaction. *J. Am. Chem. Soc.* **2013**, *135*, 16977–16987. [CrossRef] [PubMed]

Review

Research Progress of Working Electrode in Electrochemical Extraction of Lithium from Brine

Yangyang Wang [1,†], Guangya Zhang [2,3,†], Guangfeng Dong [1,4,*] and Heng Zheng [1,*]

1. School of Resource and Environmental Engineering, Wuhan University of Technology, Wuhan 430070, China
2. School of Mechanical and Electrical Engineering, Wuhan University of Technology, Wuhan 430070, China
3. SAIC GM Wuling Automobile Co., Ltd., Liuzhou 545027, China
4. SDIC Xinjiang Lop Nur Potash Co., Ltd., Hami 839000, China
* Correspondence: lbp120@126.com (G.D.); zy101801@163.com (H.Z.)
† These authors contributed equally as co-first author.

Abstract: Efficient extraction of Li from brine at a low cost is becoming a key technology to solve energy and environmental problems. Electrochemical extraction of Li has become a research hotspot due to its low energy consumption, high selectivity, and environmental friendliness. $LiMn_2O_4$, $LiFePO_4$, and $LiNi_{1/3}Co_{1/3}Mn_{1/3}O_2$ are widely used as cathode materials for the electrochemical extraction of Li but they also have some drawbacks, such as a small adsorption capacity. In this paper, the principle of electrochemical Li extraction from brine is reviewed and the research progress and analysis of the above three working electrode materials is summarized. In addition, analysis of the extraction of other rare ions from the working electrode material and the effect of micro-organisms on the working electrode material is also presented. Next, the shortcomings of working electrode materials are expounded upon and the research direction of working electrode materials in electrochemical Li extraction technology are prospected. It is hoped that this paper can provide insights and guidance for the research and application of electrochemical Li extraction from brine.

Keywords: brine; electrochemistry; lithium extraction; working electrode; micro-organism

1. Introduction

Lithium (Li) is the lightest metal element in the world, with active chemical properties, high electrical conductivity, and specific heat capacity [1]. It is widely used in batteries, ceramics, the nuclear industry, and other fields [2]. In recent years, with the development of electronic products and new energy vehicles, the market demand for lithium resources has been increasing [3]. In addition, global lithium resources are unevenly distributed, mainly in Chile, Argentina, Bolivia, China, and Australia [4]. Generally, Li exists mainly in the form of compounds in Li ore, brine, and seawater. In particular, the Li content of brine is much larger than that of Li ore, and ore lithium extraction has the disadvantages of being highly energy-intensive and polluting, with half of the global Li raw materials coming from brine extraction [5]. In addition to Li ions, brine also contains a large amount of alkali metal ions and alkaline earth metal ions (Li^+, Na^+, K^+, Mg^{2+}, Ca^{2+}, and Ba^{2+}) [6]. It is difficult to separate the lithium ions from the brine because the hydration radii and chemical properties of magnesium and lithium ions are very similar [7]. In addition, a large number of micro-organisms are present in the brine. The research shows that, in the Atacama Salt Lake in northern Chile, when the total salt concentration is 55.6%, there are still hundreds of micro-organisms, including the *Archaea halovenus*, *Halobacterium*, and haloccus, of which the most abundant is *Scutellaria* [8]. There are also a large number of halophilic micro-organisms in the Utah salt lake with a total salt concentration of more than 30%, in which the density of prokaryotes is greater than $2-3 \times 10^7$ cells/mL, mainly *Salinibacter*, *Halobacillus halophilic archaea*, unicellular green algae, and *Dunaliella* [9].

There are various methods for the extraction of brine, such as the extraction method, adsorption method, membrane separation technology, precipitation method, solar pond crystallization technology, etc. [10]. The extraction method requires a large amount of organic solvents [11]. The adsorbent in the adsorption method may cause the corrosion of equipment [12]. Membrane separation technology is expensive, and the membrane is prone to clogging [13]. The precipitation method is too time-consuming [14]. Solar pond crystallization technology has a long process cycle and is limited by geography [15]. However, the electrochemical extraction of Li has the advantages of green environmental protection [16], strong adaptability, process simplicity, and high efficiency [17], which have attracted extensive attention from researchers.

Based on the working principle of Li iron phosphate batteries, the electrochemical extraction process of Li utilizes potential-controlled electrode materials to extract Li from brine. The selection and preparation of electrode materials are one of the main factors affecting electrochemical Li extraction. Therefore, the research progress of working electrode materials in the electrochemical extraction of Li was reviewed in this paper. In addition, due to the strong corrosiveness of brine and the existence of micro-organisms, the influence of brine and micro-organisms on the working electrode material was further analyzed. Next, analysis of the shortcomings of the working electrode in the current electrochemical extraction process of Li is summarized and its follow-up research direction is prospected. It is hoped that this review will provide new ideas for the development and application of Li electrochemical extraction processes.

2. Principle of Electrochemical Extraction of Li from Brine

Typically, Li-ion battery cathode materials are employed in the electrochemical extraction process of Li [18]. The extraction of Li is achieved through the movement of Li ions between the electrode and the electrolyte during charge and discharge [19]. Li ions are reduced from the electrolyte and intercalated into the cathode material (working electrode material) during discharge; Li ions are desorbed from the cathode material into the electrolyte during charging. According to the principle of charge and discharge, the electrolyte is replaced with brine. Li ions in brine intercalate into the cathode material during discharge; Li ions enter a single recovery solution to achieve Li-ion extraction during charging. The principle of electrochemical extraction of Li from brine is shown in Figure 1. Based on the above principles, Pasta et al. [20] proposed the electrochemical ion pumping technology, using $FePO_4$ as the working electrode, which does not need to regenerate the adsorption material through chemical substances, requires a short time, minimizes the enthalpy term related to chemical bond breaking and formation, and requires less energy. The reaction formulas are given in (1) and (2) [21].

$$FePO_4 + LiCl + Ag \rightarrow LiFeO_4 + AgCl \quad (1)$$

$$LiFeO_4 + AgCl \rightarrow FeO_4 + LiCl + Ag \quad (2)$$

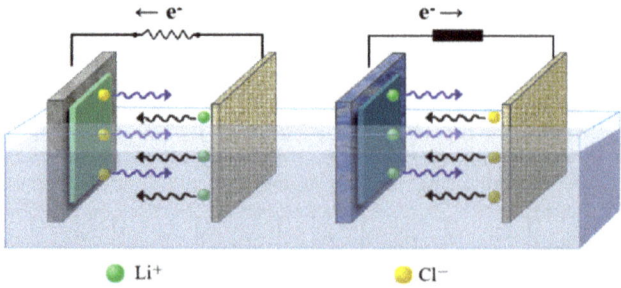

Figure 1. The principle of electrochemical extraction of Li.

Currently, $LiMn_2O_4$, $LiFePO_4$, and $LiNi_{1/3}Co_{1/3}Mn_{1/3}O_2$ are widely used as cathode materials for the electrochemical extraction of Li. The electrochemical extraction process of Li based on $LiMn_2O_4/Li_{1-x}Mn_2O_4$ has the advantages of low cost and high Li selectivity and is an effective method to alleviate the shortage of Li resources and the extraction process is shown in Figure 2 [22]. In particular, $LiMn_2O_4$ has been widely used in the extraction of Li resources due to its high selectivity for Li^+. Due to the high selectivity of electrode materials, only lithium ions are allowed to be embedded in the corresponding lattice in the extraction process and acidic or strong oxidizing eluents are not used to elute the ion sieve in the process of lithium extraction, which realizes the environmental friendliness of lithium extraction and avoids the damage of eluents to the structure of the ion sieve. In addition, the properties of cathode materials are also one of the main factors limiting the electrochemical extraction of Li. Therefore, the optimization of electrode materials and performance improvement are of great significance for the application of the electrochemical extraction of lithium. Selectivity, exchange capacity, cycling stability, etc., are the directions of electrode material optimization. In particular, selectivity needs to be prioritized.

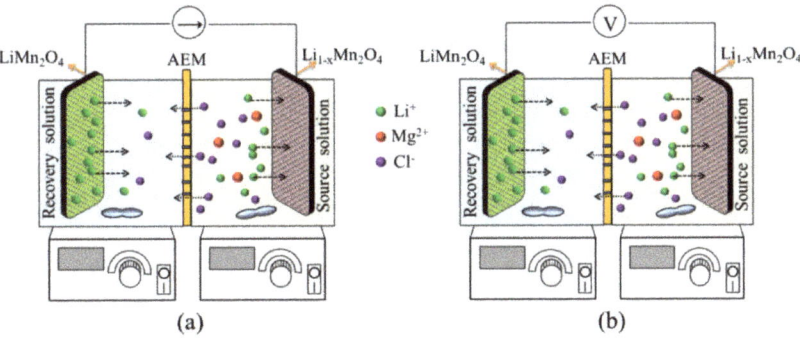

Figure 2. Schematic of lithium extraction with the driving mode of (**a**) CC and (**b**) CV [22].

3. Research Progress of Working Electrode Materials in Electrochemical Extraction of Li from Brine

At present, Li manganate, Li iron phosphate, Li cobalt oxide, etc., are commonly used as working electrode materials for the electrochemical extraction of Li, which can improve the electrochemical insertion and removal of Li ions. The above electrode materials should be most considered in the practical application process for their selectivity, exchange capacity, and cyclic stability of lithium ions. Spinel-structured Li manganate, olivine-structured Li iron phosphate, and layer-structured $LiNi_{1/3}Co_{1/3}Mn_{1/3}O_2$ have been widely studied due to their desirable properties (Figure 3). Therefore, analysis of the structure, principle, and research progress of the above three working electrode materials is summarized here.

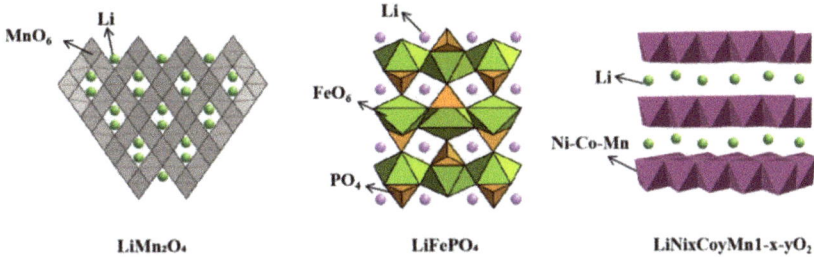

Figure 3. Three kinds of working electrodes for electrochemical extraction of Li.

3.1. Spinel Structure

LiMn$_2$O$_4$ has higher electrical conductivity than LiFePO$_4$, which might be due to the alternate arrangement of manganese and oxygen in MnO$_2$. The structure formed a channel that is favorable for the (de)intercalation of Li ions. In particular, the spinel-type structure remained unchanged during the extraction or intercalation process and the λ-MnO$_2$ formed after Li extraction was highly selective to Li [23]. However, LiMn$_2$O$_4$ exhibited poor cycling stability due to Mn leaching, which could be improved by improving its preparation method [24].

To overcome the above deficiencies, Shang et al. [25] prepared a multi-walled carbon nanotube (CNT) tandem LiMn$_2$O$_4$ (CNT-s-LMO) composite, which exhibited a favorable selectivity and extraction rate (84%) that was synergistic with the CNT-s-LMO hybrid capacitive deionization (HCDI). Furthermore, the capacity retention rate was 90% after 100 cycles [26]. In addition, spinel-type Li$_{1-x}$Ni$_{0.5}$Mn$_{1.5}$O$_4$ (LNMO) had a higher capacity than LiMn$_2$O$_4$ (the adsorption capacity can reach 1.259 mmol/g) and the working electrode does not deteriorate after 50 cycles. It can be used as a Li-ion deintercalation material for the electrochemical extraction of Li [27]. It has been reported that the λ-MnO$_2$/rGO-based CDI system exhibited favorable selectivity and high cycle stability for Li extraction from synthetic salt lake brine, which was attributed to its special intercalation structure. The structure has abundant active sites and a fast ion transport rate [28]. Similarly, the separation factors of Li$^+$/Na$^+$ and Li$^+$/M^{2+} in simulated brine are 1040.57 and 358.96 for the prepared scalable 3D porous composite electroactive membrane (λ-MnO$_2$/rGO/Ca-Alg), respectively. The excellent Li-ion extraction performance is due to the porous network structure and the potential-responsive ion pump effect in the ESIX process [29]. Xie et al. designed an electrochemical flow-through HCDI system with adequate trapping ability and stability for Li ions, and the lithium absorption capacity was as high as 18.1 mg/g, which was attributed to the trapping of Li ions in the λ-MnO$_2$ electrode via a Faraday redox reaction. Additionally, the λ-MnO$_2$ electrode exhibited excellent Li-ion selectivity when the brine contained a variety of cations, while avoiding the use of harmful acids or organic solvents [30]. Mu et al. [31] developed an electrode based on mesoporous λ-MnO$_2$/LiMn$_2$O$_4$ modification with a large specific surface area of 183 m^2/g, an extracted Li content of 75 mg/h per gram of LiMn$_2$O$_4$, and energy consumption of 23.4 Wh/mol; the electrode system provides an energy-efficient method for Li$^+$ extraction from brine. To improve the cycling stability of the electrodes, LiMn$_2$O$_4$ electrodes coated with Al$_2$O$_3$-ZrO$_2$ thin films were prepared. Due to the synergistic effect of Al$_2$O$_3$-ZrO$_2$ during charge and discharge, the chemical stability and high active sites on the electrode surface significantly improved the cycle capacity. After 30 cycles, the extraction capacity of lithium increased from 29.21% to 57.67% [32]. The reaction formulas for extracting lithium using LiMn$_2$O$_4$ are given in (3) and (4).

$$2\lambda\text{-MnO}_2 + Ag + LiCl = LiMn_2O_4 + AgCl \quad (3)$$

$$LiMn_2O_4 + AgCl = 2\lambda\text{-MnO}_2 + Ag + LiCl_4 \quad (4)$$

Electrochemical extraction of Li needs to be carried out in corrosive brines, so cathode materials with high cycling stability are required. The three-dimensional nano-structured inorganic gel framework electrode prepared by introducing polypyrrole/Al$_2$O$_3$ on the surface of LiMn$_2$O$_4$. Lithium was extracted from simulated brine with an initial capacity of 1.85 mmol/g and after 100 cycles, it showed a capacity retention rate of 85%, indicating its high cycling stability [33]. Fang et al. fabricated LiMn$_2$O$_4$@C/N-4 (LMO@CN-4) membrane electrodes with a maximum capacity of 34.57 mg/g in 40 min through in situ polymerization and high-level annealing. This might be due to the carbon encapsulation as a conductive layer that enhances charge and ion transport and prevents the bulk collapse of the crystal and the dissolution of Mn as a buffer layer [34]. In addition, Li extraction from low-concentration brine, seawater, and wastewater with low-concentration Li content should be of concern.

3.2. Olivine Structure

LiFePO$_4$ with an olivine structure is a crystal framework composed of many FeO$_6$ octahedra and PO$_4$ tetrahedrons, which realizes the insertion and extraction of Li$^+$ during the oxidation and reduction of iron [35]. LiFePO$_4$ has a higher theoretical capacity and lower Li intercalation potential than λ-MnO$_2$ (Ramasubramanian et al., 2022). LiFePO$_4$ electrode material (Ag is used as the counter electrode) exhibited high stability and Li-ion deintercalation capacity in an aqueous solution; the Li-Na ratio increased from 1:100 to 5:1, so it was selected as the working electrode for electrochemical extraction of Li [36]. The PO$_4$ tetrahedron between LiO$_6$ and FeO$_6$ in the olivine structure limited the volume change in LiFePO$_4$, which also limited the insertion and extraction of Li$^+$ during charge and discharge. The olivine-structured LiFePO$_4$ had better cycling stability due to its high lattice stability. The PPy/Al$_2$O$_3$/LMO of 3D nanocomposite inorganic gel framework structure prepared by the sol-gel method and polymerization method effectively improved the adsorption capacity and cycling stability of Li. This was attributed to the protection of the PPy/Al$_2$O$_3$ coating and the larger specific surface area [33]. In addition, LiFePO$_4$ exhibited high efficiency and stability during selective Li extraction from seawater, which was mainly achieved by the difference in electrochemical potential in the intercalation or deintercalation reaction and the diffusion-activated barrier in the FePO$_4$ framework, with molar selectivity as high as 1.8×10^4 [37]. Kim et al. [38] used FePO$_4$ electrode to recover Li from simulated artificial seawater. Inspired by mussels, they coated the electrode surface with polydopamine coating, which increased the amount of Li recovered and improved the selectivity by about 20 times. The Li$_{0.3}$FePO$_4$ electrode exhibited favorable ion selectivity, cycling stability, and adsorption capacity, showing promising application potential [39]. The reaction formulas for extracting lithium using LiFePO$_4$ are given in (5) and (6).

$$FePO_4 + Li^+ + e^- \rightarrow LiFePO_4 \quad (5)$$

$$LiFePO_4 - e^- \rightarrow FePO_4 + Li \quad (6)$$

3.3. Layered Structure

The layered structure of LiNi$_{1/3}$Co$_{1/3}$Mn$_{1/3}$O$_2$ (NCM) had the advantages of high theoretical discharge capacity, high charge-discharge rate, effective cycle stability, low cost, and low environmental toxicity [40]. NCM is the most ideal working electrode material, which is widely used in the electrochemical extraction of lithium. The synthesis method is simple and the electrochemical performance is excellent. The initial specific capacity of NCM was 193 mAh/g. After 1000 cycles at 1 C, the specific capacity is still 155 mAh/g [41]. NCM adopts a diamond-shape α-NaFeO$_2$ structure and continuous alternating [MO$_2$]$^-$ (M = Ni, Co, Mn) and Li layers, in which only Ni^{2+} and CO^{3+} are active; Mn^{4+} is conducive to maintaining the stability of crystal structure. The research showed that the NCM material in the Li electrochemical extraction system exhibited favorable selectivity and adsorption capacity and could achieve efficient Li extraction under the condition of the coexistence of various impurity ions. Compared with traditional Li extraction, it had the advantages of low energy consumption, high Li$^+$ yield, short duration, and green environmental protection [42]. In addition, compared with the organic solution, NCM exhibited a fast charge-discharge rate and adequate stability in aqueous solution. The diffusion coefficient of NCM in the aqueous solution is 1.39×10^{-10} and the charge and discharge are completed within a few seconds. After 1000 cycles, its capacity loss was only 9.1% [43]. Applied electrode material NCM to the actual brine can show high selectivity for lithium ions and can obtain Li chloride with a purity of up to 96.4% (i = ± 0.25 mA/cm) [42]. Zhao et al. developed a continuous-flow NMMO/AC hybrid supercapacitor (CF-NMMO/AC) using a depolymerized LNMMO cathode (NMMO) and an AC anode, exhibiting high capacity, high rate, and excellent cycling stability. The device consumed only 7.91 Wh/mol in simulated brine, and the extraction rate of Li$^+$ was as high as 97.2% [44]. Although NCM has a fast Li-ion intercalation and deintercalation rate and favorable cycle stability, its preparation conditions are harsh and the cost is high. In addition, the corrosiveness of brine puts forward higher requirements on the chemical stability of electrode materials. Zhao et al. prepared a graphene gauze-modified $LiNi_{0.6}Co_{0.2}Mn_{0.2}O_2$ core-shell microsphere (rGO/NCM) with high capacity, effective cycling stability, and fast (de)intercalation rate, and the extraction rate of Li^+ was up to 93% in simulated saline. This was attributed to the electron transfer pathway provided by graphene gauze instead of ion transfer between lattices, which effectively reduced the possibility of NCM lattice collapse [45]. In conclusion, NCM with excellent screening performance and a simple preparation process is the main research direction for future studies.

In view of the current shortcomings of the above three working electrode materials, new electrode materials with favorable selectivity, high adsorption capacity, and effective cycle stability could be developed by modification methods, such as electrode doping and coating. In addition, it is also a research direction to combine the excellent properties of the three. In particular, layered spinel hetero-structured Li-rich material (LSNCM) and nanocrystalline bismuth (NCBI) constituted a desalination battery with a Li recovery rate as high as 99%, which could be used for Li extraction from low-salinity brine [46]. The lithium extraction performance of the above three electrode materials is shown in Table 1.

Table 1. Summary of lithium extraction performance of three electrode materials.

Properties	$LiMn_2O_4$	$LiFePO_4$	$LiNi_{1/3}Co_{1/3}Mn_{1/3}O_2$
Initial brine [Mg/Li ratio]	147.8 [47]	134.4 [48]; 132 [49]	5.15 [42,45]
Selectivity	High Lithium ion selectivity [47]	High Lithium ion selectivity [48]	High Lithium ion selectivity [42]
Cyclic stability	After 100 cycles, the capacity retention was 91% [47]	Capacity retention exceeds 90% after 100 cycles [39]	excellent cycle stability [42,45]
Absorption capacity [mg/g]	37 [50]; 12 [47]	25 [48]; 9.13 [49]	1.56 [42]; 13.84 [45]
Purity [%]	96.2 [51]	74.3–99.98 [52]	93 [42]; 96.4 [45]
Energy [Wh/mol]	7.63 [50]; 37 [47]	2.8–29.5 [52]	2.6 [42]; 1.4 [45]
Efficiency [%]	83.3 [47]	83 [48]; 82.23 [39]; 91.11 [49]	84.4 [45]
Advantages	Highly selective, higher electrical conductivity	Adequate ion selectivity, cycling stability and adsorption capacity	High theoretical discharge capacity, high charge/discharge rate, favorable cycle stability, low cost and low environmental toxicity
Disadvantages	Poor cycle stability, Harsh preparation process, low capacity of reversible embedded lithium	High power consumption, low conductivity and tap density	Preparation conditions are harsh and the cost is high

3.4. Working Electrode Material to Extract Other Rare Ions

Electrochemical working electrode materials can extract not only lithium ions from seawater or brine but also other high-value ions, such as rubidium ions, by the same mechanism [53]. Rubidium is widely present in brines, and its coexistence with alkali metals with similar properties makes extraction more complicated [54]. Xu et al. [55] prepared lithium/rubidium imprinted layered porous silica (Li/Rb-IHPS) for the selective recovery of lithium and rubidium from aqueous solutions, which showed high absorption capacities of 166 μg/g and 141 μg/g for both lithium and rubidium, respectively. Current studies on the extraction of other high-value ions using working electrode materials are scarce, so the exploration of other rare elements using existing electrochemical lithium extraction systems is essential.

3.5. The Effects of Micro-Organisms on Working Electrode Materials

There are many micro-organisms in brine, such as halophilic bacteria, and some of them can be used as anode catalysts. Salt-loving bacteria can be used as anode catalysts in microbial fuel cells to enhance the performance of microbial fuel cells, such as starch degradation [56]. In addition, using chromium wastewater as an anion and anaerobic micro-organisms as an anode biocatalyst, hexavalent chromium (Cr^{6+}) can be reduced on an abiotic cathode by using an exogenous biofilm on the anode of a microbial fuel cell (MFC) [57]. It is found that *Halophilic archaea safranine sodium* can cause corrosion of steel materials and its impact on cathode materials is rarely reported [58]. The aerobic halophilic archaeon *Natronorubrum tibetense* can cause severe localized corrosion of Q235 carbon steel [59]. Therefore, the interactions between micro-organisms and working electrode materials should also be paid attention to. In addition, there are a large number of suspended solids and micro-organisms in the brine, which can block the electrode materials [60].

4. Conclusions and Prospects

The demand for Li and its compounds is increasing year by year with the rapid development of the new energy industry. The electrochemical extraction process of Li has attracted more and more attention due to its advantages of being green, having high efficiency, saving energy, and being safe. However, there are also a series of problems in the actual application process, such as low Li-ion migration rate, low Li-ion adsorption capacity, and poor cycle stability in the working electrode. Therefore, we should address the above problems from the following perspectives:

(1) It is necessary to develop working electrode materials with excellent comprehensive properties. Improving the electrochemical performance of the working electrode is the key to realizing engineering applications, which can start from the aspects of porosity, particle size, and synthesis route. The high saline-alkali environment of brine should also be considered. Furthermore, protective coatings can be developed to enhance the corrosion resistance of electrode materials by exploring the relationship between lattice size change and corrosion.

(2) The current research on working electrodes is heavily dependent on experimental conditions and lack of standardized methods. Therefore, it is necessary to establish a scientific evaluation system of electrode materials as soon as possible to obtain electrode materials with application prospects.

(3) The influence of micro-organisms on extraction equipment and electrode materials deserves further study due to their abundant presence in brines. Micro-organisms affect the performance of cathode materials and affect the safety, stability, and extraction efficiency of extraction equipment by corroding metal materials (such as pipes), and their mechanism of action needs to be further explored. In addition, the working electrode is prone to be blocked by micro-organisms during long-term use, so the removal of micro-organisms in brine deserves attention.

(4) Material genomics integrate high-throughput computing, high-throughput preparation, high-throughput detection, and database systems, which can greatly shorten the material development cycle. Facing the problems existing in the current working electrode for electrochemical lithium extraction, material genomics can be used to design and prepare electrode materials.

(5) Parameters such as brine flow rate and electric field distribution also have an impact on the efficiency, ion concentration, and lifetime of electrochemical lithium extraction [61]. For example, a flow-by-flow configuration with a small amount of intercalation material is not suitable for large-scale lithium extraction from brine, while a cross-flow configuration is suitable for industrial scale-up at a moderate flow rate. In addition, the lithium extraction efficiency also depends on the total current applied to the reactor. Therefore, operational parameters must be traded off to find the optimal conditions (capacity and capture rate) for the electrochemical extraction of lithium.

(6) In addition to the working electrode materials, the construction of an electrochemical lithium extraction system has an important impact on the cost, efficiency, and energy consumption of electrochemical lithium extraction. Therefore, the development of electrochemical lithium extraction can be promoted by improving the electrode system and one such improvement is the exploration of the counter electrode system. The function of a counter electrode is to form a closed circuit in the electrochemical lithium extraction system and maintain the electric neutrality of the electrochemical lithium extraction system. Common counter electrodes are Ag electrode, Pt electrode, titanium electrode, activated carbon electrode, polymer electrode, etc. Compared with traditional precious metals, the environmentally friendly material-activated carbon electrode yields significant cost benefits. Next, we should also integrate the working electrode material with the counter electrode system to enhance the performance of the electrochemical lithium extraction system.

(7) The biggest challenge of current electrochemical lithium extraction is the amplification effect in the real industrial scale-up process for which the existing working electrode materials, devices, specific operating parameters (such as DC potential, feed flow rate, the cycles of the recovered solution, etc.), energy consumption, etc., should be optimized and integrated.

In general, the electrochemical extraction of Li from brine is still the trend of Li resource acquisition and electrode materials are the bottleneck restricting its industrial application. The performance of electrode materials are one of the main research directions for future studies. In addition, the impact of other aspects of brine on electrode materials should also be comprehensively considered to sustainably extract Li resources from brine.

Author Contributions: Y.W., H.Z. and G.D. conceived the manuscript. Y.W. and G.Z. wrote the first draft of the manuscript. Y.W., G.Z., G.D. and H.Z. revised each part of the manuscript in detail. All four authors participated in the revision of the manuscript. All authors have read and agreed to the published version of the manuscript.

Funding: This research was funded by the Key Technologies and Demonstration Applications for Comprehensive Recovery and Utilization of Sulfate-type Salt Lake Tailings Resources (2021E02038).

Institutional Review Board Statement: Not applicable.

Informed Consent Statement: Not applicable.

Data Availability Statement: The study did not report any data.

Conflicts of Interest: The authors declare that the research was conducted in the absence of any commercial or financial relationships that could be construed as a potential conflict of interest.

References

1. Battistel, A.; Palagonia, M.S.; Brogioli, D.; La Mantia, F.; Trocoli, R. Electrochemical Methods for Lithium Recovery: A Comprehensive and Critical Review. *Adv. Mater.* **2020**, *32*, 1905440.
2. Hu, B.; Shang, X.; Nie, P.; Zhang, B.; Yang, J.; Liu, J. Lithium ion sieve modified three-dimensional graphene electrode for selective extraction of lithium by capacitive deionization. *J. Colloid Interface Sci.* **2022**, *612*, 392–400. [CrossRef] [PubMed]
3. Calvo, E.J. Electrochemical methods for sustainable recovery of lithium from natural brines and battery recycling. *Curr. Opin. Electrochem.* **2019**, *15*, 102–108.
4. Cubillos, C.F.; Aguilar, P.; Grágeda, M.; Dorador, C. Microbial Communities From the World's Largest Lithium Reserve, Salar de Atacama, Chile: Life at High LiCl Concentrations. *J. Geophys. Res. Biogeoences* **2018**, *123*, 3668–3681.
5. Delmas, C.; Maccario, M.; Croguennec, L.; Le Cras, F.; Weill, F. Lithium deintercalation in LiFePO4 nanoparticles via a domino-cascade model. *Nat. Mater.* **2008**, *7*, 665–671. [CrossRef] [PubMed]
6. Fabre, C.; Ourti, N.E.; Ballouard, C.; Mercadier, J.; Cauzid, J. Handheld LIBS analysis for in situ quantification of Li and detection of the trace elements (Be, Rb and Cs). *J. Geochem. Explor.* **2022**, *236*, 106979. [CrossRef]
7. Fang, J.-W.; Wang, J.; Ji, Z.-Y.; Cui, J.-L.; Guo, Z.-Y.; Liu, J.; Zhao, Y.-Y.; Yuan, J.-S. Establishment of PPy-derived carbon encapsulated $LiMn_2O_4$ film electrode and its performance for efficient Li^+ electrosorption. *Sep. Purif. Technol.* **2022**, *280*, 119726.
8. Flexer, V.; Fernando Baspineiro, C.; Ines Galli, C. Lithium recovery from brines: A vital raw material for green energies with a potential environmental impact in its mining and processing. *Sci. Total Environ.* **2018**, *639*, 1188–1204. [CrossRef]

9. Gandoman, F.H.; Jaguemont, J.; Goutam, S.; Gopalakrishnan, R.; Firouz, Y.; Kalogiannis, T.; Omar, N.; Van Mierlo, J. Concept of reliability and safety assessment of lithium-ion batteries in electric vehicles: Basics, progress, and challenges. *Appl. Energy* **2019**, *251*, 113343.
10. Luo, G.; Zhu, L.; Li, X.; Zhou, G.; Sun, J.; Chen, L.; Chao, Y.; Jiang, L.; Zhu, W. Electrochemical lithium ions pump for lithium recovery from brine by using a surface stability Al_2O_3-ZrO_2 coated $LiMn_2O_4$ electrode. *J. Energy Chem.* **2022**, *69*, 244–252. [CrossRef]
11. Guo, Z.-Y.; Ji, Z.-Y.; Wang, J.; Guo, X.-F.; Liang, J.-S. Electrochemical lithium extraction based on "rocking-chair" electrode system with high energy-efficient: The driving mode of constant current-constant voltage. *Desalination* **2022**, *533*, 115767. [CrossRef]
12. Guo, Z.Y.; Ji, Z.Y.; Wang, J.; Chen, H.Y.; Yuan, J.S. Development of electrochemical lithium extraction based on a rocking chair system of $LiMn_2O_4$/$Li_{1-x}Mn_2O_4$: Self-driven plus external voltage driven. *Sep. Purif. Technol.* **2020**, *259*, 118154. [CrossRef]
13. Hannan, M.A.; Lipu, M.S.H.; Hussain, A.; Mohamed, A. A review of lithium-ion battery state of charge estimation and management system in electric vehicle applications: Challenges and recommendations. *Renew. Sustain. Energy Rev.* **2017**, *78*, 834–854. [CrossRef]
14. He, L.; Xu, W.; Song, Y.; Luo, Y.; Liu, X.; Zhao, Z. New Insights into the Application of Lithium-Ion Battery Materials: Selective Extraction of Lithium from Brines via a Rocking-Chair Lithium-Ion Battery System. *Glob. Chall.* **2018**, *2*, 1700079. [CrossRef] [PubMed]
15. He, X.; Kaur, S.; Kostecki, R. Mining Lithium from Seawater. *Joule* **2020**, *4*, 1357–1358. [CrossRef]
16. Hong, Z.; Zhu, Q.; Liu, Y.; Wang, S.; Wu, J.; Jiang, H.; Hu, X.; Liu, K. Dependence of concentration polarization on discharge profile in electrochemical lithium extraction. *Desalination* **2022**, *527*, 115567. [CrossRef]
17. Zheng, J.; Jia, X.; Wang, C.; Zheng, M.; Dong, Q. Electrochemical Performance of the $LiNi\_(1/3)Co\_(1/3)Mn\_(1/3)O\_2$ in Aqueous Electrolyte. *J. Electrochem. Soc.* **2010**, *16*, 151.
18. Kim, J.S.; Lee, Y.H.; Choi, S.; Shin, J.; Choi, J.W. An Electrochemical Cell for Selective Lithium Capture from Seawater. *Environ. Sci. Technol.* **2015**, *49*, 9415–9422. [CrossRef]
19. Kim, N.; Su, X.; Kim, C. Electrochemical lithium recovery system through the simultaneous lithium enrichment via sustainable redox reaction. *Chem. Eng. J.* **2020**, *420*, 127715. [CrossRef]
20. Lawagon, C.P.; Nisola, G.M.; Cuevas, R.; Torrejos, R.; Kim, H.; Lee, S.P.; Chung, W.J. $Li_{1-x}Ni_{0.5}Mn_{1.5}O_4$/Ag for electrochemical lithium recovery from brine and its optimized performance via response surface methodology. *Sep. Purif. Technol.* **2019**, *212*, 416–426. [CrossRef]
21. Lawagon, C.P.; Nisola, G.M.; Cuevas, R.A.I.; Kim, H.; Lee, S.P.; Chung, W.J. $Li_{1-x}Ni_{0.33}Co_{1/3}Mn_{1/3}O_2$/Ag for electrochemical Lithium recovery from brine. *Chem. Eng. J.* **2018**, *348*, 1000–1011.
22. Li, X.; Mo, Y.; Qing, W.; Shao, S.; Tang, C.Y.; Li, J. Membrane-based technologies for lithium recovery from water lithium resources: A review. *J. Membr. Sci.* **2019**, *591*, 117317. [CrossRef]
23. Liu, C.; Li, Y.; Lin, D.; Hsu, P.C.; Chu, S. Lithium Extraction from Seawater through Pulsed Electrochemical Intercalation. *Joule* **2020**, *4*, 1459–1469. [CrossRef]
24. Liu, D.; Xu, W.; Xiong, J.; He, L.; Zhao, Z. Electrochemical system with $LiMn_2O_4$ porous electrode for lithium recovery and its kinetics. *Sep. Purif. Technol.* **2021**, *270*, 118809. [CrossRef]
25. Liu, G.; Zhao, Z.; Ghahreman, A. Novel approaches for lithium extraction from salt-lake brines: A review. *Hydrometallurgy* **2019**, *187*, 81–100. [CrossRef]
26. Mathew, M.; Kong, Q.H.; McGrory, J.; Fowler, M. Simulation of lithium ion battery replacement in a battery pack for application in electric vehicles. *J. Power Sources* **2017**, *349*, 94–104. [CrossRef]
27. Meshram, P.; Pandey, B.D.; Mankhand, T.R. Extraction of lithium from primary and secondary sources by pre-treatment, leaching and separation: A comprehensive review. *Hydrometallurgy* **2014**, *150*, 192–208. [CrossRef]
28. Mu, Y.; Zhang, C.; Zhang, W.; Wang, Y. Electrochemical lithium recovery from brine with high Mg^{2+}/Li^+ ratio using mesoporous λ-MnO_2/$LiMn_2O_4$ modified 3D graphite felt electrodes. *Desalination* **2021**, *511*, 115112.
29. Oren, A. The microbiology of red brines. *Adv. Appl. Microbiol.* **2020**, *113*, 57–110.
30. Pasta, M.; Battistel, A.; Mantia, F.L. Batteries for lithium recovery from brines. *Energy Environ. Sci.* **2012**, *5*, 9487–9491.
31. Pasta, M.; Wessells, C.D.; Cui, Y.; Mantia, F.L. A Desalination Battery. *Nano Lett.* **2012**, *12*, 839–843. [PubMed]
32. Peng, H.; Zhao, Q. A Nano-Heterogeneous Membrane for Efficient Separation of Lithium from High Magnesium/Lithium Ratio Brine. *Adv. Funct. Mater.* **2021**, *31*, 2009430.
33. Qian, H.; Zhang, D.; Lou, Y.; Li, Z.; Xu, D.; Du, C.; Li, X. Laboratory investigation of microbiologically influenced corrosion of Q235 carbon steel by halophilic archaea Natronorubrum tibetense. *Corros. Sci.* **2018**, *145*, 151–161. [CrossRef]
34. Shang, X.; Liu, J.; Hu, B.; Nie, P.; Yang, J.; Zhang, B.; Wang, Y.; Zhan, F.; Qiu, J. CNT-Strung $LiMn_2O_4$ for Lithium Extraction with High Selectivity and Stability. *Small Methods* **2022**, *6*, 2200508.
35. Sophia, A.C.; Saikant, S. Reduction of chromium(VI) with energy recovery using microbial fuel cell technology. *J. Water Process Eng.* **2016**, *11*, 39–45. [CrossRef]
36. Sun, Y.; Wang, Y.; Liu, Y.; Xiang, X. Highly Efficient Lithium Extraction from Brine with a High Sodium Content by Adsorption-Coupled Electrochemical Technology. *Acs Sustain. Chem. Eng.* **2021**, *9*, 11022–11031. [CrossRef]
37. Tian, L.; Liu, Y.; Tang, P.; Yang, Y.; Wang, X.; Chen, T.; Bai, Y.; Tiraferri, A.; Liu, B. Lithium extraction from shale gas flowback and produced water using $H_{1.33}Mn_{1.67}O_4$ adsorbent. *Resour. Conserv. Recycl.* **2022**, *185*, 106476.

38. Trócoli, R.; Battistel, A.; Mantia, F.L. Selectivity of a Lithium-Recovery Process Based on LiFePO4. *Chem. A Eur. J.* **2014**, *20*, 9888–9891.
39. Trócoli, R.; Erinmwingbovo, C.; La Mantia, F. Optimized Lithium Recovery from Brines by using an Electrochemical Ion-Pumping Process Based on λ-MnO$_2$ and Nickel Hexacyanoferrate. *ChemElectroChem* **2017**, *4*, 143–149. [CrossRef]
40. Vijay, A.; Arora, S.; Gupta, S.; Chhabra, M. Halophilic starch degrading bacteria isolated from Sambhar Lake, India, as potential anode catalyst in microbial fuel cell: A promising process for saline water treatment. *Bioresour. Technol.* **2018**, *256*, 391–398. [CrossRef]
41. Xie, N.; Li, Y.; Yuan, Y.; Gong, J.; Hu, X. Fabricating a Flow-Through Hybrid Capacitive Deionization Cell for Selective Recovery of Lithium Ions. *ACS Appl. Energy Mater.* **2021**, *4*, 13036–13043.
42. Xing, P.; Wang, C.; Chen, Y.; Ma, B. Rubidium extraction from mineral and brine resources: A review. *Hydrometallurgy* **2021**, *203*, 105644.
43. Xiong, J.; He, L.; Zhao, Z. Lithium extraction from high-sodium raw brine with Li$_{0.3}$FePO$_4$ electrode. *Desalination* **2022**, *535*, 115822.
44. Xiong, J.; Zhao, Z.; Liu, D.; He, L. Direct lithium extraction from raw brine by chemical redox method with LiFePO$_4$/FePO$_4$ materials. *Sep. Purif. Technol.* **2022**, *290*, 120789. [CrossRef]
45. Xu, X.; Zhou, Y.; Feng, Z.; Kahn, N.U.; Haq Khan, Z.U.; Tang, Y.; Sun, Y.; Wan, P.; Chen, Y.; Fan, M. A Self-Supported—MnO$_2$ Film Electrode used for Electrochemical Lithium Recovery from Brines. *ChemPlusChem* **2018**, *83*, 521–528. [CrossRef] [PubMed]
46. Xu, W.; He, L.; Zhao, Z. Lithium extraction from high Mg/Li brine via electrochemical intercalation/de-intercalation system using LiMn$_2$O$_4$ materials. *Desalination* **2021**, *503*, 114935. [CrossRef]
47. Xu, X.; Li, Y.; Yang, D.; Zheng, X.; Wang, Y.; Pan, J.; Zhang, T.; Xu, J.; Qiu, F.; Yan, Y.; et al. A facile strategy toward ion-imprinted hierarchical mesoporous material via dual-template method for simultaneous selective extraction of lithium and rubidium. *J. Clean. Prod.* **2018**, *171*, 264–274.
48. Yang, Z.; Zhang, X.; Huang, M.; Huang, J.; Fang, Z. Preparation and Rate Capability of Carbon Coated LiNi$_{1/3}$Co$_{1/3}$Mn$_{1/3}$O$_2$ as Cathode Material in Lithium Ion Batteries. *Acs Appl. Mater. Interfaces* **2017**, *9*, 12408–12415. [CrossRef] [PubMed]
49. Yang, S.; Zhang, F.; Ding, H.; He, P.; Zhou, H. Lithium Metal Extraction from Seawater. *Joule* **2018**, *2*, 1648–1651.
50. Yu, J.; Zheng, M.; Wu, Q.; Nie, Z.; Bu, L. Extracting lithium from Tibetan Dangxiong Tso Salt Lake of carbonate type by using geothermal salinity-gradient solar pond. *Sol. Energy* **2015**, *115*, 133–144. [CrossRef]
51. Zhang, L.; Li, L.; Rui, H.; Shi, D.; Peng, X.; Ji, L.; Song, X. Lithium recovery from effluent of spent lithium battery recycling process using solvent extraction. *J. Hazard. Mater.* **2020**, *398*, 122840. [CrossRef] [PubMed]
52. Zhang, Y.; Sun, W.; Xu, R.; Wang, L.; Tang, H. Lithium extraction from water lithium resources through green electrochemical-battery approaches: A comprehensive review. *J. Clean. Prod.* **2021**, *285*, 124905. [CrossRef]
53. Zhao, X.; Feng, M.; Jiao, Y.; Zhang, Y.; Sha, Z. Lithium extraction from brine in an ionic selective desalination battery. *Desalination* **2020**, *481*, 114360.
54. Zhao, X.; Jiao, Y.; Xue, P.; Feng, M.; Sha, Z. Efficiently lithium extraction from brine by using three-dimensional (3D) nanostructured hybrid inorganic-gel framework electrode. *ACS Sustain. Chem. Eng.* **2020**, *8*, 4827–4837.
55. Zhao, X.; Li, G.; Feng, M.; Wang, Y. Semi-continuous electrochemical extraction of lithium from brine using CF-NMMO/AC asymmetric hybrid capacitors. *Electrochim. Acta* **2019**, *331*, 135285. [CrossRef]
56. Zhao, X.; Yang, H.; Wang, Y.; Sha, Z. Review on the electrochemical extraction of lithium from seawater/brine. *J. Electroanal. Chem.* **2019**, *850*, 113389. [CrossRef]
57. Zhao, X.; Yang, H.; Wang, Y.; Yang, L.; Zhu, L. Lithium extraction from brine by an asymmetric hybrid capacitor composed of heterostructured lithium-rich cathode and nano-bismuth anode. *Sep. Purif. Technol.* **2021**, *274*, 119078.
58. Zhao, Y.; Zhou, E.; Xu, D.; Yang, Y.; Zhao, Y.; Zhang, T.; Gu, T.; Yang, K.; Wang, F. Laboratory investigation of microbiologically influenced corrosion of 2205 duplex stainless steel by marine Pseudomonas aeruginosa biofilm using electrochemical noise. *Corros. Sci.* **2018**, *143*, 281–291. [CrossRef]
59. Zheng, H.; Chen, X.; Yang, Y.; Li, L.; Li, G.; Guo, Z.; Feng, C. Self-Assembled LiNi$_{1/3}$Co$_{1/3}$Mn$_{1/3}$O$_2$ Nanosheet Cathode with High Electrochemical Performance. *ACS Appl. Mater. Interfaces* **2017**, *9*, 39560–39568. [CrossRef]
60. Zhang, Z.; Du, X.; Wang, Q.; Gao, F.; Jin, T.; Hao, X.; Ma, P.; Li, J.; Guan, G. A scalable three-dimensional porous λ-MnO$_2$/rGO/Ca-alginate composite electroactive film with potential-responsive ion-pumping effect for selective recovery of lithium ions. *Sep. Purif. Technol.* **2021**, *259*, 118111. [CrossRef]
61. Zubi, G.; Dufo-Lopez, R.; Carvalho, M.; Pasaoglu, G. The lithium-ion battery: State of the art and future perspectives. *Renew. Sustain. Energy Rev.* **2018**, *89*, 292–308.

Article

Molybdenum Nitride and Oxide Quantum Dot @ Nitrogen-Doped Graphene Nanocomposite Material for Rechargeable Lithium Ion Batteries

Lixia Wang [1,2,*], **Taibao Zhao** [1], **Ruiping Chen** [1], **Hua Fang** [1,2], **Yihao Yang** [1], **Yang Cao** [1,2] and **Linsen Zhang** [1,2,*]

[1] School of Material and Chemical Engineering, Zhengzhou University of Light Industry, Zhengzhou 450001, China
[2] Ceramic Materials Research Center, Zhengzhou University of Light Industry, Zhengzhou 450001, China
* Correspondence: 2014050@zzuli.edu.cn (L.W.); hnzhanglinsen@163.com (L.Z.); Fax: +86-371-86609676 (L.W.); +86-371-86609676 (L.Z.)

Abstract: A multistage architecture with molybdenum nitride and oxide quantum dots (MON-QDs) uniformly grown on nitrogen-doped graphene (MON-QD/NG) is prepared by a facile and green hydrothermal route followed by a one-step calcination process for lithium ion batteries (LIBs). Characterization tests show that the MON-QDs with diameters of 1–3 nm are homogeneously anchored on or intercalated between graphene sheets. The molybdenum nitride exists in the form of crystalline Mo_2N (face-centered cubic), while molybdenum oxide exists in the form of amorphous MoO_2 in the obtained composite. Electrochemical tests show that the MON-QD/NG calcinated at 600 °C has an excellent lithium storage performance with an initial discharge capacity of about 1753.3 mAh g^{-1} and a stable reversible capacity of 958.9 mAh g^{-1} at current density of 0.1 A g^{-1} as well as long-term cycling stability at high current density of 5 A g^{-1}. This is due to the multistage architecture, which can provide plenty of active sites, buffer volume changes of electrode and enhance electrical conductivity as well as the synergistic effect between Mo_2N and MoO_2.

Keywords: molybdenum nitride; molybdenum oxide; quantum dots; nitrogen-doped graphene; lithium ion batteries; electrochemical performance

1. Introduction

Transition metal nitrides (MN_x, M = Mo, Fe, Ni, V, W, etc.) are useful materials with numerous industrial applications, such as abrasives, cutting tools, electronics, catalysis as well as electrochemical applications. Due to the excellent metallic conductivity and low polarization loss, transition metal nitrides have been used widely as energy storage materials, e.g., LIB anode materials, in recent years. The molybdenum nitrides are promising anode materials in LIBs among the various transition metal nitrides [1–4]. For the sake of overcoming the limitations of large volume change during the charge/discharge progress, low diffusion rate of electrolyte and lithium ion as well as poor electron transport at high rate of cycles, the molybdenum nitrides were prepared into all kinds of nano-shaped particles or/and combined with other materials forming composites [3,5–17]. Zheng et al. [18] prepared a nano-complex with Mo_2N quantum dots @MoO_3@nitrogen-doped carbon (MON-NC) by a sol–gel method. Electrochemical performance tests showed that MON-NC has much higher rate performance and longer life cycle performance than that of $Mo_2N@MoO_3$ and nitrogen doped carbon (NC). Liu et al. [19] prepared a composite with the Mo_2N-coated hollow nanostructure of MoO_2. The composite has a reversible capacity of 815 mAh g^{-1} after 100 cycles at current density of 0.1 A g^{-1}. Zhang et al. [11] obtained a molybdenum nitride-doped graphene (MoN/GNS) composite material by calcining the precursor in NH_3 atmosphere. The MoN/GNS composite had good rate and cycle performance. However, these preparation methods usually contain complicated procedures and harsh synthetic conditions such as the use of templates and toxic or dangerous gases (e.g., NH_3 or H_2), which are adverse to its practical application and large-scale production. Calcinating the complex AM-HMTA precursors of ammonium molybdate ((NH_4)$_4Mo_7O_{24}$•$4H_2O$, AM) and hexamethylenetetramine ($C_6H_{12}N_4$, HMTA) under N_2/H_2 mixed atmosphere is a relatively simple method to prepare molybdenum nitride [20,21]. In our previous research, we found that the Mo_2N products contained part of MoO_2 by calcinating the AM-HMTA precursors in pure N_2 atmosphere [22]. The oxide MoO_2 is also an intensely appealing anode material for LIBs owing to its higher theoretical specific capacity (838 mAh g^{-1}) and higher density (6.5 g cm^{-3}) than those of the currently used graphite anode.

Herein, we prepared molybdenum nitride and oxide quantum dots (MON-QDs) anchored on nitrogen-doped graphene (MON-QD/NG) composite material by a facile and green hydrothermal route followed by a one-step calcination process in pure N_2 atmosphere. In the synthesis, graphene oxide (GO) can be added as an "assembled binder" to anchor these generated MON-QDs on reduced graphene oxide uniformly. The MON-QD/NG showed excellent lithium storage performance owing to the synergistic effect between 0D MON-QDs and 2D nitrogen-doped graphene nanosheets. MON-QDs can improve the electrochemical activity of electrode material by providing plenty of active sites while graphene nanosheets can inhibit the structure collapse and shorten the lithium ion diffusion pathway. Consequently, these superior characteristics endow MON-QD/NG with high lithium storage capacity, good cycle stability as well as excellent rate performance, and it is a promising anode material for LIBs.

2. Experimental Details

2.1. Preparation of Samples

Firstly, the precursors AM-HMTA and graphene oxide (GO) were prepared. The AM-HMTA was obtained by the following steps: (1) 3.5 g ammonium molybdate ((NH_4)$_4Mo_7O_{24}$•$4H_2O$, AM) and 6 g hexamethylenetetramine ($C_6H_{12}N_4$, HMTA) were dissolved in 50 mL deionized water, respectively; (2) the aqueous HMTA solution was added into the AM solution; (3) after magnetic stirring for 4 h, the mixed solution was left overnight to precipitate the white complex completely; (4) the AM-HMTA precursor was gained by filtering and then drying the white complex. The GO solution was prepared using graphite powder based on the method of modified Hummers [23]. Then, the prepared AM-HMTA was added to GO solution (4 mg mL^{-1}) with a weight ratio of 7:3 followed by stirring for 12 h to form a stable AM-HMTA/GO solution. After the above processes, the AM-HMTA/GO solution was sealed in an autoclave and heated at 180 °C for 6 h to obtain nitrogen-doped sponge, which was freeze-dried for 72 h further. Lastly, the freeze-dried sponge was calcinated at 500 °C, 600 °C and 700 °C for 2 h under N_2 atmosphere, respectively. The obtained samples were abbreviated as MON-QD/NG-500, MON-QD/NG-600 and MON-QD/NG-700 according to their calcinating temperature. Reduced graphene oxide (rGO) was also obtained via calcinating freeze-dried GO under the same conditions as that of MON-QD/NG.

2.2. Characterization

X-ray diffraction (XRD) tests were performed through a Rigaku Ultima IV instrument with Cu Kα radiation. A HORIBA Scientific LabRAM HR Evolution Raman spectrometer system was used to record the Raman spectra. X-ray photoelectron spectra (XPS) were recorded on a Perkin-Elmer PHI ESCA system. The microstructure and morphology of samples were observed by scanning electron microscope (SEM) (JSM-7001F) and transmission electron microscope (TEM) (JSM-2100). The specific surface area was measured by the Barrett–Emmett–Teller (BET) method. The pore size distribution was calculated by the Barrett–Joyner–Halenda (BJH) model.

2.3. Electrochemical Measurement

Electrochemical performances of the samples were measured through CR2016 coin type cells. The fabricated electrodes were prepared by mixing MON-QD/NG, polyvinylidene fluoride (PVDF) and carbon black in a weight ratio of 8:1:1 using N-methyl-2-pyrrolidone (NMP) as solvent. Then, the homogeneous slurry was coated onto a copper foil. Celgard 2400 polypropylene and pure lithium metal sheets were used as separator and counter electrode, respectively. The cells were assembled in an argon-filled glove box with concentration of H_2O and O_2 below 0.5 ppm. The assembled cells were allowed to stand for 24 h before electrochemical testing. Galvanostatic charge/discharge measurements were carried out in the voltage range between 0.01 and 3.00 V (vs. Li/Li$^+$) on a battery test instrument (LANHE CT2001A, Wuhan, China). Rate performance was tested at current density from 0.1 to 5 A g^{-1}. After rate performance tests, the samples directly underwent charge–discharge cycles 300 times at 5 A g^{-1}. Cyclic voltammetry (CV) tests were performed in the voltage range of 0.01–3.00 V on an electrochemical workstation (CHI660E) at different scan rates from 0.1 to 1 mV s^{-1}. Electrochemical impedance spectroscopy (EIS) was used in the frequency range of Hz to 0.01 Hz. The EIS spectra were fitted by Zview software.

3. Results and Discussion

3.1. Structural and Characterization

The crystal structures of the MON-QD/NG samples were confirmed by XRD patterns, as shown in Figure 1 The XRD pattern of rGO shows a broad diffraction peak at about 24° corresponding to the (002) diffraction plane of graphite, which may be related to the semi-graphitized nature [24,25]. Similarly, XRD patterns of MON-QD/NG samples show obvious diffraction characteristic peaks of graphene, corresponding to the rGO. Only weak diffraction peaks at 37.3°, 43.6°, 63.2° and 75.5° corresponding to the (111), (200), (220) and (311) crystal planes of the face-centered cubic (fcc) Mo$_2$N (JCPDS:25-1366) can be observed. The weak and broadened characteristic peaks of fcc Mo$_2$N may be related to the different micromorphology, such as crystal size. However, no diffraction peaks of MoO$_2$ can be detected, due to the amorphous state.

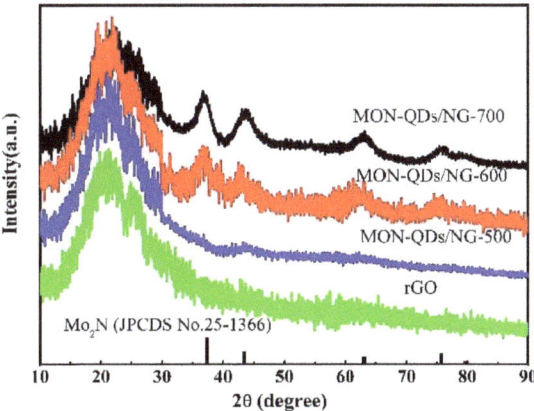

Figure 1. XRD pattern of MON-QD/NG, rGO and the standard card of face-centered cubic Mo$_2$N.

Raman tests were conducted on MON-QD/NG to further explore the structural information of the composites prepared at different temperatures, as shown in Figure 2. There are two strong peaks at about 1350 cm^{-1} and 1595 cm^{-1} of the three composites, which belong to D peak (associated with the sp^3 defective) and G peak (arises from the bond stretching of all sp^2−bonded pairs), respectively. In general, peak D represents the disordered induction peak of sp^3 carbon, while peak G represents the stretching vibration of the C−C bond, which is related to the defect or small crystal size of graphene and the graphite lattice pattern and sp² bond, respectively. The strength ratio of I_D/I_G was used to characterize the structural disorder degree, defect degree and graphitization properties of carbon materials [26]. Figure 2a shows the two characteristic peaks D, and G of graphene, and the intensity of peak D is significantly higher than that of peak G. Meanwhile, the asymmetric trailing extension of peak D extends to about 970 cm^{-1}, which is caused by the nitrogen-doped graphene [27]. The strength ratios (I_D/I_G) of MON-QD/NG-500, MON-QD/NG-600 and MON-QD/NG-700 are 1.08, 1.09 and 1.12, respectively, indicating the higher graphitization degree of MON-QD/NG with the increase in temperature. Meanwhile, there are many topological edge defects in the three composites, and these defects may come from the small size of graphite fragments and residual functional groups. A large number of studies have pointed out that atom replacement defects in and outside the graphene surface formed by nitrogen and boron atoms can improve the electrical conductivity of graphene, which is more beneficial to the lithium storage performance of the materials. In addition, from the lower Raman shift spectra (Figure 2b), it is observed that the scattering peaks at 275, 329 and 364 cm^{-1} in the three composites are related to the phonon vibration modes of MoO_2, suggesting the existence of MoO_2 in the composites [28].

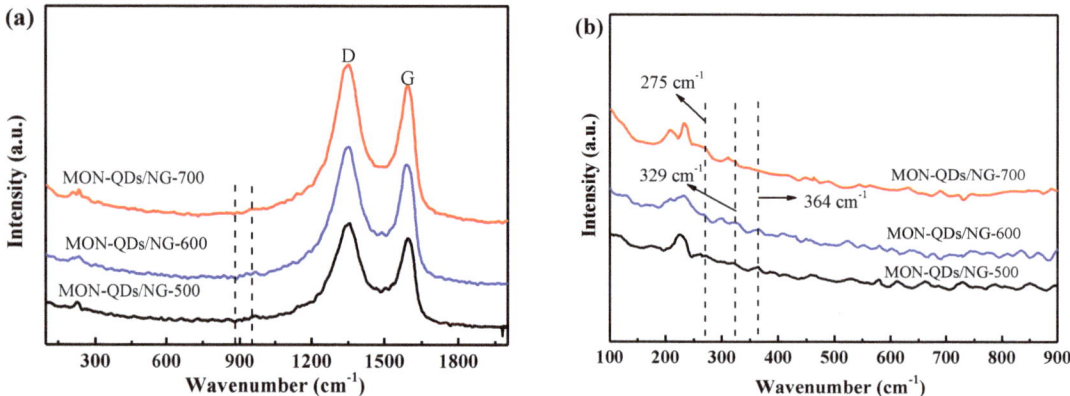

Figure 2. Raman spectra of MON-QD/NG at different calcination temperatures (a,b).

The chemical composition and electronic state of the prepared MON-QD/NG were investigated by XPS. The full spectrum of MON-QD/NG-600 is shown in Figure 3a. The sample mainly contains Mo, N, C and O elements. It can be seen from the full spectrum that the sample contains more O elements, which is caused by the existence of MoO_2 and oxidation of the surface of the sample. Figure 3b is the peak fitting result of O element, which can be fitted into three peaks: peak I at 529.9 eV is characteristic of the Mo-O bond, while peaks II at 531.2 eV and III at 532.8 eV are characteristic of the C-O-C bond and C=O bond, respectively, suggesting strong chemical interaction between MoO_2 and graphene. In the N1s spectrum (Figure 3c), peak II at 397.5 eV is the characteristic peak of the Mo-N bond, indicating the formation of Mo_2N. Peaks III, IV and V at 398.2 eV, 399.7 eV and 401.6 eV correspond to pyridine nitrogen, pyrrole nitrogen and graphene nitrogen, respectively, indicating that nitrogen-doped graphene has been successfully prepared. In Figure 3d, there are four characteristic peaks in the C1s spectrum, which are respectively attributed to carbon in different chemical states: peak I at 284.6 eV is the characteristic peak of the C-C bond, peak II at 285.5 eV and peak IV at 289.3 eV are the characteristic peaks of the C-O bond and O=C-O bond, respectively, and peak III at 287.3 eV is the characteristic peak of the C-N bond. These characteristic peaks indicate that MoO_2 and Mo_2N have strong chemical interactions with graphene, which is beneficial to charge transfer during charge and discharge. Figure 3e–g are the test results of Mo elements of MON-QD/NG-500, MON-QD/NG-600 and MON-QD/NG-700, respectively. The Mo3d spectra of the three composites can be divided into six peaks: peak I located at about 229.8 eV and peak II located at 232.0 eV can be identified as the presence of Mo^{4+}, which corresponds to MoO$_2$ phase. Peaks II and IV at 231.7 eV and 232.7 eV are characteristic peaks of the Mo-N bond in the samples, indicating the further formation of Mo$_2$N in the composites; peaks V and VI at 233.4 eV and 235.2 eV are characteristic peaks of Mo^{6+}, corresponding to the formation of MoO$_3$ phase, which is caused by the oxidation of oxygen in the air. In addition, it is found from the Mo3d spectra of three composite materials that the content of Mo$_2$N in the sample increases first and then decreases, while the content of MoO$_2$ increases all the time with the increase in calcination temperature. It can also be found that the content of Mo$_2$N in MON-QD/NG-700 composite is the lowest, while the content of MoO$_2$ is the highest among the three kinds of composites. This is because Mo element itself is easily combined with O element. Furthermore, the activity of O atoms increases with the increase in calcination temperature. However, the poor conductivity of MoO$_2$ restricts the cyclic and rate capability of the materials, which greatly affect the lithium storage properties of MON-QD/NG-700.

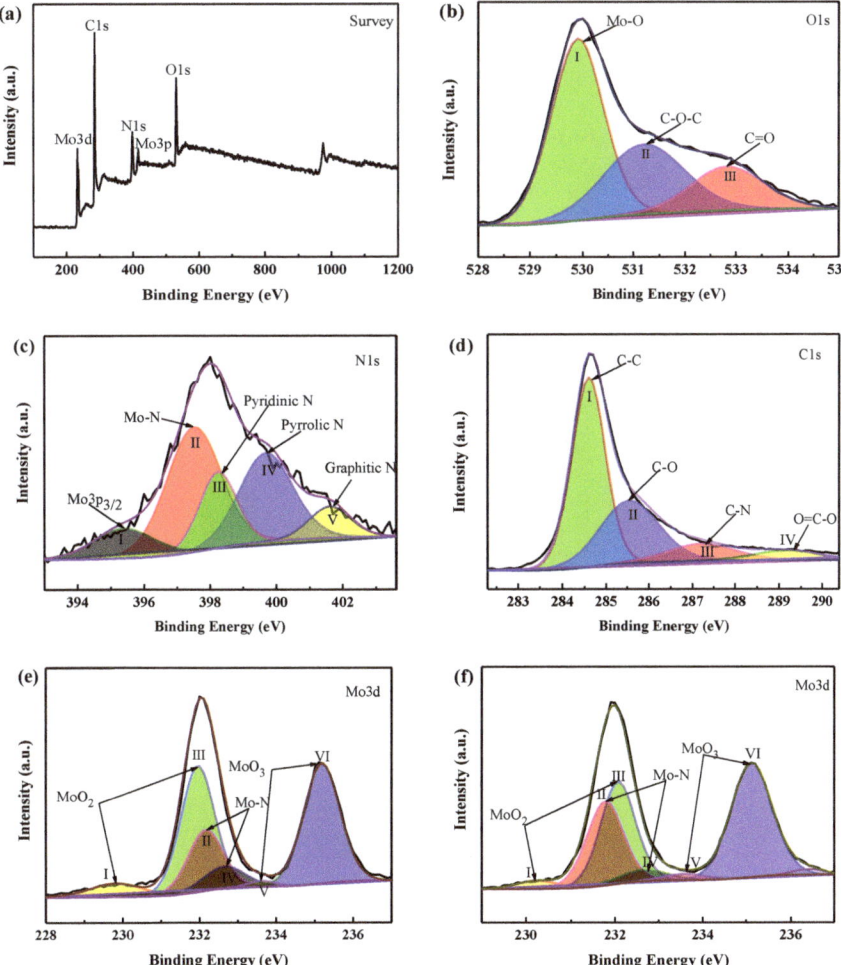

Figure 3. *Cont.*

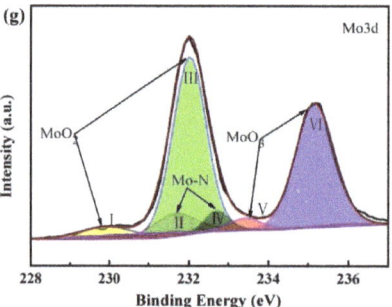

Figure 3. XPS spectra of MON-QD/NG-600 sample: (**a**) survey; (**b**) O1s; (**c**) N1s; (**d**) C1s; (**e**) Mo3d spectrum of MON-QD/NG-500; (**f**) Mo3d spectrum of MON-QD/NG-600; (**g**) Mo3d spectrum of MON-QD/NG-700.

The morphology of MON-QD/NG-600 composite was observed by SEM and TEM, as shown in Figure 4a–d. Only the typical folded lamellar morphology of graphene can be observed in the SEM image (Figure 4a). However, from the TEM (Figure 4b) and HRTEM (Figure 4c) images, it can be found that a large number of small quantum dots with size of 1–3 nm evenly distribute on the graphene lamellae. Further observation from Figure 4c revealed that Mo_2N quantum dots are evenly distributed in MoO_2. The diffraction rings in the SAED image (Figure 4d) are correspond to the (111), (200) and (220) crystal planes of Mo_2N, indicating the existence of Mo_2N quantum dots in the composite material. However, the diffraction rings of MoO_2 are not found in the SAED image, which indicates that MoO_2 quantum dots exist in an amorphous state. The morphology of quantum dots can alleviate the volume change in electrode materials in the charging and discharging process, and GO also plays the role of template in the preparation process, resulting in a great change in the morphology of composite materials.

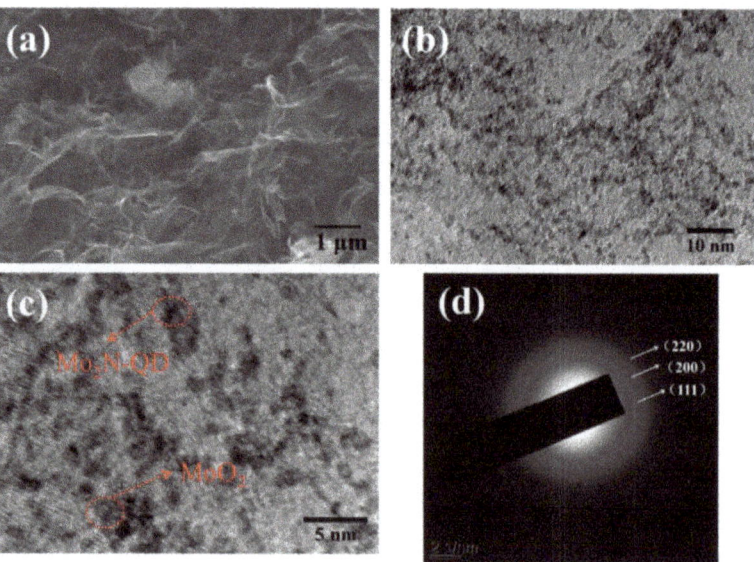

Figure 4. Morphology of MON-QD/NG-600: (**a**) SEM; (**b**) TEM; (**c**) HRTEM; (**d**) SAED.

Figure 5a,b are the adsorption and desorption isothermal curves and pore size distribution of the three MON-QD/NG samples, respectively. There are obvious hysteresis loops in the three MON-QD/NG composites with a pressure ratio of about 0.8–1.0 (Figure 5a), indicating the existence of mesoporous structures in the composites. It can also be observed from Figure 5b that a large number of mesopores with pore sizes of about 2–50 nm exist in the three composites. The specific surface area of MON-QD/NG-600 is 116.9 m^2 g^{-1} and pore volume is 0.60 cm^3 g^{-1}, which is larger than those of MON-QD/NG-500 (105 m^2 g^{-1}, 0.53 cm^3 g^{-1}) and MON-QD/NG-700 (100 m^2 g^{-1}, 0.48 cm^3 g^{-1}). The large specific surface area provides a fast channel for the migration of Li$^+$ ions and the diffusion of electrolyte, which is conductive to the rapid insertion and extraction of electrolyte ions, thereby greatly improving electrochemical performance, particularly rate performance [8,29]. Meanwhile, high pore volume can provide buffer for material volume change during the charging and discharging process, thus improving energy density.

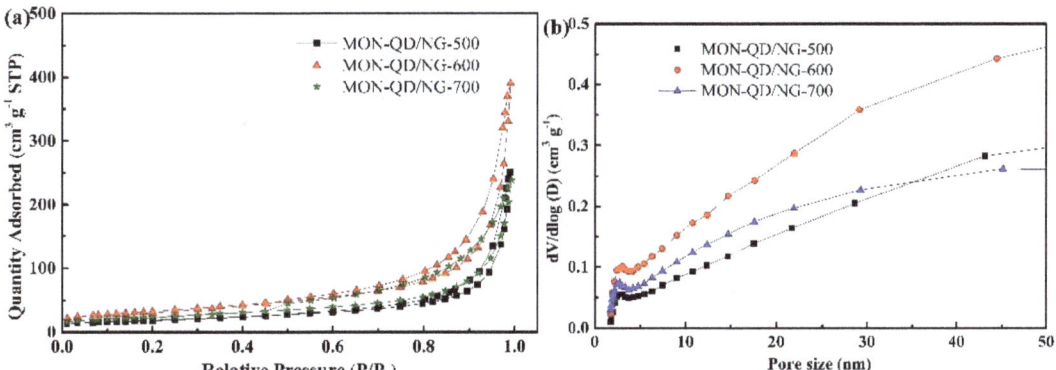

Figure 5. BET tests: (**a**) The adsorption and desorption isothermal curves of three MON-QD/NG composite materials; (**b**) pore size distribution.

3.2. Electrochemical Characterization

The cyclic voltammetry curve of MON-QD/NG-600 at 0.1 mV s^{-1} in the 0.01–3.0 V voltage range is shown in Figure 6. During the initial lithium process, the diffusion peak from about 1.65 V to 1.0 V may be Li$^+$ inserted into the amorphous MoO$_2$, accompanied by the formation of Li$_x$MoO$_2$. However, during the first cathodic process, the peak of Li$^+$ insertion into the MoO$_2$ lattice (usually above 2.2 V) is not visible, which may be related to the amorphous nature of MoO$_2$ [18,30]. When the electrode discharge voltage is lower than 1 V, the original Li$_x$MoO$_2$ reacts with Li$^+$ and gradually transforms into Mo and Li$_2$O [30]. In addition, a significant reduction peak was observed at about 0.7 V in the first CV curve, which disappeared in the subsequent CV test, probably due to the formation of the solid electrolyte interface layer (SEI) [31]. Subsequently, two oxidation peaks at 1.41 V and 1.70 V are related to the transformation process of Li$_x$MoO$_2$ and Mo to MoO$_2$. The CV curves of the second and third cycles almost coincide, indicating that the electrode material has high reversibility.

Charge and discharge curves at 0.1 A g^{-1} of MON-QD/NG-500, MON-QD/NG-600 and MON-QD/NG-700 are shown in Figures 7a, 7b and 7c, respectively. It is observed from the figures that the charge–discharge curves of the three composite materials all have obvious charge–discharge platforms. This is consistent with the redox peak in the CV curve. Additionally, the initial discharge-specific capacities of the three composites are 1445.6, 1753.3 and 1347.6 mAh g^{-1}, respectively, while the initial charging capacities are 837, 957.7 and 777.3 mAh g^{-1}, respectively. It is observed that MON-QD/NG-500 and MON-QD/NG-700 composites decay rapidly in subsequent cycles, while MON-QD/NG-600 composites exhibit excellent lithium storage performance. In order to explore the electrochemical lithium storage performance of the three electrode materials, the charge and discharge rate performance at current density from 0.1 to 5 A g^{-1} and the subsequent cycling property at 5 A g^{-1} of MON-QD/NG-500, MON-QD/NG-600 and MON-QD/NG-700 were tested, as shown in Figure 8a. In addition, charge–discharge curves at different current densities of the three materials are provided in the SI (Figure S1a–c). Obviously, MON-QD/NG-600 displays the best rate capability which has reversible capacity of 958.9, 727.4, 610.5, 476.0, 350.5, 297.2 mAh g^{-1} at the current density of 0.1, 0.2, 0.5, 1, 2, 5 A g^{-1}, respectively. Furthermore, the reversible capacity of MON-QD/NG-600 is superior to the MoN/GNS (see Table S1 in the SI) [11], Mo$_2$N nanolayer coated MoO$_2$ hollow nanostructure [19], MON-NC [18], etc.

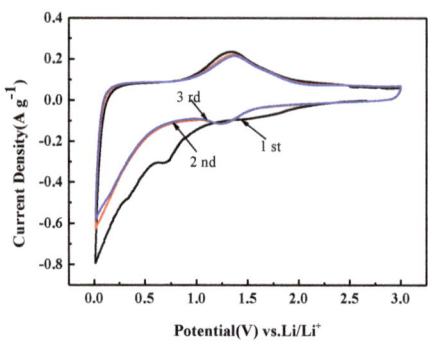

Figure 6. CV curves of MON-QD/NG-600 composite material.

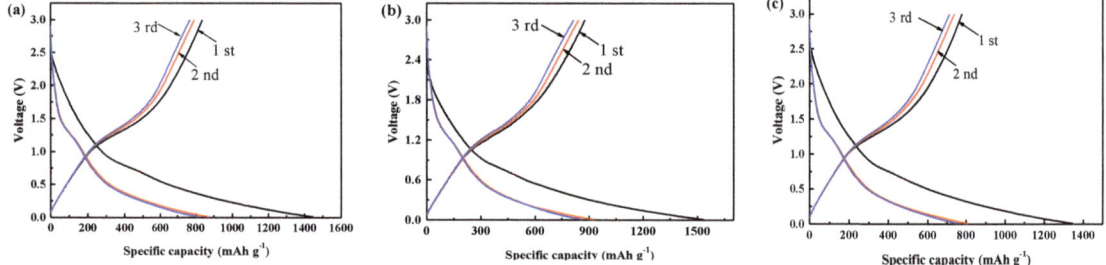

Figure 7. Charge and discharge curves of composite materials at different calcination temperatures: (**a**) MON-QD/NG-500; (**b**) MON-QD/NG-600; (**c**) MON-QD/NG-700.

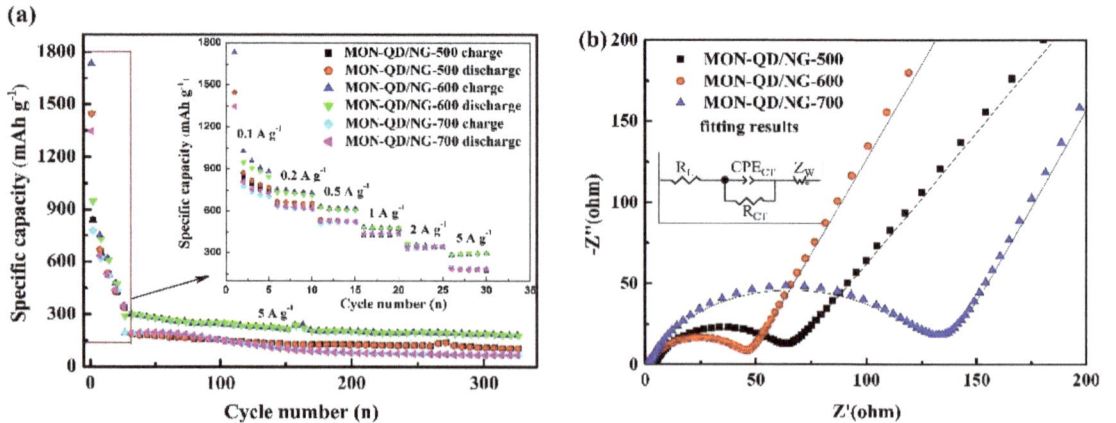

Figure 8. Electrochemical performance of electrode materials at different calcination temperatures: (**a**) rate performance and cycling stability at 5 A g^{-1}; (**b**) Nyquist plots of electrode materials.

Subsequently, the three composites were cycled 300 times at a current density of 5 A g^{-1}. Some selected charge–discharge curves (the first, 100, 200 and 300 cycles) during cycling tests at 5 A g^{-1} of MON-QD/NG-600 are also provided in the SI (as shown in Figure S2). The test results show that MON-QD/NG-600 still has a reversible capacity of about 180.1 mAh g^{-1} after about 300 cycles. However, MON-QD/NG-500 and MON-QD/NG-700 have a reversible capacity of 105.5 and 68.1 mAh g^{-1} after 300 cycles, respectively. It is quite evident that the cycling performance of MON-QD/NG-600 is obviously superior to that of MON-QD/NG-500 and MON-QD/NG-700.

The electrochemical kinetics of the three composites were analyzed by EIS spectra (as shown in Figure 8b). Each of them possesses one depressed semicircle in the high-frequency region and an inclined line in the low-frequency region, which relate to the charge transfer resistance and Li$^+$ diffusion process, respectively [32–34]. Therefore, an equivalent circuit, as shown in the illustration in Figure 8b, was selected for fitting the impedance spectra, where R_L, R_{CT}, CPE_{CT} and Z_W represent the electrolyte resistance, charge transfer resistance, double layer capacitance and Warburg impedance, respectively [35,36]. It is clearly shown that the fitted data are basically consistent with the test data. The calculated charge transfer resistances of MON-QD/NG-500, MON-QD/NG-600 and MON-QD/NG-700 are 45 Ω, 28 Ω and 78 Ω, respectively. Furthermore, the MON-QD/NG-600 electrode has a larger slope straight line, suggesting a faster Li$^+$ diffusion process. Although the graphitization degree of graphene increases and the conductivity is enhanced with the increase in temperature, the content of MoO$_2$ increases, which is not conducive to the lithium storage performance of electrode materials under high current density.

In order to gain additional understanding of the lithium storage mechanism of MON-QD/NG-600, CV curves at different scan rates from 0.1 to 1 mV s^{-1} were tested. As shown in Figure 9a, the CV curves at different scan rates exhibit a pair of redox peaks between 1.0 and 1.7 V. Obviously, the height and area of redox peak increase as the scan rate increases, which is due to electrode capacity obtained by dividing the peak area by scan rate which should be constant [37,38]. In addition, it can be observed that the oxidation peak shifted to higher potential slightly while the corresponding reduction peak shifted to a lower potential slightly, indicating that the MON-QD/NG-600 electrode showed increasingly obvious irreversible reaction at relatively high scan rates. Hence, the Randles–Ševčík equation (Equation (1)) was applied to calculate the diffusion constant D, which can well describe the relationship between square root of the scan rate v$^{1/2}$ and peak current i_p:

$$i_p = 2.69 \times 10^5 n^{3/2} A D_{Li}^{1/2} v^{1/2} \Delta C_o \tag{1}$$

where i_p is the peak current, A is the effective contact area between the electrode and electrolyte (cm^2), n is the number of electrons involved in the reaction, D_{Li} is the diffusion coefficient of Li$^+$ (cm^2 s^{-1}), v is the scan rate (V s^{-1}), ΔC_o is the change in Li$^+$ concentration in the electrode before and after the reaction (mol cm^{-3}). According to Equation (1), the Li$^+$ diffusion coefficients of the electrochemical reaction corresponding to the anodic and cathodic peaks are 4.08×10^{-10} and 4.71×10^{-10} cm^2 s^{-1}, respectively. Obviously, the Li$^+$ diffusion coefficients of anodic and cathodic reaction have the same order of magnitude, indicating the excellent reversibility of the MON-QD/NG-600 electrode. Usually, the stored charge of an electrode can be divided into three components: (1) the faradaic contribution due to Li$^+$ insertion process, (2) the faradaic contribution caused by charge transfer behavior, (3) the double layer capacitance. Generally, (1) and (2) are grouped into the same monomial, namely, the capacity-controlled process and the diffusion-controlled process. The current (i) and the scan rate (v) obey the power law in the CV curves, which can be proved by Equations (2) and (3) [39]:

$$i = av^b \tag{2}$$

$$\log(i) = b\log(v) + \log(a) \tag{3}$$

where a and b are variables. The b value can be determined by the slope of the plot log (i) versus log (v) curves [40]. b = 0.5 indicates a diffusion-controlled behavior, whereas

$b = 1$ reflects that the electrochemical reaction is controlled by a capacitance-dominated process. As shown in Figure 9b, the slopes of the anodic and cathodic peak are 0.77 and 0.81, respectively, indicating fast kinetics contributed by both behaviors. Therefore, there is a good relationship between $v^{1/2}$ and i_p. Hence, the CV curves with different scan rates can be used to quantitatively calculate the capacitive contribution by using Equation (4) [41,42]:

$$i = k_1 v + k_2 v^{1/2} \left(\text{equivalent to } \frac{i}{v^{1/2}} = k_1 v^{1/2} + k_2 \right) \quad (4)$$

where k is a constant. $k_1 v$ and $k_2 v^{1/2}$ are consistent with the contribution of the capacitance effect and the diffusion-controlled behavior, respectively [43]. The contribution of the capacitive process was also calculated. As shown in Figure 9c, the capacitive process contributes about 67.5% of the total capacity at a scan rate of 0.6 mV s^{-1}. The contribution ratios of the two processes at various scan rates were also incidentally calculated. The capacitive contribution progressively increases from 48.7% at 0.1 mV s^{-1} to a maximum value of 77.8% at 1 mV s^{-1}, as shown in Figure 9d. With the consideration of these, it shows that most of the charge stored in MON-QD/NG-600 was a capacitive process. This characteristic is highly beneficial for the fast transport of Li$^+$, resulting in high reversible rate performance and cycling performance.

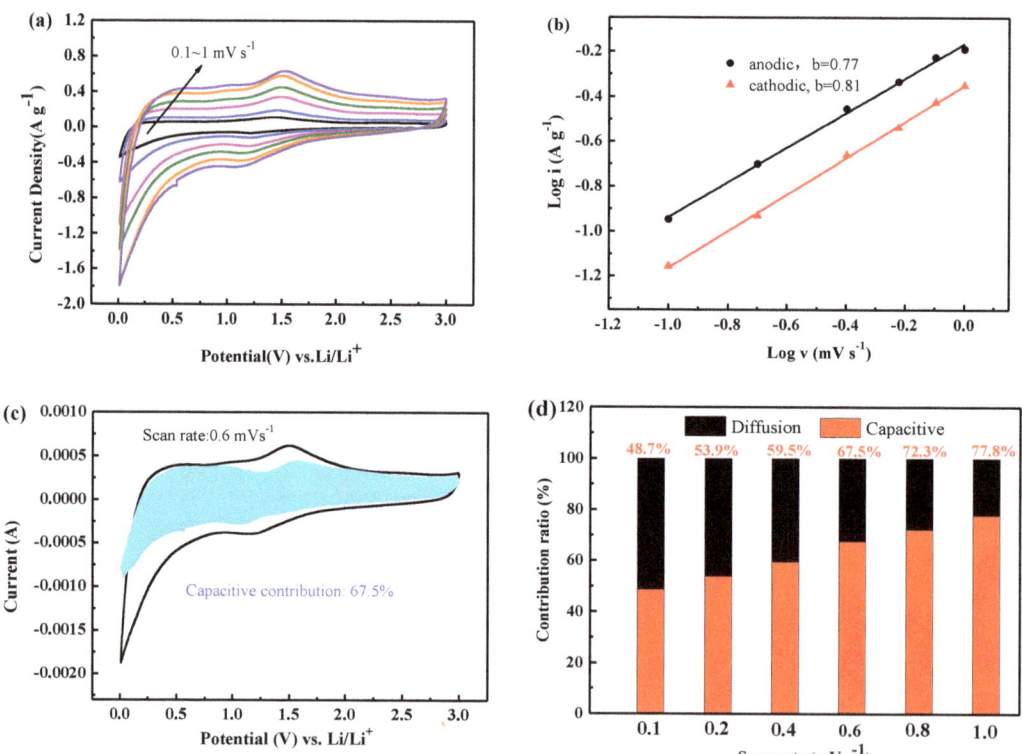

Figure 9. Kinetics analysis of the electrochemical performance toward Li$^+$ for the MON-QD/NG-600 electrode: (**a**) CV curves at scan rates from 0.1 to 1.0 mV s^{-1}; (**b**) the corresponding relationship between the scan rate and peak current; (**c**) CV curves and capacitive contribution to the total charge storage of MON-QD/NG-600 electrode at 0.6 mVs^{-1}; (**d**) contribution ratios of the capacitive and diffusion-controlled capacities at different scan rates.

4. Conclusions

MON-QD/NG was successfully synthesized through a green and facile hydrothermal route, followed by an annealing process. The electrochemical lithium storage performance test of electrode materials indicates that temperature has a great influence on the growth of Mo_2N and MoO_2. The content of MoO_2 in the composite increases with the increase in temperature. However, the poor conductivity of MoO_2 affects the electrochemical performance of the materials. Therefore, MON-QD/NG-600 has the best lithium storage performance. It has a reversible capacity of about 958.9 mAh g^{-1} at current density of 0.1 A g^{-1}, and even has a reversible capacity of about 350.5 mAh g^{-1} and 297.2 mAh g^{-1} at 2 A g^{-1} and 5 A g^{-1}, respectively. The characterization of MON-QD/NG-600 found that the prepared Mo_2N and MoO_2 exist in the shape of quantum dots with size of about 1–3 nm. The quantum dots inhibit the volume change in materials during charging and discharging, reduce the diffusion path of lithium ions as well as prevent the agglomeration between graphene sheets effectively. Moreover, the distribution of quantum dots on or between graphene sheets also provides a large number of reaction sites for electrochemical reactions. Due to the unique structure and synergy of Mo_2N and MoO_2, MON-QD/NG-600 exhibits excellent lithium storage performance in terms of stable long cycle life and superior rate capability. Therefore, the MON-QD/NG composite can be used as a promising electrode material candidate for high-performance LIBs.

Supplementary Materials: The following supporting information can be downloaded at: https://www.mdpi.com/article/10.3390/batteries9010032/s1, Figure S1: Charge-discharge curves of different current density; Figure S2: Some selected charge-discharge curves (the first, 100, 200 and 300 cycles) during cycling tests at 5 A g-1 of MON- QD/NG-600; Table S1: The comparison of molybdenum nitride related anode material for lithium ion battery [44–46].

Author Contributions: Conceptualization, L.W. and L.Z.; methodology, H.F.; software, Y.C.; formal analysis, T.Z.; investigation, R.C. and Y.Y. All authors have read and agreed to the published version of the manuscript.

Funding: This work is financially supported by Science and Technology Project of Henan Province (Nos. 212102210649, 212102210648, 222102240122).

Informed Consent Statement: Not applicable.

Data Availability Statement: Not applicable.

Conflicts of Interest: The authors declare no conflict of interest. We declare that we do not have any commercial or associative interest that represents a conflict of interest in connection with the work submitted.

References

1. Wang, L.; Zhang, K.; Pan, H.; Wang, L.; Wang, D.; Dai, W.; Qin, H.; Li, G.; Zhang, J. 2D molybdenum nitride nanosheets as anode materials for improved lithium storage. *Nanoscale* **2018**, *10*, 18936–18941. [CrossRef] [PubMed]
2. Nandi, D.K.; Sen, U.K.; Choudhury, D.; Mitra, S.; Sarkar, S.K. Atomic Layer Deposited Molybdenum Nitride Thin Film: A Promising Anode Material for Li Ion Batteries. *ACS Appl. Mater. Interfaces* **2014**, *6*, 6606–6615. [CrossRef] [PubMed]
3. Park, H.-C.; Lee, K.-H.; Lee, Y.-W.; Kim, S.-J.; Kim, D.-M.; Kim, M.-C.; Park, K.-W. Mesoporous molybdenum nitride nanobelts as an anode with improved electrochemical properties in lithium ion batteries. *J. Power Sources* **2014**, *269*, 534–541. [CrossRef]
4. Joshi, S.; Qi, W.; Puntambekar, A.; Chakrapani, V. Facile Synthesis of Large Area Two Dimensional Layers of Transition Metal Nitride and Their Use as Insertion Electrodes. *ACS Energy Lett.* **2018**, *2*, 1257–1262. [CrossRef]
5. Ji, L.; Lin, Z.; Alcoutlabi, M.; Zhang, X. Recent developments in nanostructured anode materials for rechargeable lithium-ion batteries. *Energy Environ. Sci.* **2011**, *4*, 2682–2699. [CrossRef]
6. Wu, H.; Zheng, G.; Liu, N.; Carney, T.J.; Yang, Y.; Cui, Y. Engineering Empty Space between Si Nanoparticles for Lithium-Ion Battery Anodes. *Nano Lett.* **2012**, *12*, 904–909. [CrossRef]
7. Dang, W.; Wang, W.; Yang, Y.; Wang, Y.; Huang, J.; Fang, X.; Wu, L.P.; Rong, Z.H.; Chen, X.; Li, X.; et al. One-step hydrothermal synthesis of 2D WO3 nanoplates@ graphene nanocomposite with superior anode performance for lithium ion battery. *Electrochim. Acta* **2019**, *313*, 99–108. [CrossRef]
8. Xu, H.; Zhang, H.; Fang, L.; Yang, J.; Wu, K.; Wang, Y. Hierarchical Molybdenum Nitride Nanochexes by a Textured Self-Assembly in Gas-Solid Phase for the Enhanced Application in Lithium Ion Batteries. *ACS Nano* **2015**, *9*, 6817–6825. [CrossRef]

9. Sun, Y.; Zhou, Y.; Zhu, Y.; Shen, Y.; Xie, A. In-Situ Synthesis of Petal-Like MoO$_2$@MoN/NF Heterojunction As Both an Advanced Binder-Free Anode and an Electrocatalyst for Lithium Ion Batteries and Water Splitting. *ACS Sustain. Chem. Eng.* **2019**, *7*, 9153–9163. [CrossRef]
10. Li, L.; Jia, X.N.; Zhang, Y.; Qiu, T.Y.; Hong, W.W.; Jiang, Y.L.; Zou, G.Q.; Hou, H.S.; Chen, X.C.; Ji, X.B. Li$_4$Ti$_5$O$_{12}$ quantum dot decorated carbon frameworks from carbon dots for fast lithium ion storage. *Mat. Chem. Front.* **2019**, *3*, 1761–1767. [CrossRef]
11. Zhang, B.; Cui, G.; Zhang, K.; Zhang, L.; Han, P.; Dong, S. Molybdenum nitride/nitrogen-doped graphene hybrid material for lithium storage in lithium ion batteries. *Electrochim. Acta* **2014**, *150*, 15–22. [CrossRef]
12. Zhang, Q.H.; Zhang, X.D.; He, W.; Xu, G.G.; Ren, M.M.; Liu, J.H.; Yang, X.N.; Wang, F. In situ fabrication of Na$_3$V$_2$(PO$_4$)(3) quantum dots in hard carbon nanosheets by using lignocelluloses for sodium ion batteries. *J. Mater. Sci. Technol.* **2019**, *35*, 2396–2403. [CrossRef]
13. Pan, X.; Xi, B.; Lu, H.; Zhang, Z.; An, X.; Liu, J.; Feng, J.; Xiong, S. Molybdenum Oxynitride Atomic Nanoclusters Bonded in Nanosheets of N-Doped Carbon Hierarchical Microspheres for Efficient Sodium Storage. *Nanomicro Lett.* **2022**, *14*, 163. [CrossRef] [PubMed]
14. Gayathri, V.; Peter, I.J. Graphene Quantum Dots Supported Mo$_3$S$_4$ as a Promising Candidate for Pt-Free Counter Electrode in Dye-Sensitized Solar Cell and Supercapacitor Applications. *ECS J. Solid State Sci. Technol.* **2021**, *10*, 091002. [CrossRef]
15. Kumar, R.; Bhuvana, T.; Sharma, A. Ammonolysis synthesis of nickel molybdenum nitride nanostructures for high-performance asymmetric supercapacitors. *New J. Chem.* **2020**, *44*, 14067–14074. [CrossRef]
16. He, T.; Zhang, W.; Manasa, P.; Ran, F. Quantum dots of molybdenum nitride embedded in continuously distributed polyaniline as novel electrode material for supercapacitor. *J. Alloys Compd.* **2020**, *812*, 152138. [CrossRef]
17. Borah, D.J.; Mostako, A.T.T.; Borgogoi, A.T.; Saikia, P.K.; Malakar, A. Modified top-down approach for synthesis of molybdenum oxide quantum dots: Sonication induced chemical etching of thin films. *RSC Adv.* **2020**, *10*, 3105–3114. [CrossRef]
18. Zheng, C.; Luo, N.; Huang, S.; Wu, W.; Huang, H.; Wei, M. Nanocomposite of Mo$_2$N Quantum Dots@MoO$_3$@Nitrogen-Doped Carbon as a High-Performance Anode for Lithium-Ion Batteries. *ACS Sustain. Chem. Eng.* **2019**, *7*, 10198–10206. [CrossRef]
19. Liu, J.; Tang, S.; Lu, Y.; Cai, G.; Liang, S.; Wang, W.; Chen, X. Synthesis of Mo$_2$N nanolayer coated MoO2 hollow nanostructures as high-performance anode materials for lithium-ion batteries. *Energy Environ. Sci.* **2013**, *6*, 2691–2697. [CrossRef]
20. Afanasiev, P. New single source route to the molybdenum nitride Mo(2)N. *Inorg. Chem.* **2002**, *41*, 5317–5319. [CrossRef]
21. Akhmetzyanova, U.; Pelíšková, L.; Skuhrovcová, L.; Tišler, Z.; Opanasenko, M. Molybdenum Nitrides, Carbides and Phosphides as Highly Efficient Catalysts for the (hydro)Deoxygenation Reaction. *ChemistrySelect* **2019**, *4*, 8453. [CrossRef]
22. Wang, L.; Zhang, Z.; Li, L.; Zhang, L.; Fang, H.; Song, Y.; Li, X. Preparation and lithium storage properties of Mo$_2$N quantum dots@nitrogen-doped graphene composite. *CIESC J.* **2020**, *71*, 5854–5862. [CrossRef]
23. Marcano, D.C.; Kosynkin, D.V.; Berlin, J.M.; Alexander, S.; Zhengzong, S.; Alexander, S.; Alemany, L.B.; Wei, L.; Tour, J.M. Improved synthesis of graphene oxide. *ACS Nano* **2010**, *4*, 4806. [CrossRef]
24. Zheng, C.; Liu, M.; Chen, W.; Zeng, L.; Wei, M. An in situ formed Se/CMK-3 composite for rechargeable lithium-ion batteries with long-term cycling performance. *J. Mater. Chem. A* **2016**, *4*, 13646–13651. [CrossRef]
25. Liu, X.B.; Amiinu, I.S.; Liu, S.J.; Pu, Z.H.; Li, W.Q.; Ye, B.; Tan, D.M.; Mu, S.C. H$_2$O$_2$-Assisted Synthesis of Porous N-Doped Graphene/Molybdenum Nitride Composites with Boosted Oxygen Reduction Reaction. *Adv. Mater. Interfaces* **2017**, *4*, 8. [CrossRef]
26. Liang, J.; Zhao, J.; Li, Y.; Lee, K.-T.; Liu, C.; Lin, H.; Cheng, Q.; Lan, Q.; Wu, L.; Tang, S.; et al. In situ SiO$_2$ etching strategy to prepare rice husk-derived porous carbons for supercapacitor application. *J. Taiwan Inst. Chem. Eng.* **2017**, *81*, 383–390. [CrossRef]
27. Maldonado, S.; Morin, S.; Stevenson, K.J. Structure, composition, and chemical reactivity of carbon nanotubes by selective nitrogen doping. *Carbon* **2006**, *44*, 1429–1437. [CrossRef]
28. Zhang, Q.; Li, X.; Ma, Q.; Zhang, Q.; Bai, H.; Yi, W.; Liu, J.; Han, J.; Xi, G. A metallic molybdenum dioxide with high stability for surface enhanced Raman spectroscopy. *Nat. Commun.* **2017**, *8*, 14903. [CrossRef]
29. Peng, T.; Tan, Z.; Zhang, M.; Li, L.; Wang, Y.; Guan, L.; Tan, X.; Pan, L.; Fang, H.; Wu, M. Facile and cost-effective manipulation of hierarchical carbon nanosheets for pseudocapacitive lithium/potassium storage. *Carbon* **2020**, *165*, 296–305. [CrossRef]
30. Wang, C.; Jiang, J.; Ruan, Q.; Ao, S.; Ostrikov, K.; Zhang, W.; Lu, J.; Li, Y.Y. Construction of MoO$_2$ Quantum Dot-Graphene and MoS$_2$ Nanoparticle-Graphene Nanoarchitectures toward Ultrahigh Lithium Storage Capability. *ACS Appl. Mater. Interfaces* **2017**, *9*, 28441–28450. [CrossRef]
31. Shi, W.; Zhu, J.; Rui, X.; Cao, X.; Chen, C.; Zhang, H.; Hng, H.H.; Yan, Q. Controlled synthesis of carbon-coated cobalt sulfide nanostructures in oil phase with enhanced li storage performances. *ACS Appl Mater. Interfaces* **2012**, *4*, 2999–3006. [CrossRef]
32. Liang, Y.; Chen, Y.; Ke, X.; Zhang, Z.; Wu, W.; Lin, G.; Zhou, Z.; Shi, Z. Coupling of triporosity and strong Au-Li interaction to enable dendrite-free lithium plating/stripping for long-life lithium metal anodes. *J. Mater. Chem. A* **2020**, *8*, 18094–18105. [CrossRef]
33. Liu, G.; Yang, Y.; Lu, X.; Qi, F.; Liang, Y.; Trukhanov, A.; Wu, Y.; Sun, Z.; Lu, X. Fully Active Bimetallic Phosphide Zn0.5Ge0.5P: A Novel High-Performance Anode for Na-Ion Batteries Coupled with Diglyme-Based Electrolyte. *ACS Appl. Mater. Interfaces* **2022**, *14*, 31803–31813. [CrossRef]
34. Ali, Y.; Lee, S. An integrated experimental and modeling study of the effect of solid electrolyte interphase formation and Cu dissolution on CuCo$_2$O$_4$-based Li-ion batteries. *Int. J. Energy Res.* **2021**, *46*, 3017–3033. [CrossRef]

35. Wang, F.-M.; Rick, J. Synergy of Nyquist and Bode electrochemical impedance spectroscopy studies to commercial type lithium ion batteries. *Solid State Ion.* **2014**, *268*, 31–34. [CrossRef]
36. Ogihara, N.; Kawauchi, S.; Okuda, C.; Itou, Y.; Takeuchi, Y.; Ukyo, Y. Theoretical and Experimental Analysis of Porous Electrodes for Lithium-Ion Batteries by Electrochemical Impedance Spectroscopy Using a Symmetric Cell. *J. Electrochem. Soc.* **2012**, *159*, A1034–A1039. [CrossRef]
37. Rui, X.H.; Li, C.; Liu, J.; Cheng, T.; Chen, C.H. The Li3V2(PO4)(3)/C composites with high-rate capability prepared by a maltose-based sot-gel route. *Electrochim. Acta* **2010**, *55*, 6761–6767. [CrossRef]
38. Zhu, L.; Zhang, M.; Yang, L.; Zhou, K.; Wang, Y.; Sun, D.; Tang, Y.; Wang, H. Engineering hierarchical structure and surface of Na4MnV(PO4)3 for ultrafast sodium storage by a scalable ball milling approach. *Nano Energy* **2022**, *99*, 107396. [CrossRef]
39. Wang, J.; Polleux, J.; Lim, J.; Dunn, B. Pseudocapacitive Contributions to Electrochemical Energy Storage in TiO_2 (Anatase) Nanoparticles. *J. Phys. Chem. C* **2007**, *111*, 14925–14931. [CrossRef]
40. Li, H.; Lang, J.; Lei, S.; Chen, J.; Wang, K.; Liu, L.; Zhang, T.; Liu, W.; Yan, X. A High-Performance Sodium-Ion Hybrid Capacitor Constructed by Metal-Organic Framework-Derived Anode and Cathode Materials. *Adv. Funct. Mater.* **2018**, *28*, 1800757. [CrossRef]
41. Das, S.R.; Majumder, S.B.; Katiyar, R.S. Kinetic analysis of the Li+ ion intercalation behavior of solution derived nano-crystalline lithium manganate thin films. *J. Power Sources* **2005**, *139*, 261–268. [CrossRef]
42. Veronica, A.; Jérémy, C.; Lowe, M.A.; Jong Woung, K.; Pierre-Louis, T.; Tolbert, S.H.; Abruña, H.D.; Patrice, S.; Bruce, D. High-rate electrochemical energy storage through Li+ intercalation pseudocapacitance. *Nat. Mater.* **2013**, *12*, 518.
43. Zhu, T.; Wang, Y.; Li, Y.; Cai, R.; Zhang, J.; Yu, C.; Wu, J.; Cui, J.; Zhang, Y.; Ajayan, P.M.; et al. CoO Quantum Dots Anchored on Reduced Graphene Oxide Aerogels for Lithium-Ion Storage. *ACS Appl. Nano Mater.* **2020**, *3*, 10369–10379. [CrossRef]
44. Wang, L.; Li, L.; Guo, X.; Zhang, J.; Liu, P.; Yan, J. Porous Mo_2N nano-column array thin film electrode for lithium ion storage. *Results Phys.* **2019**, *15*, 102715. [CrossRef]
45. Ji, W.; Shen, R.; Yang, R.; Yu, G.; Guo, X.; Peng, L.; Ding, W. Partially nitrided molybdenum trioxide with promoted performance as an anode material for lithium-ion batteries. *J. Mater. Chem. A* **2014**, *2*, 699–704. [CrossRef]
46. Zhou, D.; Wu, H.; Wei, Z.; Han, B.H. Graphene-molybdenum oxynitride porous material with improved cyclic stability and rate capability for rechargeable lithium ion batteries. *Phys. Chem. Chem. Phys.* **2013**, *15*, 16898–16906. [CrossRef]

Disclaimer/Publisher's Note: The statements, opinions and data contained in all publications are solely those of the individual author(s) and contributor(s) and not of MDPI and/or the editor(s). MDPI and/or the editor(s) disclaim responsibility for any injury to people or property resulting from any ideas, methods, instructions or products referred to in the content.

Review

Review and Stress Analysis on the Lithiation Onset of Amorphous Silicon Films

Kai Zhang [1], Erwin Hüger [2,3,*], Yong Li [4], Harald Schmidt [2,3] and Fuqian Yang [5,*]

[1] School of Aerospace Engineering and Applied Mechanics, Tongji University, Shanghai 200092, China
[2] Clausthal Centre of Material Technology, Clausthal University of Technology, D-38678 Clausthal-Zellerfeld, Germany
[3] Solid State Kinetics Group, Institute of Metallurgy, Clausthal University of Technology, D-38678 Clausthal-Zellerfeld, Germany
[4] Jiangsu Key Laboratory of Engineering Mechanics, School of Civil Engineering, Southeast University, Nanjing 210096, China
[5] Materials Program, Department of Chemical and Materials Engineering, University of Kentucky, Lexington, KY 40506, USA
* Correspondence: erwin.hueger@tu-clausthal.de (E.H.); fuqian.yang@uky.edu (F.Y.); Tel.: +49-5323724919 (E.H.); +1-8592572994 (F.Y.)

Abstract: This work aims to review and understand the behavior of the electrochemical lithiation onset of amorphous silicon (a-Si) films as electrochemically active material for new generation lithium-ion batteries. The article includes (i) a review on the lithiation onset of silicon films and (ii) a mechanochemical model with numerical results on the depth-resolved mechanical stress during the lithiation onset of silicon films. Recent experimental studies have revealed that the electrochemical lithiation onset of a-Si films involves the formation of a Li-poor phase ($Li_{0.3}Si$ alloy) and the propagation of a reaction front in the films. The literature review performed reveals peculiarities in the lithiation onset of a-Si films, such as (i) the build-up of the highest mechanical stress (up to 1.2 GPa) during lithiation, (ii) a linear increase in the mechanical stress with lithiation which mimics the characteristics of linear elastic deformation, (iii) only a minute volume increase during Li incorporation, which is lower than expected from the number of Li ions entering the silicon electrode, (iv) the largest heat generation appearing during cycling with only a minor degree of parasitic heat contribution, and (v) an unexpected enhanced brittleness. The literature review points to the important role of mechanical stresses in the formation of the Li-poor phase and the propagation of the reaction front. Consequently, a mechanochemical model consisting of two stages for the lithiation onset of a-Si film is developed. The numerical results calculated from the mechanochemical model are in good accord with the corresponding experimental data for the variations in the volumetric change with state of charge and for the moving speed of the reaction front for the lithiation of an a-Si film of 230 nm thickness under a total C-rate of C/18. An increase in the total C-rate increases the moving speed of the reaction front, and a Li-rich phase is likely formed prior to the end of the growth of the Li-poor phase at a high total C-rate. The stress-induced phase formation of the Li-poor phase likely occurs during the lithiation onset of silicon electrodes in lithium-ion battery.

Keywords: lithiation onset; silicon electrode; diffusion; stress; phase formation; reaction front

1. Introduction

The necessity of incorporating silicon in lithium-ion batteries (LIBs) to achieve high-capacity charge storage has been discussed in many publications, e.g., in references [1–17], but the lithiation process of silicon is scarcely discussed in the literature, as stated in a recent letter (reference [1]). Briefly, silicon, which is well known to be technologically well established, benign, and naturally abundant, possesses the highest gravimetric (3579 mAhg^{-1}

and volumetric (8334 mAhcm^{-3}) capacities, next to Li metal, for storing Li ions. Consequently, there are great efforts to implement silicon as an anode material in next-generation LIBs, although the endeavours remain unsatisfactory. The desire to store large amounts of charge (Li ions) by using large amounts of silicon represents concomitantly its disadvantage due to the rapid development of electrode damage (pulverization of active material) within the first cycle. Thus, it is of interest to understand the lithiation process of silicon even in the first cycle.

This work aims to review and understand the behavior of the lithiation onset of amorphous silicon (a-Si) films. Literature data are addressed, with a focus on some peculiarities in the properties of silicon electrodes during the lithiation onset which have not been reviewed until yet. Recent experimental studies have revealed that the lithiation onset of silicon films involves the formation of a Li-poor phase ($Li_{0.3}Si$ alloy) and the propagation of a reaction front in the films (Figure 1e–g), instead of the formation of a layer between the solid electrode and the liquid electrolyte (solid electrolyte interphase, SEI, layer) (Figure 1a,b) widely reported in the literature, as discussed in reference [1]. These results challenge the conventional consensus that the lithiation onset of silicon films is due to the formation of solid electrolyte interphase and to the reduction of native surface oxide (Figure 1a,b). Figure 1 summarizes the two different points of view concerning the lithiation onset of a-Si films with explanations given in the figure caption. To address this issue, we propose a mechanochemical model consisting of two stages for the lithiation onset of a-Si film. The numerical results are in good accord with the corresponding experimental data for the variations in the radial stress and volumetric change with state of charge and for the moving speed of the reaction front for the lithiation of an a-Si film. The analysis suggests the presence of the stress-induced phase formation of a Li-poor phase for the lithiation onset of a-Si and likely resolves the debate associated with the lithiation onset, which plays an important role in the structural integrity of Si-based anodes and the applications of Si-based LIB, as outlined in the following.

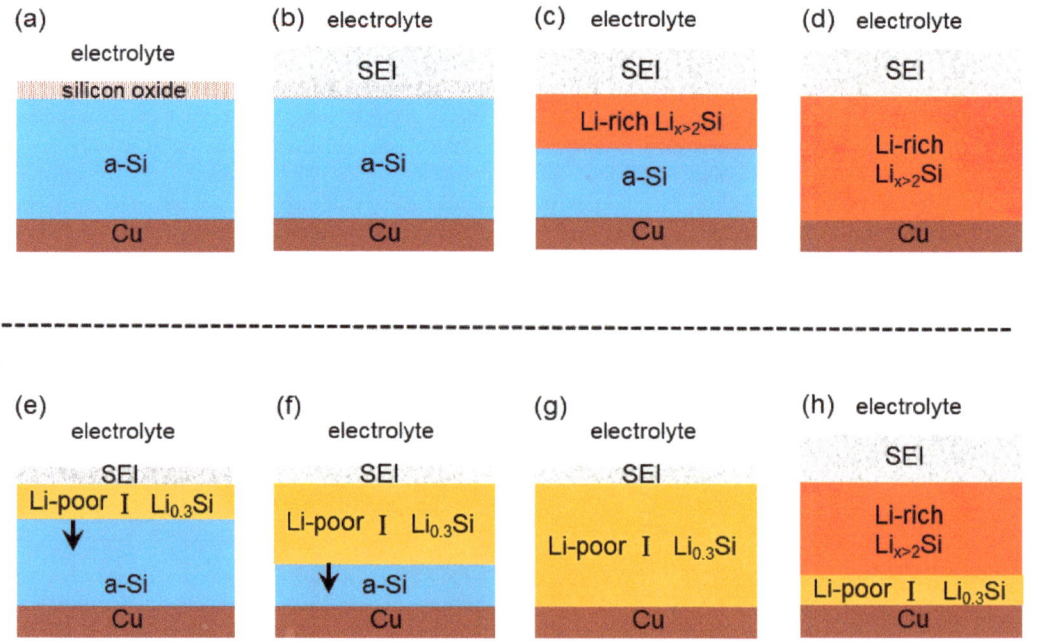

Figure 1. Schematic of two possible lithiation mechanisms (with and without a Li-poor silicide phase) of an a-Si film in the first lithiation process, summarizing the introduction section of this work. Shown

are cross sections of the Si film (blue) during the lithiation of the first cycle. The copper current collector is marked with brown. (**a–d**) Schematic of the lithiation process of the a-Si film according to the conventional consensus without the formation of the Li-poor phase. (**a**) Virgin silicon film electrode, i.e., before lithiation. The native surface oxide is approximately 2 nm thick. The copper current collector is very thick (well above 1 µm). The thickness of the a-Si film can be 1 µm or less. (**b**) Lithiation onset corresponding up to 10% state of charge (SOC) where firstly a solid electrolyte interphase (SEI) layer (marked with marbled gray) grows on un-lithiated silicon. The thickness of the SEI layer is considered to be some tens of nanometers (e.g., ~30 nm in reference [1]). (**c**) Further lithiation produces only a Li-rich lithium silicide (Li_xSi with $2 < x < 3.75$) underneath the SEI layer. (**d**) At full lithiation, the Li-rich layer reaches the current collector. (**e–h**) Schematic of the lithiation process in the presence of a Li-poor phase (yellow) as discussed in the present work. (**e**) The lithiation onset may produce an SEI layer (e.g., the reduction of the native surface oxide layer (panel (**a**)), but of lower extent than in panel (**b**). (**e–g**) A Li-poor region ($Li_{0.3}Si$) builds up during the lithiation onset which corresponds to the Li^+ uptake marked with roman number I in Figure 2. The lithiation process I continues until all silicon is converted into a Li-poor phase (panel (**g**)) at approximatively 10% SOC. The main part of the Li^+-uptake I in Figure 2 is not consumed for the SEI formation but for the lithiation of amorphous silicon [1]. (**h**) The rest of the Li-poor phase remains sandwiched between the current collector and a Li-rich phase. The SEI layer becomes thicker at the end of the lithiation process.

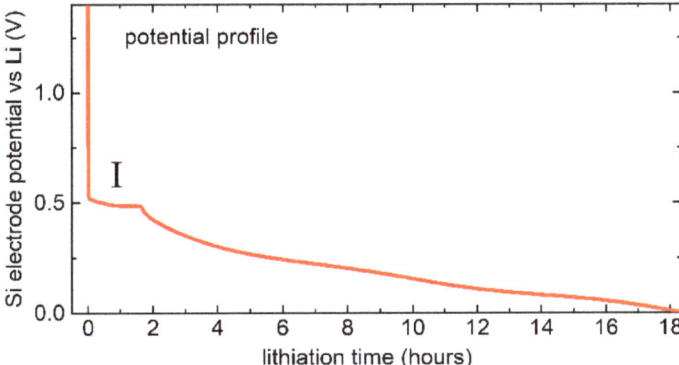

Figure 2. Variation in Si electrode potential with lithiation time during the first lithiation process of an a-Si thin film of 230 nm thickness at a constant current density of 11 µA/cm^2. Unpublished experimental data collected by the authors of the present work.

The authors of this work chose to work on this topic because they consider that it contributes to an understanding of the lithiation process of silicon, which plays an important role in determining the performance and reliability of Si-based LIBs during operation. The appearance of the Li-poor phase (Figure 1) may be proper and important for the LIB operation. It is more likely to accept that the formation of the Li-poor phase avoids structural disruption. Pure silicon is in proximity to the Li-poor Li_xSi phase (x ~ 0.3) (Figure 1e,f) and not directly to a Li-rich Li_xSi phase (x > 2) (Figure 1c), and may be preferable for the LIB operation.

Moreover, the appearance of the Li-poor phase has an influence not only on the bonding between different phases inside the silicon at different lithiated stages (Figure 1c,e,f) but also on the bonding of the active material (silicon) to the current collector (Figure 1), which is mainly copper for negative electrodes. The latter, i.e., the bonding of the active material to the current collector, is of importance for the electrode's integrity during LIB operation. On this issue, let us mention that experiments have shown that the Li-poor (x ~ 0.3) zone remains close to the copper current collector if the electrode is, electrochemically as possible, fully lithiated (Figure 1h), and also after several electrochemical cycles [1]. From that point of view, let us examine the literature on the bonding of the active material to the current collector. Before lithiation, the bonding of silicon to the current collector (copper) (Figure 1a) is experimentally and theoretically well known to be reasonably strong. The lithiation process inevitably increases the Li content in silicon in the vicinity of the current collector (Figure 1d,g,h). The ab initio molecular dynamics calculations at finite temperatures within the framework of density functional theory as implemented in the VASP code [18] predict that the bonding strength of the Li_xSi active material to the current collector decreases with the Li content in silicon. The reason for the weakened bonding at the a-Si/Cu interface represents a change in interfacial bonding upon lithiation [18]. According to the calculated interfacial work of separation versus Li content in the silicon electrode [18], the bonding strength of the Li poor silicide phase (x ~ 0.3) to copper (Figure 1g,h) is only about 2% lower than that for pure silicon bonded to copper (situation depicted in Figure 1a,e,f), but decreases for higher Li contents (situation depicted in Figure 1d), e.g., being seven times lower (i.e., 15%) for Li_xSi with x = 1. More drastic is the situation for the critical shear strength during sliding at the Si/Cu interface [18]. On the one hand, the interfacial work of separation (Li_xSi/Cu) is reduced by only ~15% upon full lithiation (x ~ 3.75) (situation depicted in Figure 1d) compared to a Li poor phase (x ~ 0.3) (situation depicted in Figure 1g,h); on the other hand, the critical shear strength is reduced by an order of magnitude for the situation presented in Figure 1d compared to that in Figure 1g,h. Similarly, the net charge on copper [18] increases by one order of magnitude upon full lithiation (situation depicted in Figure 1d) compared to a Li-poor phase (situation depicted in Figure 1g,h). The calculated derivative of the interface energy along the sliding direction [18] shows a peaked profile (i.e., a great fluctuation) vs sliding distance for the Li-poor phase (for the situation depicted in Figure 1g,h), whereas the Li-rich phase possesses a smooth profile (for the situation depicted in Figure 1d). This allows for the "frictionless" sliding of the Li-rich phase at the Li_xSi/copper interface but not for the Li-poor-phase at the Li_xSi/copper interface, which likely suggests a separation of the Li-rich phase (Figure 1d) from the copper current collector similar to the behavior of tin-based LIBs [19], but not the Li-poor phase (Figure 1g,h). The inhibition of "frictionless" sliding is crucial to prevent LIB failure by the loss of electrical contact between the active material and the current collector. As mentioned, experiments have shown that a Li-poor (x ~ 0.3) layer remains close to the copper current collector also at full lithiation (Figure 1h). Consequently, it is more favorable to form the Li-poor phase first, and to maintain the Li-poor phase at the interface to the current collector (Figure 1g,h) during LIB cycling. In that context, this work is focused on the analysis of this initial lithiation process, i.e., on the apparition of the Li-poor phase, which is, as mentioned, scarcely discussed in the literature, and has, consequently, remained below public awareness.

The article is organized as follows. Section 2 presents a literature survey on the lithiation onset of a-silicon. Section 2.1 describes the lithiation onset of a-silicon. Section 2.2 presents literature data on some peculiarities in the properties of silicon electrodes during the lithiation onset which have not been reviewed until yet. Section 2.3 elaborates some explanatory reasons for the phase front mechanism in the lithiation onset. Section 3 presents numerical investigations (modeling). Section 3.1 introduces a mechanochemical model aiming to numerically investigate the lithiation onset of a-silicon. Section 3.2 presents the numerical results on the depth-resolved mechanical stress during the lithiation onset of a-silicon. The last section summarizes the findings of this work.

2. Literature Survey

2.1. Description of the Lithiation Onset of Amorphous Silicon

The lithiation onset of a-Si thin films (i.e., up to 10% SOC, and a gravimetric capacity of ~350 mA/g) is characterized by a predominant Li^+ uptake, which takes place at a nearly constant potential of about 0.5 V in respect to lithium reference electrode [1], i.e., the flat plateau marked with roman number I in the potential profile of Figure 2. For information on the experimental procedure, see Appendix A.1. The flat plateau I corresponds to a predominant Li^+ uptake peak marked also with roman number I in the voltammograms (cyclic voltammetry, CV), and differential charge (dQ/dV plot) and capacity (dC/dV plot) curves presented in a recent letter [1]. The dC/dV curves are obtained by simply dividing the ordinate (dQ) of the dQ/dV plot by the electrode active mass, which is taken to be constant during cycling. Hence, the shape of dQ/dV curves is identical to that of dC/dV curves, and both are similar to voltammograms (for more details see references [1] and [2]). Most works in the literature usually attribute the lithiation onset to the formation of the SEI layer and to the reduction of native surface oxide [1].

The plateau I in Figure 2 was also observed for different liquid electrolytes (see reference [3]). It also appears when artificial solid-state SEI layers, i.e., a 60 nm thick oxide (Li_3PO_4) layer, are sputter-deposited on the top of a-Si thin layers [3]. The potential profiles presented by Wu et al. in reference [3] reveal, in addition to the flat plateau around 0.5 V, an additional flat plateau around 1 V. The latter is attributed to the formation of the SEI through the decomposition of carbonate-like solvents and lithium salts. The plateau around 1 V is rendered absent by capping the a-Si film electrode with an artificial SEI layer [3], but the plateau around 0.5 V remains present with a minor potential shift to approximatively 0.4 V. The origin of the plateau around 0.5 V (Figure 2) is not discussed in the work of Wu et al. [3]. In the communication of high-performance all-solid-state cells fabricated with silicon electrodes by Phan et al. [4], voltammograms are presented for the lithiation of sputter-deposited a-Si films of 400 nm in thickness, which were cycled with liquid electrolytes and, separately, with solid-state electrolytes. In both cases, i.e., for liquid and solid-state electrolytes, the lithiation onset is characterized by two predominant Li^+ uptakes, a smaller one around 0.5 V and a sharper one around 0.3 V, the latter denoted with the roman number I in reference [4]. The latter is attributed to the reduction in silicon oxide existing at the interface between silicon and the solid-state electrolyte. The Li^+ uptake positioned at 0.5 V is attributed to an irreversible reaction process independent of the nature of the substrate and electrolyte [4], pointing to an origin different from the formation of SEI layer.

Concerning the SEI formation, many reports find that it does not occur exclusively at the lithiation onset of silicon, but over the whole potential range, predominantly at lower potentials [5–17], where the carbonate-based electrolytes are unstable [20]. Investigations on the lithiation process of a-Si oxide films [21] and particles [22,23] have also found a predominant Li^+ uptake peak at the lithiation onset similar to that for the lithiation of a-Si films (corresponding to the plateau I in Figure 2). For Si_xO electrodes, the lithiation onset is ascribed to the Li bonding on defects inside the Si_xO electrodes [21–23], such as reactive oxygen available for Li bonding. Recent experimental results indicate that the Li^+ uptake in region I may also appear due to an intrinsic property of silicon lithiation [1], and it cannot be attributed to the SEI formation but to an in-depth lithiation of silicon [1,24]. Details on the lithiation onset of a-Si thin films are given in reference [1]. Briefly, the migration of Li in the silicon film experiences a two-phase lithiation process where a lithiated zone with a "constant" Li-concentration of x ~ 0.3 (Li-poor phase) is sharply delimited from non-lithiated silicon (see Figure 1e–g, and references [1,24]). The reaction front is planar and parallel to the film surface and propagates through the entire silicon film during the lithiation, as schematically shown in Figure 1 of this work.

The analysis of operando neutron reflectometry reveals that clean and well-ordered crystalline (100) oriented silicon wafers are initially also lithiated with the presence of a Li-poor region via a reaction front delimiting the Li-poor region from pure silicon [25]. This result indicates the appearance of the Li-poor silicide ($Li_{0.3}Si$) as an intrinsic property during the lithiation onset of pure and "defect-free" silicon. However, Hüger et al. [1] demonstrate that the modification of Li^+-uptake I, i.e., the appearance of the Li-poor phase, can be tailored by a change in the SEI layer and/or a change in the mechanical stress during the silicon lithiation. Cycling experiments with multiple pure silicon layers of each layer capped by a thin carbon layer are presented in the supplementary material of reference [1]. Two different Si/C multilayer (ML) electrodes were cycled with constant current. The liquid electrolyte was in direct contact with a silicon layer and a carbon layer for the first

ML ([Si(14 nm)/C(16 nm)] × 10 ML) and the second ML [C(16 nm)/Si(14 nm)] × 10 ML, respectively. For both MLs, Li$^+$ uptake I is not present [1], indicating that the appearance of Li$^+$ uptake I can be tailored. The experiments reveal that the absence of Li$^+$ uptake I is not dependent on the individual surface layer of the ML material (carbon or silicon) which is in direct contact with the electrolyte, but potentially on another factor. A possible difference of the mechanical stress inside the ML-film electrode in respect to that in single silicon-film electrodes may be responsible for the lack of the Li$^+$ uptake I in the MLs [1]. Consequently, the aim of this work represents, besides the literature survey, the calculation of the lithiation-induced depth-resolved mechanical stress on the phase boundary between a growing Li$_{0.3}$Si phase which consumes a clean and "defect-free" silicon film. Before doing so in Section 3, elaborations on the changes in the physical properties of silicon during the lithiation onset and on the reasons for the appearance of the phase front delimiting Li$_{0.3}$Si from a-silicon are presented.

2.2. Changes in Physical Properties of Silicon during the Lithiation Onset

Reviewing the experiments performed during the lithiation onset of silicon, one finds that the predominant Li$^+$ uptake I (which builds up the Li$_{0.3}$Si phase) is connected with the appearance of peculiarities which are absent at other states of charge. They are (i) the build-up of the highest mechanical stress (up to 1.2 GPa) during silicon lithiation, (ii) a linear increase in the mechanical stress with lithiation which mimics the characteristics of linear elastic deformation, (iii) only a minute volume increase during Li incorporation which is lower than expected from the number of Li ions entering the silicon electrode, (iv) the largest heat generation appearing during cycling with only a minor degree of parasitic heat contribution, and (v) an unexpected enhanced brittleness.

Figure 3 presents the potential profile measured during the first lithiation process of an a-Si film of 230 nm thick (Figure 3a) in direct comparison with sketches of corresponding mechanical stress (Figure 3b) and volume change of silicon (film thickness) (Figure 3c) during the first lithiation process of a-Si thin films. Note that although Figure 3a represents experimental data collected by the authors of the present work, Figure 3b represents only an outline (sketch) of a typical mechanical stress curve during the lithiation of an a-Si film published in the literature, e.g., in reference [26] and reference [27]. Additionally, Figure 3c of this work represents only an outline of the volume change during lithiation as measured operando with neutron reflectometry (reference [28]) and microscopy (reference [29]).

The in situ measured mechanical stresses during lithiation of a-Si films with thicknesses of 50 nm [26,30], 100 nm [30,31], 150 nm [30], 200 nm [31], 250 nm [27,30,32], 300 nm [33] and 325 nm [34] reported in the literature all exhibit similar behavior during lithiation and de-lithiation. Note also that the electrochemical Li$^+$ uptake and release is similar for different film thicknesses, suggesting the negligible effect of film thickness on the lithiation processes [1]. The substrate (current collector) confines the in-plane expansion and contraction of the active film during lithiation and delithiation, respectively. Consequently, lithiation-induced compressive stress appears and linearly increases up to ~1.2 GPa at ~10% SOC (x ~ 0.3 in Li$_x$Si) during lithiation (Figure 3b). The rapidly linear increase likely suggests a lack of plasticity during the initial lithiation process, corresponding to the predominant Li$^+$ uptake I in Figures 2 and 3a. Furthermore, the increase in the film volume (i.e., film thickness) during Li$^+$ uptake I is much less than that predicted from the inserted amount of Li (Figure 3c). Beyond 10% SOC, the volume expansion of the silicon film caused by the lithiation is in consistence with the trend predicted theoretically with the increase in the inserted Li amount (Figure 3c), which involves plastic deformation (Figure 3b).

An unexpected result was also obtained in the measurement of heat flow during lithiation process I. Housel et al. [35] performed isothermal microcalorimetry to detect the heat flow generated during the lithiation and delithiation of a silicon electrode produced from crystalline silicon powder. The highest entropic heat flow (~50 mW/g) appeared for up to 10% SOC of the first cycle, which corresponds to lithiation process I in Figures 2 and 3a. Beyond 10% SOC, the entropic heat flow decreased markedly [35]. The contribution of parasitic heat, which indicates SEI growth, was obtained from a comparison between the total heat flow and the sum of polarization heat flow and entropic heat flow [35]. Unexpectedly, during the first lithiation, the lowest dissipated heat flow associated with parasitic reactions appeared between 1% and 10% SOC, and, hence, in the region of lithiation process I.

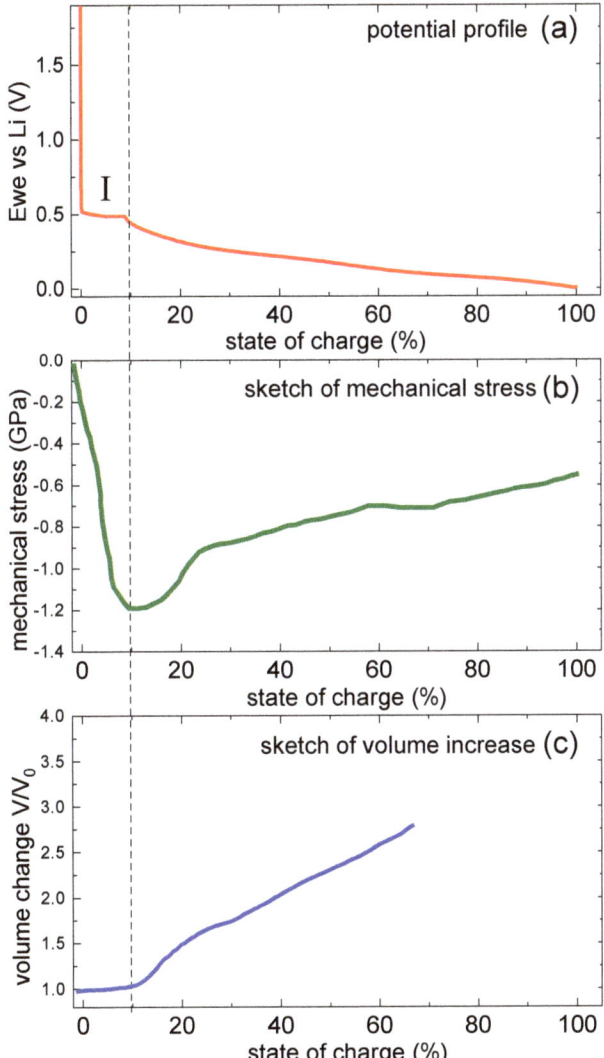

Figure 3. (**a**) Potential profile during the initial lithiation of an a-Si thin film of 230 nm in direct comparison with outlines (sketches) of (**b**) a typical mechanical stress curve during the lithiation of an a-Si film, and of (**c**) the typical volume increase curve during the lithiation of an a-Si film. All the values are plotted versus state of charge (SOC). 100% SOC corresponds to $Li_{3.75}Si$ (x = 3.75). V_0 represents the silicon film volume before lithiation. The dashed line represents the end of the predominant Li$^+$ uptake marked with roman number I in Figures 2 and 3a.

For a-Si films, nanoindentation measurements [34] reveal the dependence of the fracture behavior on the Li concentration of lithiated silicon. During initial lithiation, the fracture toughness decreases first from unlithiated a-silicon to a minimum value for a lithiated SOC corresponding to $Li_{0.31}Si$ (~10% SOC). For a higher Li content, the fracture toughness becomes higher than that of pure silicon [34], which is as expected. Lithiated silicon has been found to exhibit tensile deformation in contrary to the brittle behaviour of pure silicon [34]. The increase in fracture toughness indicates a brittle-to-ductile transition in lithiated silicon with the increase in Li concentration, which is as expected. This behaviour is in line with an increase in ductility with the insertion of Li due to the introduction of a more metallic character of atomic bonding with increased Li concentration. The atomistic mechanism for the brittle-to-ductile transition in amorphous Li_xSi is thought to be the decrease in strong covalent Si bonds (Si–Si) and the concomitant increase in delocalized (metallic) Li bonds (Li-Si) [34,36], giving rise to an alteration in the dominant atomic-level processes of deformation and fracture with increasing Li concentration. Thus, the ductility should increase with the increase in Li incorporation into silicon, but this is not the case for Li concentrations of up to x ~ 0.31 [34]. The Li insertion firstly decreases the fracture toughness up to x ~ 0.31, indicating lithiation-induced embrittlement [34]. There is an unexpected embrittlement increase for Li insertion up to x ~ 0.31, which corresponds to Li^+ uptake process I. This means an increase in atomic bonds and in the cohesive strength during lithiation process I. This is in line with the measurement of only a minute volume change in contrast to that expected from the amount of Li inserted for SOC up to 10% (Figure 3c). The smaller volume increase may result in a higher atomic density in amorphous $Li_{0.3}Si$ in respect to pure a-silicon, which is indicative of stronger cohesive strength in $Li_{0.3}Si$ than in pure silicon.

2.3. Reasons for the Appearance of the Phase Front Delimiting $Li_{0.3}Si$ from a-Silicon

The following elaboration gives some explanations as to (i) why the lithiation onset of silicon involves a phase front which delimits the Li-poor phase from unreacted silicon, and (ii) why the Li-poor phase possesses the particular Li concentration corresponding to x ~ 0.3 in Li_xSi.

The appearance of the Li-poor silicide ($Li_{0.3}Si$) as an intrinsic property during the lithiation onset of pure and "defect-free" silicon is unexpected, since no stable Li_xSi silicide phases appear for x < 1 at ambient pressure in the Li-Si phase diagram [37–42]. The first principle calculation predicts that mechanical pressure above 8 GPa [41] is necessary for the Li-poor silicide with x ~ 0.3 ($LiSi_3$) to become stable. Such a high mechanical pressure has not been measured during the lithiation of a-Si thin films [26,27,30–34]. The mechanical pressure needed to produce the energy barriers around the x ~ 0.3 phase in order to change its instability into metastability is probably lower for an amorphous network. Nanometric zones of crystalline $LiSi_3$ with dimensions below 2 nm and without a long-range order may also appear, mimicking an amorphous network, and may need a lower mechanical pressure for the occurrence of the $LiSi_3$ phase compared to bulk crystalline $LiSi_3$. It should be noted that the first principle calculation was performed at 0 K and that the energy barrier (mechanical pressure) for the presence of the stable Li-poor silicide is probably less than 8 GPa at a higher temperature (room temperature).

The analysis of the experimental data in reference [1] shows that the Li-poor phase, which builds up during the Li^+ uptake I, is not formed from an invariant reaction because it does not always occur at the same potential [1]. This might hint at a metastable reaction that is irreversible upon the first cycle [1]. A metastable state points to the existence of energy barriers between the metastable state and the global minimum of energy. Even if the metastable state appears and is stable at room temperature, temperatures higher than room temperature often cause the transition of the metastable phase into a phase of a lower energy minimum, or it is not even developed at that temperature. Consequently, it is obvious that the amorphous and metastable Li-poor phase formed at room temperature during the Li^+ uptake I is not found in the Li-Si phase diagram [37–42].

The presence of a reaction front associated with the Li-poor phase at the lithiation onset of silicon (Figure 1e–g) is likely attributed to the coupling between lithiation-induced stress (chemical stress) and lithium diffusion. The lithium diffusion is controlled simply by the gradient of Li concentration if no stress gradient is present. Under the action of stress, the driving force for the lithium diffusion is the gradient of chemical potential, which consists of the contributions of the gradient of Li concentration and the stress gradient [43]. The compositional change in a silicon electrode (the formation of an intermetallic compound) corresponding to Li$^+$ uptake I under a stress is calculated to be more than the Li concentration in the silicon electrode at an SOC of 10% (x ~ 0.3 in Li$_x$Si), i.e., it amounts to $\Delta x \sim -0.5$ as obtained from the curve for stress-induced composition changes in silicon films presented in reference [26]. This suggests that the stress gradient produced in Li$^+$ uptake I hinders Li migration into silicon. That is to say, the chemical stress introduced by the lithium diffusion has the tendency to reject a significant amount of lithium ($\Delta x \sim -0.5$) from the silicon electrode as it would be maximal at Li$^+$ uptake I (x ~ 0.3). This process should continue for lithiation up to x ~ 0.5. Therefore, there are two opposite processes involving the lithiation onset of a silicon electrode—one is the electrochemical lithiation under the gradient of Li concentration, which drives lithium through the Li-poor phase and into the un-lithiated silicon to produce a Li$_x$Si phase, and the other is a stress-limited process with the chemical stress trying to hold the silicon electrode free of lithium during the lithiation. The interaction between these two processes may lead to a two-phase lithiation mechanism with the formation of a sharp interface (reaction front) delimiting non-lithiated silicon from lithiated silicon with x < 0.5 as depicted in Figure 1e–g.

From experimental results obtained in Li isotope exchange experiments and secondary ion mass spectrometry (SIMS) depth profiling [44–46], one can estimate the Li flux (Li permeability) at room temperature through sputter-deposited a-Si thin films (e.g., Table 2 in reference [45]). The Li flux (Li permeability) in a-Si thin layers of ~100 nm is estimated to be ~10^{-38} m^{2}s^{-1}, which is 17 orders of magnitude lower than ~10^{-21} m^{2}s^{-1} in the amorphous Li$_{0.3}$Si phase, using the kinetical data given in references [44–46] to determine Li permeability. This means that the high Li flux through the Li-poor phase toward un-lithiated silicon does not permeate inside the un-lithiated silicon, but actually stops at the interface between the Li-poor phase and the un-lithiated silicon. At that interface, silicon is converted into the Li-poor phase, as schematically shown in Figure 1e–g. Further discussion on Li diffusivity and Li solubility at room temperature in un-lithiated a-silicon and in the Li$_{0.3}$Si phase is given in the next section. However, the question still remains open as to why the Li poor phase possesses the particular Li content of x ~ 0.3.

First principle calculations found that the formation energy of Li$_x$Si phases becomes negative only for $x \geq 0.3$ [47,48]. This means that the formation of a Li$_x$Si phase is energetically favorable only for $x \geq 0.3$. Thus, firstly, the total energy calculation predicts that pure silicon does not allow Li with a concentration of x < 0.3 to be incorporated into silicon above its extremely low Li solubility limit [45]. Consequently, the lithiation of silicon starts by the formation of Li$_{0.3}$Si. Secondly, as mentioned above, stresses due to lithiation have the tendency to repel Li atoms from silicon for x < 0.5. Finally, the Li permeation (Li flux) in the Li$_{0.3}$Si phase is, as stated above, 17 orders of magnitude higher than that in pure silicon. Thus, there are three factors—(i) extremely low Li permeability in pure silicon and relative high Li permeability in Li$_{0.3}$Si phase, (ii) Li repulsion from pure silicon due to mechanical stress, and (iii) energy gain by the formation of Li$_x$Si only for $x \geq 0.3$—which likely contribute to why initially a two-phase lithiation mechanism appears which sharply delimits pure silicon from a Li$_x$Si phase of x ~ 0.3 (Figure 1e–g). In order to investigate the peculiar lithiation onset of a-Si films, we present and apply in this work a mechanochemical model to analyze the stress evolution during the lithiation onset of an a-Si film for a planar phase-front mechanism (Figure 4) by modelling the interaction between diffusion and stress in the lithiation onset of Si-based electrodes.

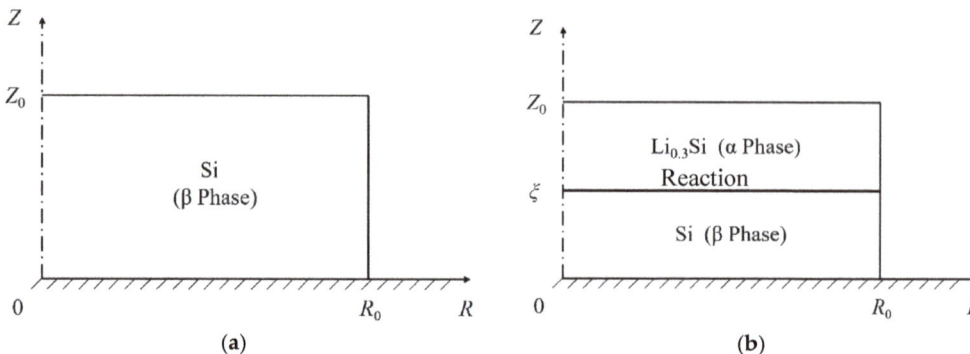

Figure 4. Schematic of Li diffusion in an a-Si thin film electrode at (**a**) stage 0, and (**b**) stage I with the formation of Li-poor phase $Li_{0.3}Si$ (α phase) and a phase boundary (reaction front), ξ. In a cylindrical coordinate (R, Θ, Z), lithium diffuses into the thin film from the surface of $Z = Z_0$ and the thin film is fixed on the substrate at $Z = 0$. The pristine un-lithiated silicon is denoted as β phase.

3. Numerical Investigation

A mechanochemical model is introduced in this work and used to analyze the stress evolution during the lithiation onset of an a-Si film. Numerical simulation was performed by taking into consideration the experimental results of (i) the phase-front lithiation mechanism of a Li-poor phase consuming unreacted silicon, and (ii) a Li-poor phase with a Li concentration corresponding to $x \sim 0.3$ in Li_xSi. The appearance of the planar phase-front lithiation mechanism, and the Li concentration of the Li-poor phase, are not a result of the simulation, but instead represent the ingredients of the simulation.

There are studies on the stress analysis of the lithiation of silicon, e.g., references [34,49,50], but none of them have focused on the subject presented in this work, i.e., the phase front mechanism appearing in the lithiation plateau I in Figure 2. Currently, most modeling analyses have been based on the concept of diffusion-induced stress (with and without the contribution of stress-limited diffusion) and the concentration dependence of mechanical properties in the framework of elasticity [51–53], elastoplasticity [54,55] or viscoplasticity [56–58]. To address the lithiation of electrode materials with the possible presence of the reaction front, phase-field models have also been developed to study stress evolution in electrode materials during electrochemical cycling [59,60]. It should be noted that there are few works addressing the formation of new compounds in the analysis [61–63], even though Yang [64] had formulated the formulas with the contributions of chemical reactions to the mass transport and the stress evolution.

3.1. Mechanochemical Model

The analysis in this work focuses on the stress evolution and the motion of the reaction front for the growth of the Li-poor phase in an a-Si film, which is based on the theory of linear elasticity and the coupling between diffusion and stress [43]. A moving boundary condition is used to describe the propagation of the reaction front. The temporal evolution of the stresses is illustrated, and a comparison between the numerical results and those reported in the literature is performed.

According to the lithiation experiment (i.e., Figure 2), we consider a silicon disk of 200 nm in thickness and 2000 nm in radius instead of ~1 cm used in experiments by Hüger et al. [1,65] Note that the results obtained with the radius of 2000 nm are the same as those with the radius of ~1 cm for the region far away from the disk edge (see Figure A1 in Appendix A.2 for more detail). A cylindrical coordinate system, (R, Θ, Z), shown in Figure 4, is used, and the constitutive equations of the disk are [66]

$$\sigma = \frac{\nu E}{(1+\nu)(1-2\nu)}[Tr(\varepsilon) - \Omega C]\mathbf{I} + \frac{E}{1+\nu}(\varepsilon - \frac{1}{3}\Omega C\mathbf{I}) \quad (1)$$

where σ, ε and $\mathbf{I}$ are the stress, strain and unit tensors, respectively, Ω is the partial molar volume, C is the Li-concentration, and E and ν are Young's modulus and Poisson's ratio of silicon, respectively. The normal strain components and stress components are ε_i and σ_i ($i = R, \Theta, Z$), respectively. Note that the contribution of surface elasticity [67] is not considered in this model since the analysis is focused on the planar surface and interface.

With axisymmetric characteristics, the strain components are calculated from nonzero radial and axial displacements, (u, w), as

$$\varepsilon_R = \frac{\partial u}{\partial R}, \; \varepsilon_\Theta = \frac{u}{R}, \; \varepsilon_Z = \frac{\partial w}{\partial Z}, \; \varepsilon_{ZR} = \frac{1}{2}\left(\frac{\partial u}{\partial Z} + \frac{\partial w}{\partial R}\right). \quad (2)$$

The equilibrium equations without body force are

$$\left.\begin{array}{l}\frac{\partial \sigma_R}{\partial R} + \frac{\partial \tau_{ZR}}{\partial Z} + \frac{\sigma_R - \sigma_\Theta}{R} = 0 \\ \frac{\partial \sigma_Z}{\partial Z} + \frac{\partial \tau_{RZ}}{\partial R} + \frac{\tau_{ZR}}{R} = 0\end{array}\right\}. \quad (3)$$

The boundary conditions are

$$\sigma_R|_{R=R_0} = 0, \; \sigma_Z|_{Z=Z_0} = 0, \; \tau_{RZ}|_{Z=Z_0} = 0, \; w|_{Z=0} = 0 \text{ and } u|_{R=0} = 0. \quad (4)$$

For the diffusion of lithium into the Si thin film involving the formation of a $Li_{0.3}Si$ phase, we divide the diffusion process into two stages, as shown in Figure 5: stage 0 is the diffusion of lithium in a single-phase system before the lithium concentration reaches the critical value for the formation of a new phase, and stage I is the diffusion of lithium in a bi-phase system with the new phase.

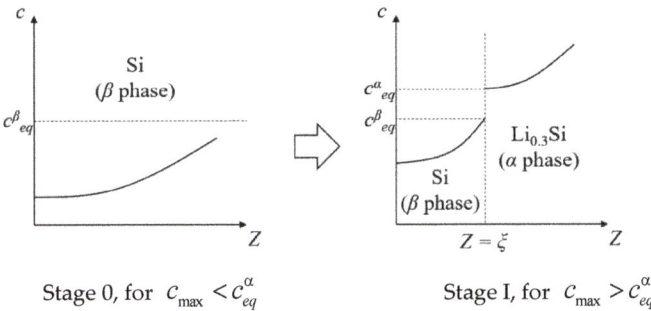

Figure 5. Schematic of the Li concentration distribution at two different stages during the lithiation of the Si thin film.

At the onset of the lithiation, there is only β phase (Si). Increasing the lithiation time leads to the diffusion of lithium into the Si thin film and to the increase in Li concentration. When the Li concentration at the surface of the Si thin film reaches the equilibrium concentration, c^α_{eq}, for the formation of α phase ($Li_{0.3}Si$), the α phase is formed and stage I starts. In stage I, there are two phases of α and β, which are separated by an interface, ξ (Figure 5b). The Li concentrations in each phase at the interface are c^α_{eq} and c^β_{eq}, respectively. The interface moves towards the surface of the rigid substrate (current collector, at $Z = 0$) and causes the change in lithium concentration from c^β_{eq} to c^α_{eq}.

For the diffusion of Li, the diffusional flux, **J**, is calculated as

$$\mathbf{J} = -MC_{max}c\,\text{grad}\mu \quad \text{and} \quad \mu = \mu_0 + RT\ln c - \sigma_h\Omega, \tag{5}$$

with M (=D/R_gT) as the mobility, c (=C/C_{max}) as the normalized concentration, C_{max} as the maximum stoichiometric concentration of Li, μ as the chemical potential of Li, D as the diffusion coefficient, R_g as the gas constant, T as the absolute temperature, and σ_h (=$[\sigma_R + \sigma_\Theta + \sigma_Z]/3$) as hydrostatic stress. The diffusion equation is

$$\begin{cases} \dfrac{\partial c^\alpha}{\partial t} - \dfrac{\partial}{\partial Z}\left(D^\alpha\left(\dfrac{\partial c^\alpha}{\partial Z} - \dfrac{\Omega c^\alpha}{R_gT}\dfrac{\partial \sigma_h}{\partial Z}\right)\right) = 0, & Z \in (\xi, Z_0] \\ \dfrac{\partial c^\beta}{\partial t} - \dfrac{\partial}{\partial Z}\left(D^\beta\left(\dfrac{\partial c^\beta}{\partial Z} - \dfrac{\Omega c^\beta}{R_gT}\dfrac{\partial \sigma_h}{\partial Z}\right)\right) = 0, & Z \in [0, \xi) \end{cases}, \tag{6}$$

where the superscripts of α and β represent the corresponding parameters in the α and β phases, respectively. Note that we consider the dependence of Li concentration only on variable Z for the disk thickness much less than the disk radius.

At stage 0, lithium diffuses into the thin film from the surface of $Z = Z_0$, as shown in Figure 2a. The initially un-lithiated film gives

$$c^\beta(R, Z)\big|_{t=0} = 0. \tag{7}$$

The galvanostatic operation of the film [66] follows

$$-D^\beta\left(\dfrac{\partial c^\beta}{\partial Z} - \dfrac{\Omega c^\beta}{R_gT}\dfrac{\partial \sigma_h}{\partial Z}\right)\bigg|_{Z=Z_0} = J_0, \tag{8}$$

and the impermeable condition of the substrate yields

$$\left(\dfrac{\partial c^\beta}{\partial Z} - \dfrac{\Omega c^\beta}{R_gT}\dfrac{\partial \sigma_h}{\partial Z}\right)\bigg|_{Z=0} = 0. \tag{9}$$

At stage I, the phase formation occurs after the maximum normalized concentration reaches the critical value. A phase interface of ξ is formed, delimiting the Si thin film into Li$_{0.3}$Si alloy (α phase) and Si-solution (β phase), as shown in Figures 4b and 5b. A continuous influx of Li causes the motion of ξ from $Z = Z_0$ to $Z = 0$. The Li concentrations on the either side of the interface are

$$c^\alpha\big|_{Z=\xi^+} = c^\alpha_{eq}, \quad c^\beta\big|_{Z=\xi^-} = c^\beta_{eq}. \tag{10}$$

The moving speed of the interface is determined by the following equation

$$(c^\alpha_{eq} - c^\beta_{eq})\dfrac{d\xi}{dt} = D^\alpha\left(\dfrac{\partial c^\alpha}{\partial Z} - \dfrac{\Omega c^\alpha}{R_gT}\dfrac{\partial \sigma_h}{\partial Z}\right) - D^\beta\left(\dfrac{\partial c^\beta}{\partial Z} - \dfrac{\Omega c^\beta}{R_gT}\dfrac{\partial \sigma_h}{\partial Z}\right). \tag{11}$$

The boundary condition at the surface of the thin film is

$$-D^\alpha\left(\dfrac{\partial c^\alpha}{\partial Z} - \dfrac{\Omega c^\alpha}{R_gT}\dfrac{\partial \sigma_h}{\partial Z}\right)\bigg|_{Z=Z_0} = J_0. \tag{12}$$

The boundary condition at the bottom of the thin film is the same as Equation (9).

The numerical method was used to solve the coupled equations given above via the partial differential equation (PDE) module of COMSOL Multiphysics. There are 1000 nodes along the Z direction and 5000 nodes along the R direction for the finite element model.

Consider the presence of the $Li_{0.3}Si$ phase as the phase formation introduced by Li diffusion. The normalized equilibrium concentration for the $Li_{0.3}Si$ phase is determined as the ratio of the Li fraction in the compound to the maximum Li fraction in Si as

$$\hat{c}^\alpha_{eq} = \frac{\hat{c}_{Li_{0.3}Si}}{\hat{c}_{Li_{3.75}Si}} = \frac{0.3}{3.75} = 0.08. \tag{13}$$

The equilibrium concentration of lithium for the un-lithiated phase should be equal to the Li solubility limit of the a-Si film. Li solubility in silicon is very low (see reference [45]). Consequently, it was possible to measure Li solubility only at temperatures where the solubility attains higher values than at room temperature. The Li solubility of ~100 nm a-Si thin film sputter-deposited within the same laboratory, and using the same ion beam coater as in the present work, was experimentally determined from SIMS depth profiling for samples annealed at temperatures in between 513 K and 773 K [45]. At the lowest temperature, i.e., at 513 K, Li solubility amounts to about one Li atom per 10^5 silicon atoms, (directly notified as 10^{-5}). The experimentally determined temperature dependence of Li solubility follows an Arrhenius law [45]. The extrapolation of the Li solubility to room temperature gives a Li solubility of ~10^{-9} (Li atoms per Si atoms, [46]). Hence, experimental results reveal that the equilibrium concentration for the un-lithiated phase should be equal to ~10^{-9}, and in that manner the equilibrium concentration for the un-lithiated phase can be assumed to be zero.

The Li diffusivity of a-silicon is actually unknown. The newest experiments to electrochemically measure Li diffusivity at room temperature by GITT (Galvanostatic Intermittent Titration Technique) [68] reveal the Li chemical diffusivity as well as the Li tracer diffusivity at the onset of the lithiation process to be almost independent of Li concentration in Li_xSi for x < 0.35. This is consistent with the fact that a phase (i.e., the Li-poor phase) with a constant Li concentration of x ~ 0.3 grows into the a-Si film electrode (references [1,24,68]). The Li tracer diffusivity was measured by GITT to be around 10^{-20} m$^2 \cdot$s^{-1} [68]. Tracer diffusivities at temperatures in between 573 K and 773 K were measured by SIMS in a $Li_{0.25}Si$ alloy produced by co-sputtering of lithium-metal and silicon-wafer sputter-targets (reference [46]). The Li diffusivity obtained by extrapolation to room temperature is of about 10^{-20} m^{2}s^{-1} for the composition of $Li_{0.25}Si$, in accord with the Li tracer diffusivities obtained by GITT [68]. Consequently, the Li diffusion coefficient in the α phase (i.e., $Li_{0.3}Si$) was taken in the present model to be 1.5×10^{-20} m$^2 \cdot$s^{-1}. Non-electrochemical-based experiments performed at elevated temperatures up to 773 K (reference [45]) indicate that the Li tracer diffusion coefficient in ~100 nm a-Si thin films extrapolated to room temperature attains an inexplicable 10 orders of magnitude lower diffusivity, ~10^{-30} m$^2 \cdot$s^{-1} if the Li diffusivity was estimated from Li permeability values, or ~ 10^{-29} m$^2 \cdot$s^{-1} if the Li diffusivity was determined by the lag-time method (reference [45]). This is in accordance with the value of about 10^{-29} m$^2 \cdot$s^{-1} extrapolated to room temperature from tracer diffusivities (determined from SIMS experiments) at higher temperatures measured by Strauß et al. [46] for the alloy of $Li_{0.02}Si$ which represents almost pure silicon. The estimated Li diffusivity of only ~10^{-30} m$^2 \cdot$s^{-1} at room temperature in un-lithiated silicon is inexplicably low, and in the opinion of the authors it may be unrealistically low. It may be possible that the correlation between diffusivity and temperature changes at low temperatures down to room temperature, due to a possible change in Li diffusion and solubility mechanism at low temperatures. Consequently, we consider in the present model a higher Li diffusion coefficient in the β phase, i.e., that of the α phase (1.5×10^{-20} m$^2 \cdot$s^{-1}). The effects of the difference of Li diffusivities between the α phase ($Li_{0.3}Si$) and β phase (a-Si) on the spatial evolution of the interface position are analyzed in Figure A2 of Appendix A.2.

Consider that the silicon thin film is lithiated under a total C-rate of C/14. The corresponding Li influx at the surface is

$$J_0 = \frac{\text{Volume} \cdot \text{C} - \text{rate}}{\text{Surface area} \cdot 3600} = \frac{Z_0/14}{3600} = 3.97 \times 10^{-12} \text{ mol} \cdot \text{m}^{-2} \cdot \text{s}^{-1}. \tag{14}$$

The film thickness, Z_0, is 200×10^{-9} m. Finally, Table 1 lists the parameters used in the numerical calculation.

Table 1. Material properties and parameters used in the numerical calculation.

Symbol	Parameter	Numerical Value
D^α	Diffusion coefficient in α phase	1.5×10^{-20} m$^2\cdot$s^{-1}
D^β	Diffusion coefficient in β phase	1.5×10^{-20} m$^2\cdot$s^{-1}
Z_0	Initial thickness of the silicon thin film	200×10^{-9} m
$\hat{c}_{eq}^\alpha$	Normalized equilibrium value of α phase	0.08
$\hat{c}_{eq}^\beta$	Normalized equilibrium value of β phase	0
E	Elastic modulus	92 GPa [69]
ν	Poisson's ratio	0.28 [70]
Ω	Partial molar volume	8.18×10^{-6} m$^3\cdot$mol^{-1} [71]
C_{max}	Maximum concentration	3.67×10^{-5} mol$\cdot$m^{-3} [33]
R	Gas constant	8.314 J$\cdot$mol$^{-1}\cdot$K^{-1}
T	Absolut temperature	300 K

3.2. Numerical Results and Discussion

Figure 6 shows temporal evolution of the normalized concentration and radial stress at the center of the free surface. The Li concentration at the surface of the Si thin film increases with the increase in diffusion time (Figure 6a), revealing more lithium in the Si film with increasing the diffusion time. It takes only ~35 s for the Li concentration at the surface of the Si film to reach the equilibrium concentration of $\hat{c}_{eq}^\alpha = 0.08$. This result suggests that stage 0 ends at ~35 s with the onset of stage I. However, the radial stress near the surface is compressive, and the magnitude of the radial stress increases sharply during this short time from 0 to about 7.4 GPa (corresponding to a hydrostatic stress of −4.7 GPa and a shear stress of 3.7 GPa) (Figure 6b), slightly less than the theoretically predicted 8 GPa for the stabilization of Li-poor silicide with x ~ 0.3 [41]. The latter, i.e., the 8 GPa threshold to initiate the formation of the Li-poor phase, may be, as previously mentioned, lower in amorphous silicon and at a finite (room) temperature. This result indirectly confirms the stress-induced phase formation of $Li_{0.3}Si$ at the end of stage 0. Note that the large radial stress is only present around the top surface of the Si film, and most of the Si film remains in a nearly stress-free state, as revealed by the spatial distributions of the radial stress and the Li concentration in Figure 6c,d, respectively.

Figure 7 depicts spatial variations in normalized concentration and radial stress along the axisymmetric axis under different total C-rates at the end of stage 0. A large amount of Li accumulates near the top surface of the thin film (Figure 7a). The larger the C-rate, the sharper the drop in Li-concentration and the narrower the diffusion zone, as expected. According to Figure 7b, the larger the C-rate, the narrower the compressive-stress zone at the surface. Note that the un-lithiated Si experiences tensile stress, whose magnitude increases with the decrease in the C-rate at the end of stage 0. Moreover, the magnitude of the radial stress at the surface of the a-Si film ($Z/Z_0 = 1$) is about 7.4 GPa at the end of stage 0, independent of the total C-rate used for the lithiation.

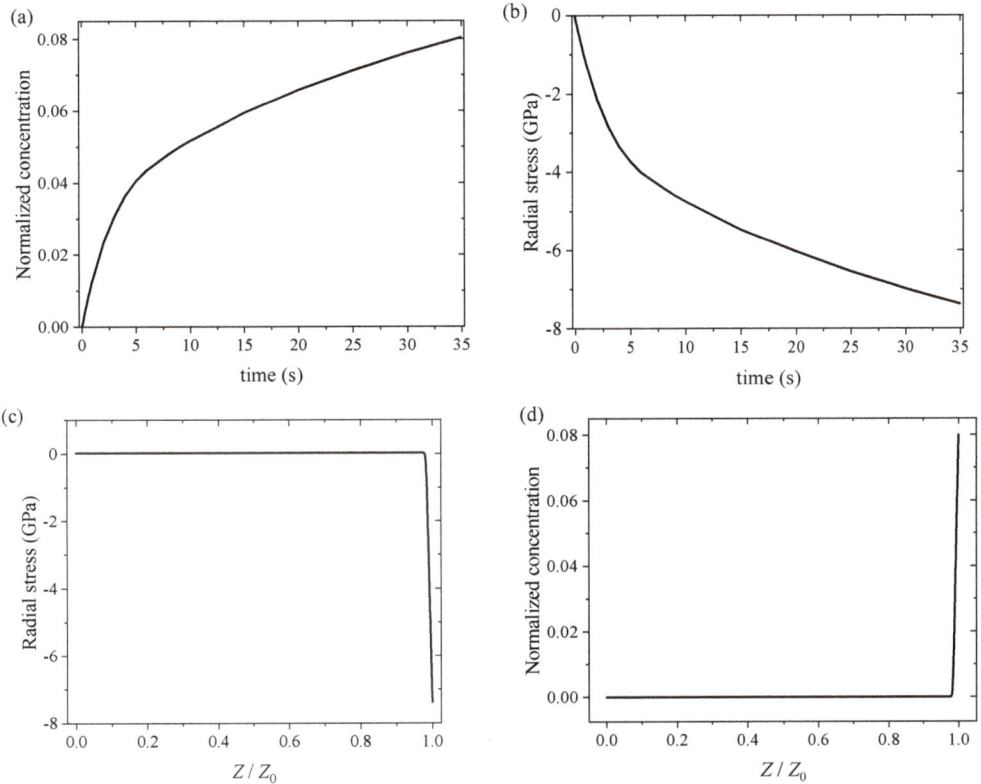

Figure 6. Temporal evolutions of (**a**) the normalized concentration and (**b**) the radial stress at the center of the top surface of the Si thin film at stage 0; spatial distribution of (**c**) the radial stress and (**d**) the normalized concentration along the film thickness at the end of stage 0.

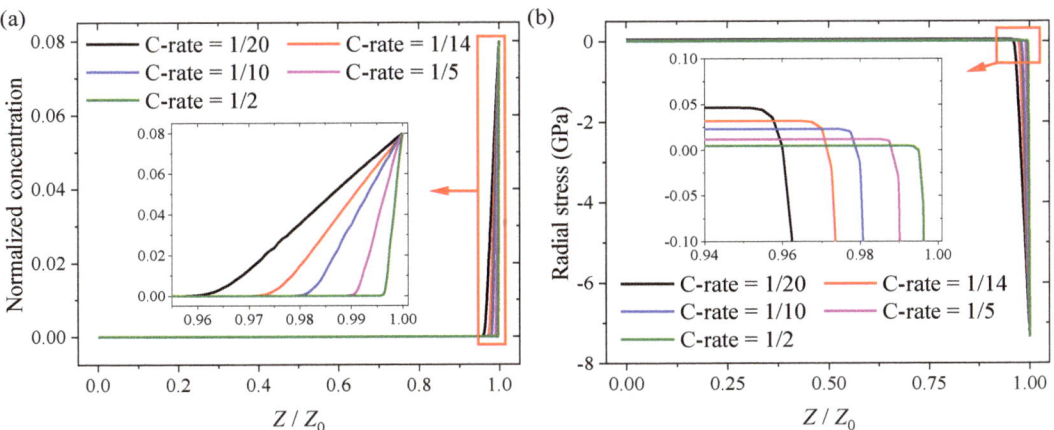

Figure 7. Spatial variations of (**a**) normalized Li concentration, and (**b**) radial stress along the thickness direction at the axisymmetric axis under different total C-rates at the end of stage 0.

It should be pointed out that such a high stress can cause the separation of the Si film from the substrate, wrinkling, surface cracks or phase formation to relieve the stress. Note that the local stress near the free surface can hardly be measured with the commonly used techniques, such as MOS (multibeam optical sensor), since the MOS system records the curvature of the film, while the local stress in such a narrow region rarely affects the curvature. An ocular inspection of silicon film electrodes lithiated at different states up to the end of stage I, and disassembled from the electrochemical cell inside a glovebox filled with protective argon gas overpressure, revealed a smooth surface without wrinkling and cracks. The smoothness of the lithiated Si-film electrode surface remains intact after being washed with propylene carbonate, isopropanol and even acetonitrile. This ocular observation was affirmed by light microscopy performed in the high-vacuum of a secondary ion mass spectroscopy (SIMS) device. Interestingly, the electrode surface remains smooth after the electrode has been stored over a long time (months) inside vacuum or protective argon gas. This experimental observation suggests that phase formation up to the end of lithiation process I does not cause structural damage and is likely responsible for the stress relief.

Figure 8 displays spatial variations in the normalized concentration and radial stress along the thickness at the axisymmetric axis for different positions of the interface at stage I for a total C-rate of 1/14 C, in which the strain energy generated in stage 0 is used for the formation of the $Li_{0.3}Si$ phase. It is evident that there are two distinct regions in the Si thin film—one with a Li-poor phase and the other with un-lithiated Si. The Li concentration in the Li-poor phase increases as the interface approaches the substrate (Figure 8a), as expected from the diffusion theory. More lithium atoms are needed to form the $Li_{0.3}Si$ phase, which requires Li diffusion through the formed $Li_{0.3}Si$ layer. It should be noted that the spatial distribution of Li concentration in the Li-poor layer is different from the constant normalized Li concentration of 0.08 experimentally observed by Hüger et al. [1,65], suggesting that there are other factors likely controlling the migration of Li in the Li-poor layer. On the other hand, the constant normalized Li-concentration of 0.08 over the whole film depth was reported from SIMS depth profiling measurements [1,24,65]. However, the sputtering process affiliated with the SIMS measurements may mask a possible low gradient in Li concentration with a decrease in Li concentration from the electrode surface toward the $Li_{0.3}Si/Si$ phase front. Note that such a decrease in Li concentration may be expected to be necessary for Li diffusion from the electrode surface toward the $Li_{0.3}Si/Si$ phase front.

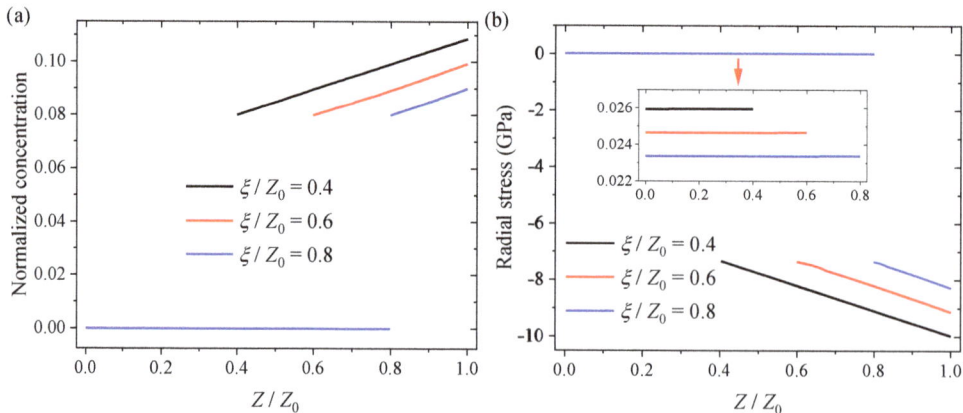

Figure 8. Spatial variations of (**a**) normalized concentration, and (**b**) radial stress along the thickness direction at the axisymmetric axis for different positions of the interface at stage I. (Total C-rate: 1/14 C).

According to Figure 8b, the $Li_{0.3}Si$ layer experiences compressive stress and the un-lithiated Si layer experiences tensile stress. The magnitude of the compressive stress is much larger than that of the tensile stress.

As pointed out above, the strain energy generated in stage 0 is likely relaxed through the formation of the $Li_{0.3}Si$ phase, i.e., there are likely no residual stresses at the onset of stage I. Thus, we can assume that the Si thin film is at a stress-free state at the onset of stage I and the strain energy associated with the radial stress of -7.4 GPa during the lithiation is used for the formation of the $Li_{0.3}Si$ phase. The following analysis is based on the stress-free state of the Si thin film at the onset of stage I and the relaxation of the stresses corresponding to the radial stress of -7.4 GPa due to the formation of the $Li_{0.3}Si$ phase.

Figure 9a presents the variation in radial stress at the center of the top surface of the Si thin film with SOC for a total C-rate of 1/14 C. For comparison, the experimental results given by Pharr et al. [33] are included in the figure. Note that the stress measured experimentally by MOS is an average stress along the thickness direction. With the ultrathin Si film, we directly compare the numerical results of the stress at the top surface with the experimental results. The magnitude of the radial stress increases from 0 to ~1.2 GPa when the SOC is increased from 0 to ~8%. Such a trend is qualitatively in accord with the experimental results.

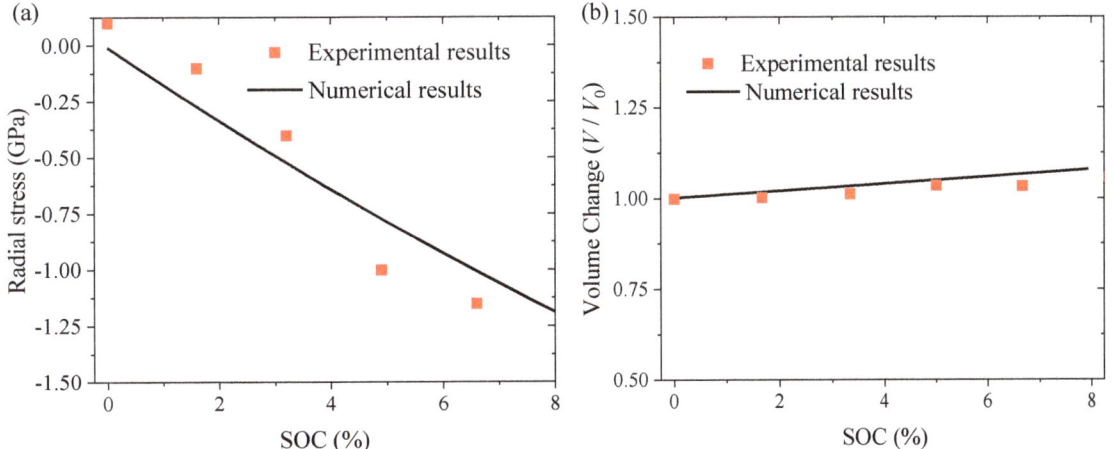

Figure 9. Variations of (**a**) radial stress at the center of the top surface, and (**b**) the volumetric change of the Si thin film with SOC at stage I. (Total C-rate: C/14).

Figure 9b shows the volumetric change of the Si thin film with SOC. The volumetric change in the Si thin film is a linearly increasing function of SOC. The experimental results given by Uxa et al. [24] are also included in the figure. It is evident that the numerical results are consistent with the experimental results. Note that the linear dependence of the volumetric change of the Si thin film on SOC is consistent with the constitutive relations of Equation (1).

Figure 10a shows the temporal evolution of the interface position separating the Li-poor phase ($Li_{0.3}Si$) from the un-lithiated Si phase under different C-rates. The interface approaches the substrate with the increase in lithiation time and is a nearly linear function of the lithiation time. From Figure 10a, we calculate the moving speed of the phase-interface between the Li-poor phase and the un-lithiated silicon. Figure 10b depicts the variation in the moving speed of the phase-interface with the total C-rate. It is evident that the moving speed of the phase-interface (reaction front) is a nonlinearly increasing function of the total C-rate. To prevent the occurrence of the stage II lithiation before the end of stage I, one can use a small C-rate to pre-lithiate Si electrodes to form a lithium-poor phase. A description of the stage II lithiation (Li-rich phase) is given in references [1,24].

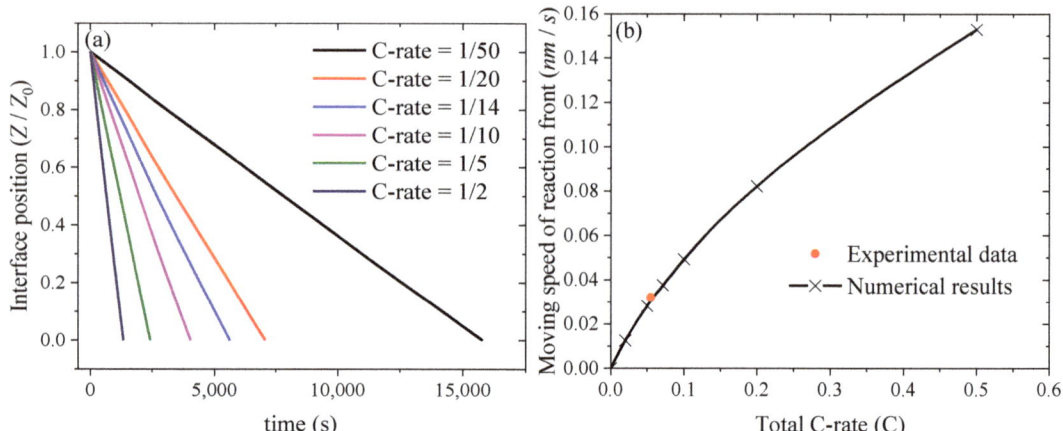

Figure 10. (**a**) Temporal variation in the phase-interface position under different total C-rates, and (**b**) variation of the moving speed of the reaction front with total C-rate, and a Si film thickness of 230 nm.

The measured potential profile presented in Figure 2 shows that full lithiation with a constant current of 11 µA/cm^2 takes about 18 h. Hence, the total C-rate amounts to C/18 (0.055 C). Figure 2 shows also that the duration of Li$^+$ uptake I amounts to approximatively 2 h. This means that the Li$_{0.3}$Si/Si phase front reaches the current collector, i.e., the Li$_{0.3}$Si phase consumes all unreacted silicon, in about two hours. Consequently, a phase-front velocity of 0.032 nm/s is obtained by dividing the film thickness of 230 nm with the time interval of two hours. The obtained experimental result of (0.055 C, 0.032 nm/s) is plotted as a red dot in Figure 10b. The numerical result is in good accord with the experimental result, supporting the mechanochemical model used in this work.

The comparison between the numerical results shown in Figures 9 and 10b and the corresponding experimental results and the agreement of the numerical result of the moving speed of the reaction front with the experimental result reveal that the mechanochemical model used in this work can effectively describe the diffusion of Li in the a-Si film with the formation of only one new phase of Li$_{0.3}$Si during the lithiation onset.

The experimental and numerical results presented and discussed in this work suggest the stress-induced phase formation of a Li-poor phase for the lithiation onset of a-Si. With this finding, one may assess some explanations for two important issues: (i) the main difference between the lithiation onset of a-Si and nanocrystalline silicon, and (ii) their roles in the lithiation onset of a-Si given that a-Si has a huge internal stress and high defect density in comparison to c-Si. According to the insights obtained from the experimental results, we can address these two issues as follows. At first view, there seems to be no major difference between the lithiation onset of c-Si (crystalline silicon) and a-Si concerning the formation of the Li-poor phase. The higher internal stress and defect density of a-Si compared to those of c-Si may lead to a higher Li content in the Li-poor zone in the case of a-Si compared to that of c-Si. The appearance of the Li-poor phase and its Li content in polycrystalline silicon may be influenced by the preferential texture (preferential orientation) and by a change in mechanical stress in polycrystalline silicon, e.g., due to the presence of additives and binders. Note that silicon is cubic structure, which exhibits isotropic behavior in diffusion and anisotropic behavior in elastic deformation. The anisotropic behavior in elastic deformation can lead to a stress state different from that of isotropic deformation for a-silicon under the same cycling conditions.

As mentioned in the literature survey presented in Section 2, operando NR investigations [25] also found the appearance of the Li-poor phase in the lithiation process of bulk c-silicon, i.e., in that of 10 mm thick (100) oriented silicon wafer. The Li content of the Li-poor phase was estimated by fitting NR simulations on NR experimental data, and was found to be $Li_{0.1}Si$ (x ~ 0.1), which is lower than x = 0.3. Operando NR experiments performed later by another group [72] also found two distinct lithium silicide zones with high and low Li concentrations, initially separated by a sharp interface which broadens with cycling. The Li-poor zone was always in between the Li-rich zone and unlithiated silicon, i.e., the Li-poor zone was in direct contact with unlithiated silicon [72] (similar to the situations presented in Figure 1e–g). From the neutron scattering length densities (SLD) obtained from their NR data [72], the ratio of Li to Si in the Li-poor zone is evaluated to be between x ~ 0.05 and x ~ 0.35. Note that, although NR is sensitive to Li incorporation into silicon, NR is generally unable to exactly distinguish between low Li contents such as x = 0.1 and x = 0.3. In contrast to NR, and especially in contrast to a large range of often applied measurement techniques including electron spectroscopy and microscopy, SIMS is able to detect low traces of Li in silicon. Lithium has a high ionization cross-section for Li^+-ions. Hence, the Li^+ SIMS signal is of an extremely high intensity, as proved in our laboratory, especially for SIMS operated with O_2^+ primary ion beams. Li SIMS signals of more than 1 million counts per second can be measured for the Li-poor phase. Hence, future SIMS investigations on the lithiation process of bulk c-silicon (e.g., Si wafers) are of great interest. Moreover, the experiments on silicon wafers [25,72] were performed only on (100) oriented wafers. Mechanical stresses may be dependent on crystal orientations. Consequently, SIMS experiments and calculations on the lithiation of silicon wafers with different orientations may be of interest to examine the appearance of the stress-induced phase formation of the Li-poor phase in single Si crystals.

Polycrystalline films often possess a textured structure. A possible preferential orientation of polycrystalline silicon film, e.g., produced by heating treatments, may impede the lithiation process, as has been observed for the case of $LiNi_xMn_yCo_zO_2$ (x + y + z = 1, NMC) active materials of positive electrodes (e.g., ref. [73]), due to reduced Li kinetics and changed mechanical stress. Polycrystalline (e.g., silicon) electrodes are mostly produced by crystalline powder embedded in carbon-like additives and binders, which also exercises mechanical stress and different phenomena (e.g., Schottky contacts) on the Si particles before and during lithiation and delithiation. In that manner, the material in contact with silicon may influence the diffusion of lithium and the appearance of the Li-poor phase. For the case of silicon wafers, there are reports about experiments performed with a thin surface layer of aluminium oxide to suppress the SEI layer formation [72]. The Li-poor phase also appeared in that case, further showing that the Li-poor phase is an intrinsic product during the lithiation of silicon, and that, in that example, it is not dependent on the SEI layer. On the other hand, this may not always be the case. As mentioned in the second section, Li^+ uptake I, which produces the Li-poor phase, is not detectable during the lithiation of amorphous Si/C ML, as presented in the supplementary material of reference [1]. The mechanical stress inside Si/C ML may interact with the chemical stress introduced by the lithiation process, interfering in the buildup of the Li-poor phase. Alternatively, the Li-poor phase may be built up in the Si/C ML but may possess a lower Li content of e.g., x ~ 0.1, which needs a much lower Li^+-uptake than that of x ~ 0.3, and in that case does not recognizably remain in the potential profile of the lithiation process. Its examination constitutes a prime subject for SIMS investigation due to the mentioned high sensitivity of SIMS to very low Li contents in silicon. Additionally, different individual layer thicknesses inside the Si/C ML may produce different mechanical stresses and different Schottky barrier (space charge) strengths before and during ML lithiation, which may promote or interfere in the formation of the Li-poor phase and also affects its Li content. Hence, future SIMS investigations on the lithiation process of multilayers (e.g., Si/C MLs) are also of interest. The examination of the Si/C MLs may also be of interest for polycrystalline silicon electrodes produced by c-Si powder embedded in carbon-like additives and binders.

Finally, it is of interest to examine whether the stress-induced phase formation of the Li-poor phase represents a universal phenomenon, i.e., if a stress-induced phase formation of the Li-poor phase appears also in other well desired LIB electrode materials such as in (amorphous and crystalline) germanium.

4. Conclusions

In summary, the lithiation onset of a-Si films was reviewed, and a mechanochemical model for the lithiation onset of a-Si film was developed. The electrochemical lithiation onset of silicon films involving the formation of a Li-poor phase ($Li_{0.3}Si$ alloy) and the propagation of a reaction front in the films delimiting the $Li_{0.3}Si$ alloy from un-lithiated silicon, as revealed experimentally, constitute the foundation for a deeper examination of the lithiation onset of a-silicon. For that, (i) a literature review was performed to describe (i.1) the lithiation onset, (i.2) the changes in material properties at the lithiation onset and (i.3) the reason for the appearance of the reaction front, after which (ii) numerical calculations of depth-resolved mechanical stress during the lithiation onset were performed.

The literature survey elaborated the following five peculiar changes in the behavior of the active electrode material during the lithiation onset of a-silicon: (i) the build-up of the highest mechanical stress (up to 1.2 GPa) during lithiation, (ii) a linear increase in the mechanical stress with lithiation which mimics the characteristics of linear elastic deformation, (iii) only a minute volume increase during Li incorporation which is lower than that expected from the amount of Li ions entering the silicon electrode, (iv) the largest heat generation appearing during cycling with only a minor degree of parasitic heat contribution, and (v) an unexpected enhanced brittleness.

A further literature review was performed to examine (i) why the lithiation onset of silicon occurs via a phase front which delimits the Li-poor phase from unreacted silicon, and (ii) why the Li-poor phase possesses the particular Li concentration corresponding to $x \sim 0.3$ in Li_xSi. The following three factors—(i) extremely low Li permeability in pure silicon and relative high Li permeability in $Li_{0.3}Si$, (ii) Li repulsion from pure silicon due to mechanical stress, and (iii) energy gain by Li_xSi formation only for $x \geq 0.3$—were found to qualitatively explain why initially a two-phase lithiation mechanism appears which sharply delimits pure silicon from a Li-poor Li_xSi phase with $x \sim 0.3$.

The literature review shows that the appearance of the Li-poor silicide ($Li_{0.3}Si$) as an intrinsic property during the lithiation onset of pure and "defect-free" silicon is unexpected, since no stable Li_xSi silicide phases appear for $x < 1$ at ambient pressure in the Li-Si phase diagram. The first principle calculation predicts that mechanical pressure above 8 GPa is necessary for the Li-poor silicide with $x \sim 0.3$ ($LiSi_3$) to become stable at 0 K. This points to the fact that the peculiar lithiation onset of a-silicon may be quantitatively explained by the depth-resolved mechanical stress appeared during the lithiation onset. Consequently, a mechanochemical model was elaborated and the lithiation onset was simulated on the stress evolution and the motion of the reaction front for the growth of the Li-poor phase in an a-Si film, based on the theory of linear elasticity and the coupling between diffusion and stress. A moving boundary condition was used to describe the propagation of the reaction front. The temporal evolution of the stresses was illustrated, and the comparison between the numerical results and those reported in the literature was performed.

The mechanochemical model consists of two stages, with stage 0 involving the diffusion of lithium in the Si film without the formation of the Li-poor phase and stage I involving the formation of the Li-poor phase with the propagation of the reaction front at the end of stage 0. The magnitude of the compressive radial stress at the end of stage 0 can reach around 8 GPa, likely corresponding to the formation of the $Li_{0.3}Si$ phase under high stress. No experimental observations of cracking and wrinkling for stress relief points to a stress relaxation by the phase formation up to the end of lithiation stage I. Increasing the total C-rate reduces the layer thickness of the high compressive-radial stress around the surface of the Si film at the end of stage 0.

For the lithiation of the Si film at stage I, the numerical results for the variations of the radial stress and volumetric change with SOC and for the moving speed of the reaction front for the lithiation under 0.055 C (C/18) are in good accord with the corresponding experimental data. These results support the mechanochemical model developed in this work, revealing the important role of the interaction of diffusion and stress in the lithiation onset of Si-based electrodes used in lithium-ion batteries and in the stress-induced phase formation of the $Li_{0.3}Si$ phase.

Author Contributions: Conceptualization, K.Z., E.H., Y.L. and F.Y.; methodology, K.Z., E.H. and F.Y.; validation, K.Z., E.H., Y.L. and F.Y.; formal analysis, K.Z., E.H., Y.L. and F.Y.; investigation, K.Z., E.H. and F.Y.; resources, E.H., H.S. and F.Y.; data curation, K.Z., E.H. and Y.L.; writing—K.Z., E.H., Y.L. and F.Y.; writing—review and editing, E.H. and F.Y.; supervision, E.H. and F.Y.; project administration, E.H., H.S. and F.Y. All authors have read and agreed to the published version of the manuscript.

Funding: This research received no external funding.

Data Availability Statement: Data are available upon request.

Conflicts of Interest: The authors declare no conflict of interest.

Appendix A

Appendix A.1 Experimental Procedure

For the lithiation and delithiation processes, a working electrode possessing a-Si film, a Li–metal reference electrode and a Li-metal counter electrode was used, together with a liquid carbon-based electrolyte (propylene carbonate) with a 1 M $LiClO_4$ salt. A 1-inch copper disk with a thickness of 1 mm was used as a current collector and a substrate for the ion-beam deposition of a-Si electrochemical active material of 230 nm thickness. The ion beam coater IBC 681 (Gatan Inc., Pleasanton, CA, USA) was used for the film deposition, as described in reference [45]. Grazing incidence X-ray diffraction, performed with a Bruker D8 DISCOVER diffractometer, found the deposited silicon films to be amorphous. The electrochemical lithiation and delithiation were performed at room temperature on a Biologic SP150 potentiostat with the EC-lab software.

Appendix A.2 Additional Figures and Information on the Mechanochemical Simulation

Figure A2a shows temporal variations of the interface position in the thin film electrode for D^β (Li diffusivity in a-Si) in a range of 1.5×10^{-20} to 1.5×10^{-30} m^2/s and D^α (Li diffusivity in $Li_{0.3}Si$) being 1.5×10^{-20} m^2/s. It is evident that there is no observable difference for the spatial position of the interface at the same lithiation time. The Li diffusivity in the a-Si has a negligible effect on the migration rate of the interface. Figure A2b shows temporal variations of the interface position in the thin film electrode for D^α in a range of 1.5×10^{-20} to 1.5×10^{-25} m^2/s and D^β being 1.5×10^{-20} m^2/s. For the lithiation time less than 10 s, there is no observable difference for the spatial position of the interface at the same lithiation time. For the lithiation time larger than 10 s, the larger the diffusivity of the α phase, the closer the interface to the current collector at the same lithiation. This result suggests that the larger the diffusivity of the α phase, the larger is the migration rate of the interface. The Li diffusivity in the Li-poor phase has a dominant effect on the migration rate of the interface.

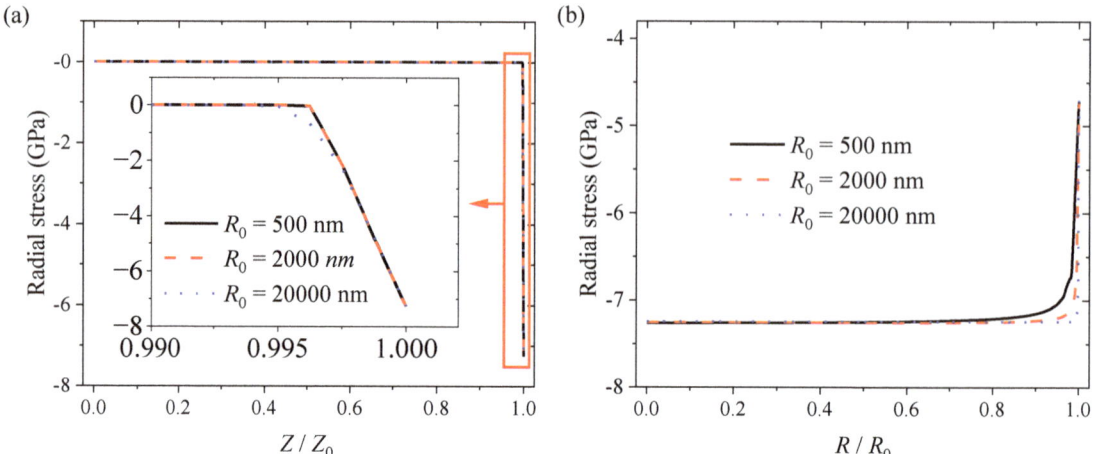

Figure A1. Variations in the radial stress along (**a**) the film thickness at axisymmetric axis and (**b**) the radial direction on the top surface at end of stage 0. Results of the disk with different initial radii of 500, 2000 and 200,000 nm are depicted. The numerical results show that there are no observable differences between the corresponding stresses in the region far away from the disk edge. Here, $Z = 0$ corresponds to the interface between the a-Si and the current collector and $Z = Z_0$ corresponds to the interface between the a-silicon electrode and electrolyte.

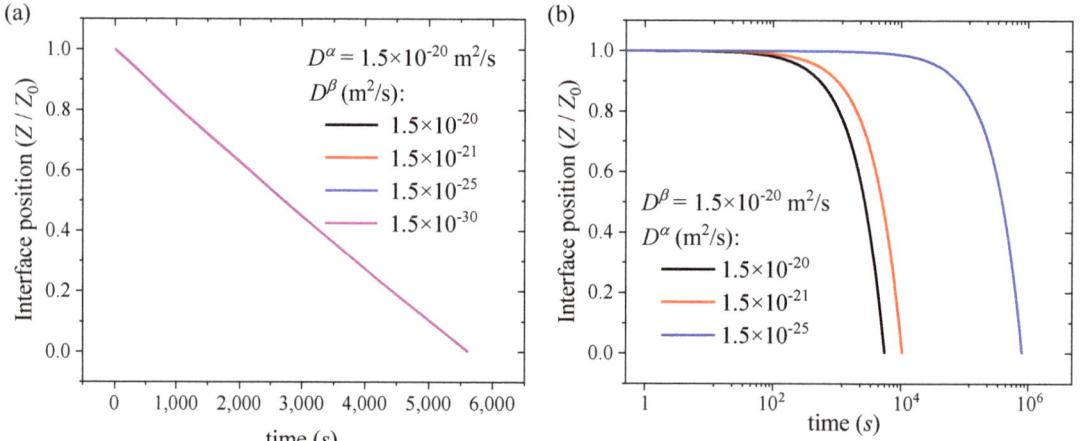

Figure A2. Temporal variation in the interface position between pure silicon (β phase) and the Li-poor phase (α phase) in the thin film electrode for different values of (**a**) D^β (Li diffusivity in a-Si-β phase) and D^α (Li diffusivity in Li$_{0.3}$Si-α phase) $=1.5 \times 10^{-20}$ m^2/s, and (**b**) D^α and $D^\beta = 1.5 \times 10^{-20}$ m^2/s.

References

1. Hüger, E.; Uxa, D.; Yang, F.; Schmidt, H. The lithiation onset of amorphous silicon thin-film electrodes. *Appl. Phys. Lett.* **2022**, *121*, 133901. [CrossRef]
2. Hüger, E.; Jin, C.; Schmidt, H. Electrochemical investigation of ion-beam sputter-deposited carbon thin films for Li-ion batteries. *J. Appl. Electrochem.* **2022**, *52*, 1715–1732. [CrossRef]
3. Wu, B.; Chen, C.; Danilov, D.L.; Jiang, M.; Raijmakers, L.H.; Eichel, R.d.-A.; Notten, P.H. Influence of the SEI formation on the stability and lithium diffusion in Si electrodes. *ACS Omega* **2022**, *7*, 32740–32748. [CrossRef]
4. Phan, V.P.; Pecquenard, B.; Le Cras, F. High-Performance all-solid-state cells fabricated with silicon electrodes. *Adv. Funct. Mater.* **2012**, *22*, 2580–2584. [CrossRef]

5. Cao, C.; Abate, I.I.; Sivonxay, E.; Shyam, B.; Jia, C.; Moritz, B.; Devereaux, T.P.; Persson, K.A.; Steinrück, H.-G.; Toney, M.F. Solid electrolyte interphase on native oxide-terminated silicon anodes for Li-ion batteries. *Joule* **2019**, *3*, 762–781. [CrossRef]
6. Browning, K.L.; Browning, J.F.; Doucet, M.; Yamada, N.L.; Liu, G.; Veith, G.M. Role of conductive binder to direct solid–electrolyte interphase formation over silicon anodes. *Phys. Chem. Chem. Phys.* **2019**, *21*, 17356–17365. [CrossRef]
7. Philippe, B.; Dedryvère, R.M.; Allouche, J.; Lindgren, F.; Gorgoi, M.; Rensmo, H.K.; Gonbeau, D.; Edström, K. Nanosilicon electrodes for lithium-ion batteries: Interfacial mechanisms studied by hard and soft X-ray photoelectron spectroscopy. *Chem. Mater.* **2012**, *24*, 1107–1115. [CrossRef]
8. Schroder, K.W.; Celio, H.; Webb, L.J.; Stevenson, K.J. Examining solid electrolyte interphase formation on crystalline silicon electrodes: Influence of electrochemical preparation and ambient exposure conditions. *J. Phys. Chem. C* **2012**, *116*, 19737–19747. [CrossRef]
9. Yin, Y.; Arca, E.; Wang, L.; Yang, G.; Schnabel, M.; Cao, L.; Xiao, C.; Zhou, H.; Liu, P.; Nanda, J. Nonpassivated silicon anode surface. *ACS Appl. Mater. Interfaces* **2020**, *12*, 26593–26600. [CrossRef]
10. Schnabel, M.; Harvey, S.P.; Arca, E.; Stetson, C.; Teeter, G.; Ban, C.; Stradins, P. Surface SiO_2 thickness controls uniform-to-localized transition in lithiation of silicon anodes for lithium-ion batteries. *ACS Appl. Mater. Interfaces* **2020**, *12*, 27017–27028. [CrossRef]
11. Schnabel, M.; Arca, E.; Ha, Y.; Stetson, C.; Teeter, G.; Han, S.-D.; Stradins, P. Enhanced interfacial stability of Si anodes for Li-ion batteries via surface SiO_2 coating. *ACS Appl. Energy Mater.* **2020**, *3*, 8842–8849. [CrossRef]
12. Hasa, I.; Haregewoin, A.M.; Zhang, L.; Tsai, W.-Y.; Guo, J.; Veith, G.M.; Ross, P.N.; Kostecki, R. Electrochemical reactivity and passivation of silicon thin-film electrodes in organic carbonate electrolytes. *ACS Appl. Mater. Interfaces* **2020**, *12*, 40879–40890. [CrossRef]
13. Stetson, C.; Schnabel, M.; Li, Z.; Harvey, S.P.; Jiang, C.-S.; Norman, A.; DeCaluwe, S.C.; Al-Jassim, M.; Burrell, A. Microscopic observation of solid electrolyte interphase bilayer inversion on silicon oxide. *ACS Energy Lett.* **2020**, *5*, 3657–3662. [CrossRef]
14. Huang, W.; Wang, J.; Braun, M.R.; Zhang, Z.; Li, Y.; Boyle, D.T.; McIntyre, P.C.; Cui, Y. Dynamic structure and chemistry of the silicon solid-electrolyte interphase visualized by cryogenic electron microscopy. *Matter* **2019**, *1*, 1232–1245. [CrossRef]
15. Steinrück, H.-G.; Cao, C.; Veith, G.M.; Toney, M.F. Toward quantifying capacity losses due to solid electrolyte interphase evolution in silicon thin film batteries. *J. Chem. Phys.* **2020**, *152*, 084702. [CrossRef] [PubMed]
16. Kohler, T.; Hadjixenophontos, E.; Joshi, Y.; Wang, K.; Schmitz, G. Reversible oxide formation during cycling of Si anodes. *Nano Energy* **2021**, *84*, 105886. [CrossRef]
17. Cao, C.; Steinrück, H.-G.; Shyam, B.; Stone, K.H.; Toney, M.F. In situ study of silicon electrode lithiation with X-ray reflectivity. *Nano Lett.* **2016**, *16*, 7394–7401. [CrossRef] [PubMed]
18. Stournara, M.E.; Xiao, X.C.; Qi, Y.; Johari, P.; Lu, P.; Sheldon, B.W.; Gao, H.J.; Shenoy, V.B. Li segregation induces structure and strength changes at the amorphous Si/Cu interface. *Nano Lett.* **2013**, *13*, 4759–4768. [CrossRef]
19. Guo, M.Q.; Meng, W.J.; Zhang, X.G.; Bai, Z.C.; Wang, G.W.; Wang, Z.H.; Yang, F.Q. Structural degradation of Cu current collector during electrochemical cycling of Sn-based lithium-ion batteries. *J. Electron. Mater.* **2019**, *48*, 7543–7550. [CrossRef]
20. Yang, Y.; Zhao, J.B. Wadsley-roth crystallographic shear structure niobium-based oxides: Promising anode materials for high-safety lithium-ion batteries. *Adv. Sci.* **2021**, *8*, 2004855. [CrossRef]
21. Al-Maghrabi, M.; Suzuki, J.; Sanderson, R.; Chevrier, V.; Dunlap, R.; Dahn, J. Combinatorial studies of $Si_{1-x}O_x$ as a potential negative electrode material for Li-ion battery applications. *J. Electrochem. Soc.* **2013**, *160*, A1587. [CrossRef]
22. Cao, Y.; Bennett, J.C.; Dunlap, R.; Obrovac, M. A simple synthesis route for high-capacity SiO_x anode materials with tunable oxygen content for lithium-ion batteries. *Chem. Mater.* **2018**, *30*, 7418–7422. [CrossRef]
23. Cao, Y.; Dunlap, R.; Obrovac, M. Electrochemistry and thermal behavior of SiO_x made by reactive gas milling. *J. Electrochem. Soc.* **2020**, *167*, 110501. [CrossRef]
24. Uxa, D.; Jerliu, B.; Hüger, E.; Dörrer, L.; Horisberger, M.; Stahn, J.; Schmidt, H. On the lithiation mechanism of amorphous silicon electrodes in Li-ion batteries. *J. Phys. Chem. C* **2019**, *123*, 22027–22039. [CrossRef]
25. Seidlhofer, B.-K.; Jerliu, B.; Trapp, M.; Hüger, E.; Risse, S.; Cubitt, R.; Schmidt, H.; Steitz, R.; Ballauff, M. Lithiation of crystalline silicon as analyzed by operando neutron reflectivity. *ACS Nano* **2016**, *10*, 7458–7466. [CrossRef]
26. Sheldon, B.W.; Soni, S.K.; Xiao, X.; Qi, Y. Stress contributions to solution thermodynamics in Li-Si alloys. *Electrochem. Solid-State Lett.* **2011**, *15*, A9. [CrossRef]
27. Sethuraman, V.A.; Srinivasan, V.; Bower, A.F.; Guduru, P.R. In situ measurements of stress-potential coupling in lithiated silicon. *J. Electrochem. Soc.* **2010**, *157*, A1253. [CrossRef]
28. Jerliu, B.; Hüger, E.; Dorrer, L.; Seidlhofer, B.-K.; Steitz, R.; Oberst, V.; Geckle, U.; Bruns, M.; Schmidt, H. Volume expansion during lithiation of amorphous silicon thin film electrodes studied by in-operando neutron reflectometry. *J. Phys. Chem. C* **2014**, *118*, 9395–9399. [CrossRef]
29. Beaulieu, L.; Hatchard, T.; Bonakdarpour, A.; Fleischauer, M.; Dahn, J. Reaction of Li with alloy thin films studied by in situ AFM. *J. Electrochem. Soc.* **2003**, *150*, A1457. [CrossRef]
30. Soni, S.K.; Sheldon, B.W.; Xiao, X.; Tokranov, A. Thickness effects on the lithiation of amorphous silicon thin films. *Scr. Mater.* **2011**, *64*, 307–310. [CrossRef]
31. Nadimpalli, S.P.; Sethuraman, V.A.; Bucci, G.; Srinivasan, V.; Bower, A.F.; Guduru, P.R. On plastic deformation and fracture in Si films during electrochemical lithiation/delithiation cycling. *J. Electrochem. Soc.* **2013**, *160*, A1885. [CrossRef]

32. Sethuraman, V.A.; Chon, M.J.; Shimshak, M.; Srinivasan, V.; Guduru, P.R. In situ measurements of stress evolution in silicon thin films during electrochemical lithiation and delithiation. *J. Power Sources* **2010**, *195*, 5062–5066. [CrossRef]
33. Pharr, M.; Suo, Z.; Vlassak, J.J. Measurements of the fracture energy of lithiated silicon electrodes of Li-ion batteries. *Nano Lett.* **2013**, *13*, 5570–5577. [CrossRef]
34. Wang, X.; Fan, F.; Wang, J.; Wang, H.; Tao, S.; Yang, A.; Liu, Y.; Beng Chew, H.; Mao, S.X.; Zhu, T.; et al. High damage tolerance of electrochemically lithiated silicon. *Nat. Commun.* **2015**, *6*, 8417. [CrossRef] [PubMed]
35. Housel, L.M.; Li, W.Z.; Quilty, C.D.; Vila, M.N.; Wang, L.; Tang, C.R.; Bock, D.C.; Wu, Q.Y.; Tong, X.; Head, A.R.; et al. Insights into reactivity of silicon negative electrodes: Analysis using isothermal microcalorimetry. *ACS Appl. Mater. Interfaces* **2019**, *11*, 37567–37577. [CrossRef]
36. Yang, F.Q. Cycling-induced structural damage/degradation of electrode materials—Microscopic viewpoint. *Nanotechnology* **2022**, *33*, 065405. [CrossRef] [PubMed]
37. Wang, P.; Kozlov, A.; Thomas, D.; Mertens, F.; Schmid-Fetzer, R. Thermodynamic analysis of the Li–Si phase equilibria from 0 K to liquidus temperatures. *Intermetallics* **2013**, *42*, 137–145. [CrossRef]
38. Zeilinger, M.; Kurylyshyn, I.M.; Haussermann, U.; Fässler, T.F. Revision of the Li–Si phase diagram: Discovery and single-crystal X-ray structure determination of the high-temperature phase $Li_{4.11}Si$. *Chem. Mater.* **2013**, *25*, 4623–4632. [CrossRef]
39. Morris, A.J.; Grey, C.; Pickard, C.J. Thermodynamically stable lithium silicides and germanides from density functional theory calculations. *Phys. Rev. B* **2014**, *90*, 054111. [CrossRef]
40. Braga, M.H.; Dębski, A.; Gąsior, W. Li–Si phase diagram: Enthalpy of mixing, thermodynamic stability, and coherent assessment. *J. Alloys Compd.* **2014**, *616*, 581–593. [CrossRef]
41. Zhang, S.; Wang, Y.; Yang, G.; Ma, Y. Silicon framework-based lithium silicides at high pressures. *ACS Appl. Mater. Interfaces* **2016**, *8*, 16761–16767. [CrossRef]
42. Chiang, H.-H.; Lu, J.-M.; Kuo, C.-L. First-principles study of the structural and dynamic properties of the liquid and amorphous Li–Si alloys. *J. Chem. Phys.* **2016**, *144*, 034502. [CrossRef] [PubMed]
43. Yang, F. Interaction between diffusion and chemical stresses. *Mater. Sci. Eng. A* **2005**, *409*, 153–159. [CrossRef]
44. Strauß, F.; Hüger, E.; Julin, J.; Munnik, F.; Schmidt, H. Lithium diffusion in ion-beam sputter-deposited lithium–silicon layers. *J. Phys. Chem. C* **2020**, *124*, 8616–8623. [CrossRef]
45. Hüger, E.; Dörrer, L.; Schmidt, H. Permeation, solubility, diffusion and segregation of lithium in amorphous silicon layers. *Chem. Mater.* **2018**, *30*, 3254–3264. [CrossRef]
46. Strauß, F.; Dörrer, L.; Bruns, M.; Schmidt, H. Lithium tracer diffusion in amorphous Li_xSi for low Li concentrations. *J. Phys. Chem. C* **2018**, *122*, 6508–6513. [CrossRef]
47. Chevrier, V.L.; Dahn, J.R. First principles model of amorphous silicon lithiation. *J. Electrochem. Soc.* **2009**, *156*, A454–A458. [CrossRef]
48. Artrith, N.; Urban, A.; Ceder, G. Constructing first-principles phase diagrams of amorphous Li_xSi using machine-learning-assisted sampling with an evolutionary algorithm. *J. Chem. Phys.* **2018**, *148*, 241711. [CrossRef] [PubMed]
49. Bucci, G.; Swamy, T.; Bishop, S.; Sheldon, B.W.; Chiang, Y.M.; Carter, W.C. The effect of stress on battery-electrode capacity. *J. Electrochem. Soc.* **2017**, *164*, A645–A654. [CrossRef]
50. Stournara, M.E.; Kumar, R.; Qi, Y.; Sheldon, B.W. Ab initio diffuse-interface model for lithiated electrode interface evolution. *Phys. Rev. E* **2016**, *94*, 012802. [CrossRef]
51. Hirai, K.; Ichitsubo, T.; Uda, T.; Miyazaki, A.; Yagi, S.; Matsubara, E. Effects of volume strain due to Li–Sn compound formation on electrode potential in lithium-ion batteries. *Acta Mater.* **2008**, *56*, 1539–1545. [CrossRef]
52. Li, Y.; Zhang, K.; Zheng, B.; Yang, F. Effect of local deformation on the coupling between diffusion and stress in lithium-ion battery. *Int. J. Solids Struct.* **2016**, *87*, 81–89. [CrossRef]
53. Zhang, F.; Wang, J.; Liu, S.; Du, Y. Effects of the volume changes and elastic-strain energies on the phase transition in the Li-Sn battery. *J. Power Sources* **2016**, *330*, 111–119. [CrossRef]
54. Selvi, G.T.; Jha, S.K. Anomalous interfacial stress generation and role of elasto-plasticity in mechanical failure of Si-based thin film anodes of Li-ion batteries. *Bull. Mater. Sci.* **2022**, *45*, 23. [CrossRef]
55. Li, J.; Fang, Q.; Wu, H.; Liu, Y.; Wen, P. Investigation into diffusion induced plastic deformation behavior in hollow lithium ion battery electrode revealed by analytical model and atomistic simulation. *Electrochim. Acta* **2015**, *178*, 597–607. [CrossRef]
56. Li, Y.; Zhang, J.; Zhang, K.; Zheng, B.; Yang, F. A defect-based viscoplastic model for large-deformed thin film electrode of lithium-ion battery. *Int. J. Plast.* **2019**, *115*, 293–306. [CrossRef]
57. Bagheri, A.; Arghavani, J.; Naghdabadi, R.; Brassart, L. A theory for coupled lithium insertion and viscoplastic flow in amorphous anode materials for Li-ion batteries. *Mech. Mater.* **2021**, *152*, 103663. [CrossRef]
58. Zhang, K.; Zhou, J.; Yang, F.; Zhang, Y.; Pan, Y.; Zheng, B.; Li, Y.; Yang, F. A stress-control charging method with multi-stage currents for silicon-based lithium-ion batteries: Theoretical analysis and experimental validation. *J. Energy Storage* **2022**, *56*, 105985. [CrossRef]
59. Zuo, P.; Zhao, Y.-P. A phase field model coupling lithium diffusion and stress evolution with crack propagation and application in lithium ion batteries. *Phys. Chem. Chem. Phys.* **2015**, *17*, 287–297. [CrossRef] [PubMed]
60. Deng, J.; Wagner, G.J.; Muller, R.P. Phase field modeling of solid electrolyte interface formation in lithium ion batteries. *J. Electrochem. Soc.* **2013**, *160*, A487. [CrossRef]

1. Liu, Z.; Cai, R.; Chen, B.; Liu, T.; Zhou, J. Effect of electrochemical reaction on diffusion-induced stress in hollow spherical lithium-ion battery electrode. *Ionics* **2017**, *23*, 617–625. [CrossRef]
2. Zhang, T.; Guo, Z.; Wang, Y.; Zhu, J. Effect of reversible electrochemical reaction on Li diffusion and stresses in cylindrical Li-ion battery electrodes. *J. Appl. Phys.* **2014**, *115*, 083504. [CrossRef]
3. Suo, Y.; Yang, F. Transient analysis of diffusion-induced stress: Effect of solid reaction. *Acta Mech.* **2019**, *230*, 993–1002. [CrossRef]
4. Yang, F. Effect of local solid reaction on diffusion-induced stress. *J. Appl. Phys.* **2010**, *107*, 103516. [CrossRef]
5. Hüger, E.; Jerliu, B.; Dörrer, L.; Bruns, M.; Borchardt, G.; Schmidt, H. A secondary ion mass spectrometry study on the mechanisms of amorphous silicon electrode lithiation in Li-ion batteries. *Z. Phys. Chem.* **2015**, *229*, 1375–1385. [CrossRef]
6. Li, J.C.-M. Physical chemistry of some microstructural phenomena. *Metall. Trans. A* **1978**, *9*, 1353–1380.
7. Zang, J.-L.; Zhao, Y.-P. A diffusion and curvature dependent surface elastic model with application to stress analysis of anode in lithium ion battery. *Int. J. Eng. Sci.* **2012**, *61*, 156–170. [CrossRef]
8. Uxa, D.; Hüger, E.; Schmidt, H. Li diffusion in thin-film Li_ySi electrodes: Galvanostatic intermittent titration technique and tracer diffusion experiments. *J. Phys. Chem. C* **2020**, *124*, 27894–27899. [CrossRef]
9. Hertzberg, B.; Benson, J.; Yushin, G. Ex-situ depth-sensing indentation measurements of electrochemically produced Si–Li alloy films. *Electrochem. Commun.* **2011**, *13*, 818–821. [CrossRef]
10. Kuznetsov, V.; Zinn, A.-H.; Zampardi, G.; Borhani-Haghighi, S.; La Mantia, F.; Ludwig, A.; Schuhmann, W.; Ventosa, E. Wet nanoindentation of the solid electrolyte interphase on thin film Si electrodes. *ACS Appl. Mater. Interfaces* **2015**, *7*, 23554–23563. [CrossRef] [PubMed]
11. Gao, F.; Hong, W. Phase-field model for the two-phase lithiation of silicon. *J. Mech. Phys. Solids* **2016**, *94*, 18–32. [CrossRef]
12. Ronneburg, A.; Silvi, L.; Cooper, J.; Harbauer, K.; Ballauff, M.; Risse, S. Solid Electrolyte Interphase Layer Formation during Lithiation of Single-Crystal Silicon Electrodes with a Protective Aluminum Oxide Coating. *ACS Appl. Mater. Interfaces* **2021**, *13*, 21241–21249. [CrossRef] [PubMed]
13. Strafela, M.; Fischer, J.; Music, D.; Chang, K.; Schneider, J.; Leiste, H.; Rinke, M.; Bergfeldt, T.; Seifert, H.; Ulrich, S. Dependence of the constitution, microstructure and electrochemical behaviour of magnetron sputtered Li–Ni–Mn–Co–O thin film cathodes for lithium-ion batteries on the working gas pressure and annealing conditions. *Int. J. Mater. Res.* **2017**, *108*, 879–886. [CrossRef]

Disclaimer/Publisher's Note: The statements, opinions and data contained in all publications are solely those of the individual author(s) and contributor(s) and not of MDPI and/or the editor(s). MDPI and/or the editor(s) disclaim responsibility for any injury to people or property resulting from any ideas, methods, instructions or products referred to in the content.

Review

Conductive Metal–Organic Frameworks for Rechargeable Lithium Batteries

Fengjun Deng [†], Yuhang Zhang [†] and Yingjian Yu *

College of Physics Science and Technology, Kunming University, Kunming 650214, China
* Correspondence: yuyingjiankmu@163.com
† These authors contributed equally to this work.

Abstract: Currently, rechargeable lithium batteries are representative of high-energy-density battery systems. Nevertheless, the development of rechargeable lithium batteries is confined by numerous problems, such as anode volume expansion, dendrite growth of lithium metal, separator interface compatibility, and instability of cathode interface, leading to capacity fade and performance degradation of batteries. Since the 21st century, metal–organic frameworks (MOFs) have attracted much attention in energy-related applications owing to their ideal specific surface areas, adjustable pore structures, and targeted design functions. The insulating characteristics of traditional MOFs restrict their application in the field of electrochemistry energy storage. Recently, some teams have broken this bottleneck through the design and synthesis of electron- and proton-conductive MOFs (c-MOFs), indicating excellent charge transport properties, while the chemical and structural advantages of MOFs are still maintained. In this review, we profile the utilization of c-MOFs in several rechargeable lithium batteries such as lithium-ion batteries, Li–S batteries, and Li–air batteries. The preparation methods, conductive mechanisms, experimental and theoretical research of c-MOFs are systematically elucidated and summarized. Finally, in the field of electrochemical energy storage and conversion, challenges and opportunities can coexist.

Keywords: conductive metal–organic frameworks; lithium-ion batteries; Li–S batteries; Li–air batteries

Citation: Deng, F.; Zhang, Y.; Yu, Y. Conductive Metal–Organic Frameworks for Rechargeable Lithium Batteries. *Batteries* **2023**, *9*, 109. https://doi.org/10.3390/batteries9020109

Academic Editors: Johan E. ten Elshof and Marco Giorgetti

Received: 23 November 2022
Revised: 16 January 2023
Accepted: 31 January 2023
Published: 3 February 2023

Copyright: © 2023 by the authors. Licensee MDPI, Basel, Switzerland. This article is an open access article distributed under the terms and conditions of the Creative Commons Attribution (CC BY) license (https://creativecommons.org/licenses/by/4.0/).

1. Introduction

High-energy density and long working lifetime are the permanent pursuits for rechargeable batteries [1–6]. Currently, rechargeable lithium batteries, particularly lithium-ion batteries (LIBs), have been commercialized on a large scale, ranging from small electronic devices such as power banks and cameras to large mobile devices such as electric vehicles and aircraft [7–10]. Although LIB technology is considered one of the most promising energy storage systems owing to its high energy density and good lifespan [11], LIBs require optimization in several aspects, such as the heat resistance of LIB diaphragms at high temperatures, the life cycle of LIBs at low temperatures, and the recovery of lithium from discarded LIBs to protect the environment [12–14]. Batteries such as Li–S and Li–air have gained popularity. Li–S batteries are environmentally friendly and safe, but they have several inherent problems such as the shuttle effect and volume expansion during operations [15,16]. Li–air batteries exhibit ultra-high energy density; however, the battery lifespan is still unsatisfactory [17,18]. To develop good rechargeable lithium batteries, research on advanced materials is essential.

Metal–organic frameworks (MOFs) are coordination polymers that combine inorganic metal ions or metal clusters as junction points and organic ligands as connection bridges. MOFs are characterized by large specific surface areas, permanent pores, tunable pore diameters, and functions that can be modified according to needs [19,20]. As a result, MOFs have received considerable attention for their applications in gas storage, sensors, medicine, agronomy, and catalysis [21–27]. In recent years, MOF-based materials have been widely used for electrochemical energy storage systems. For example, Wang et al. optimized the lithium plating/stripping behavior through the carbonization of ZIF–67 [28]. Carbonized ZIF–67 as the anode for a lithium metal battery exhibited high coulombic efficiency (CE) and stable cycling performance at ultra-high current density according to experiments and theoretical calculations. Chen et al. prepared a novel MOF gel electrolyte that suppressed the formation of Li dendrites and also exhibited excellent performance at high temperatures [29]. However, most current MOFs exhibit poor electrical conductivity, which hinders their development in energy-storage systems [30–32]. Conductive metal–organic frameworks (c-MOFs) materials are a new type of material and have attracted much attention in recent years. c-MOFs are synthesized based on MOFs through the targeted design of their conductivity. Their excellent conductive ability makes them widely applicable in the field of energy storage, such as lithium batteries, fuel cells, supercapacitors, etc. The application of c-MOFs to improve the performance of rechargeable batteries opens a new path for energy storage research. Figure 1 unveils several common metal ions (Fe, Co, Ni, Cu, Zn, Pd, Ag, Pt, etc.) and organic ligands (BHT, THQ, TABTO, HHTP, etc.), which can form one-dimensional, two-dimensional, or three-dimensional MOFs. Meanwhile, some c-MOFs with metal elements belonging to different periods are visualized in Figure 1. The application of c-MOFs in sensors, supercapacitors, and water decomposition indicates that c-MOFs have pragmatic value [33–37]. The application of c-MOFs in rechargeable lithium batteries ameliorated the current challenges of lithium batteries through the promotion of charge transfer (Figure 2). From 2018 to date, there have been numerous examples of the application of c-MOFs in rechargeable lithium batteries, including the $Ni_3(HITP)_2$ diaphragm applied in Li–S batteries, Cu–BHT (butylated hydroxytoluene)/reduced graphene oxide (rGO) applied as an anodein LIBs, Cu–1,4–benzoquinone (THQ) improving the properties of Li–air batteries (Figure 3). A few excellent reviews on the development and applications of c-MOFs have been recorded [38–41]; however, the application of c-MOFs in rechargeable lithium batteries has not been specifically and systematically described. This review elucidates the conductive mechanism, the synthesis method, and experimental and theoretical research of c-MOFs in the new generation of rechargeable lithium batteries.

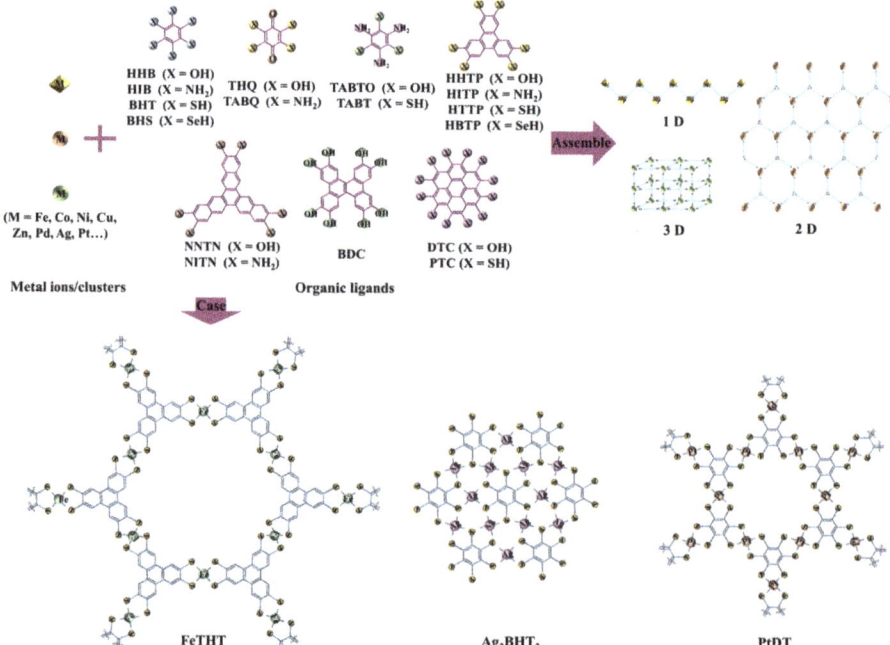

Figure 1. Schematic diagram of MOFs structure and some examples of c-MOFs.

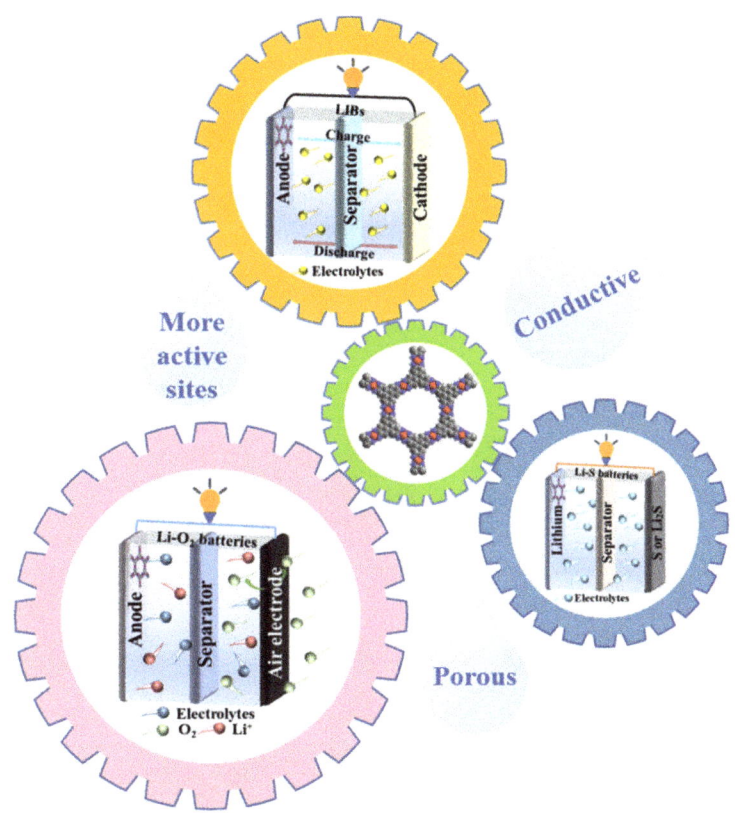

Figure 2. Application of c-MOFs in rechargeable lithium batteries.

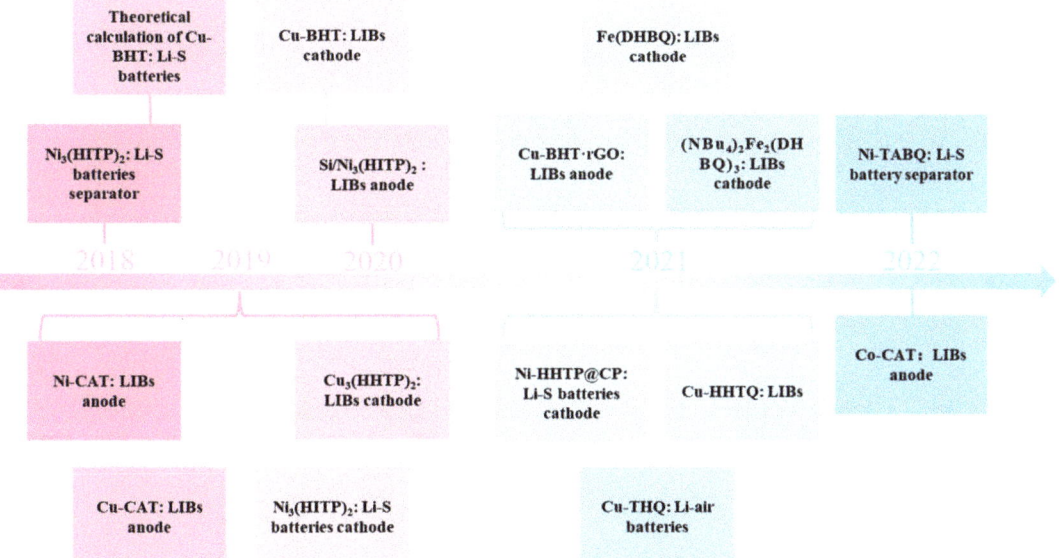

Figure 3. Time axis of c-MOFs application in lithium batteries.

2. Conduction Mechanism

MOFs with high electrical conductivity have great potential for practical applications in energy storage and microelectronic devices [38]. In this section, the conduction mechanism of c-MOFs is introduced from the aspect of electrical conductivity. Generally, electrons enter the adjacent electronic orbits under an electromotive force to generate a conduction electron, while with ionization potentials, ions move away from electronic orbits in molecular gaps under ionization energy to generate conductive ions [42]. Electrical conductivity (σ) is a vital parameter for evaluating material conductivity. The calculation is expressed as follows:

$$\sigma = e(\mu_e n_e + \mu_h n_h) \quad (1)$$

where e is the electron, h represents the hole, μ indicates the carrier mobility, and n denotes the concentration of the carrier. As the mobility or carrier concentration increases, the conductivity of c-MOF increases, and the conductivity performance would be higher (Equation (1)). To build a new type of MOF with high conductivity, a high carrier concentration and good charge mobility are required. In MOFs, both organic ligands and metal ions provide charge carriers. Organic ligands exhibit two main functions such as promoting charge transfer and providing unpaired free radicals, whereas metals require holes or high-energy electrons [42]. Numerous factors affect the conductivity of MOFs in practical applications, such as temperature and crystallinity, which change the conductivity [43,44]. For highly ordered crystalline MOFs, the conduction behavior is elucidated in the energy band theory. The energy band is divided into a conduction band, valence band, and band gap. As the temperature reaches absolute zero, the electrons occupy the positions in the valence band, while the conduction band is empty. The energy levels of electrons in isolated atoms are discrete, and the outermost electrons fill the Fermi level. For the metal conductor, the conduction and valance bands overlap, and the band gap is equal to zero. The Fermi level in the overlapped part results in a high electron concentration, and the metal exhibits good conductivity. For semiconductors and insulators, the Fermi level lies in the band gap. The band gap of semiconductors is 0~3 eV, and the band gap of insulators is greater than 4 eV. At a certain temperature, electrons could absorb a certain amount of energy to transfer from the valence band to the conduction band. Meanwhile, a hole forms in the valence band and acts as a charge carrier. The carrier concentration is obtained from Equation (2):

$$n = n_0 \exp\left(\frac{-E_a}{KT}\right) \quad (2)$$

where n_0 is a prefactor, E_a represents the activation energy, K represents Boltzmann's constant, and T represents the absolute temperature [41]. Activation energy is the potential barrier energy required to overcome the thermal excitation of electrons. The activation energy determines the charge carrier concentration (Equation (2)). The charge transport mechanisms of c-MOFs include hopping transport and band transport [45]. Both mechanisms rely on a high spatial or energetic overlap and a low barrier between the organic ligand and the symmetric orbital of the metal ion. For hopping transport, carriers are localized in discrete energy levels and transition between in situ and adjacent sites under thermal activations. The transition probability can be obtained from Equation (3):

$$p = \exp\left(-\alpha R - \frac{E}{KT}\right) \quad (3)$$

where E is the energy density between adjacent points, T denotes the absolute temperature, K represents Boltzmann's constant, R indicates distance, and α is a constant. According to Equations (2) and (3), temperature has a great influence on carrier concentration and transition probability. The higher the temperature, the higher the carrier concentration and the transition probability. Therefore, conduction mechanisms would have great influences on the application of c-MOFs at different temperatures. It can be suggested that understanding the various conduction mechanisms would guide the application of c-MOFs in lithium batteries at different temperatures. In contrast, the band transport of charge carriers is delocalized. In the band transport mechanism, carrier mobility depends on the effective mass and the frequency of carrier scattering events. The equation is expressed as follows:

$$\mu = \frac{e\tau}{m^*} \tag{4}$$

where m^* is the effective mass of the carrier, τ represents the scattering time in the two collisions, and e denotes the carrier charge. The longer the scattering time or the smaller the carrier effective mass, the greater the mobility and conductivity of the MOFs (Equation (4)). Therefore, upon the design of the synthesis strategy of c-MOFs, the doping of impurities, or the existence of holes should be avoided to reduce the loss of scattering time. To obtain a small carrier effective mass, highly symmetric lattices, simple cells, and good dispersion of the energy band are required.

The hopping transport mechanism and the band transport mechanism are realized in the chemical construction of electronic c-MOFs through bonding, space, and guest molecules. The transport mechanisms for proton c-MOFs are realized via the Grotthuss mechanism [46,47]. Figure 4 depicts the three conductive mechanisms. The transport mechanism through a bond is dependent on a covalent bond formation. The energetic and steric overlap of organic ligands and metal ions facilitates charge transport [41,45]. In 2009, Takaishi reported Cu[Cu(pdt)$_2$] as a typical example of the bond transport. The conductivity of Cu[Cu(pdt)$_2$] was 6×10^{-4} S·cm^{-1} at 300 K, which was used as a high-power density porous electrode [48]. The charge transfer through space is mainly owing to the close packing and overlapping orbitals between adjacent ligands in rigid MOFs (such as π–π stacking between organic ligands), which is a non-covalent interaction. This solves the problem of fewer bonds between adjacent electroactive units and promotes charge transfer [49]. In 2018, Carol Hua et al. demonstrated an example of the transfer of electric charge through space. In Zn(II) frames containing cofacial thiazolo[5,4-d]thiazole units, aromatic stacking interactions result in mixed valence states during electrochemical or chemical reduction [50]. The conduction via guest molecules introduces several guest molecules into the pore structure of MOFs to regulate the internal redox activity. Redox pairing and effective orbital overlap between the frame and guest molecules play a vital role in generating efficient charge transfer. A typical example was the immersion of MOFs Cu$_3$(BTC)$_2$ in methylene chloride saturated with tetracyanoquinodmethane by Yoon et al., which increased the conductivity of MOFs from 10^{-8} to 0.07 S·cm^{-1} [51]. The German chemist Theodor von Grotthuss contributed greatly to the development of the proton conduction mechanism. Hydrogen bonding is essential in this mechanism and is the key to elucidating proton transport. As a non-covalent interaction, the hydrogen bond is a weak bond formed between the hydrogen atom on the X–H (X is an atom more electronegative than hydrogen) of the molecular fragment and the same molecule or other molecules [52–54]. At room temperature, hydrogen bonds can form or break in response to thermal fluctuations, and that causes the transfer of protons in the Grotthuss mechanism.

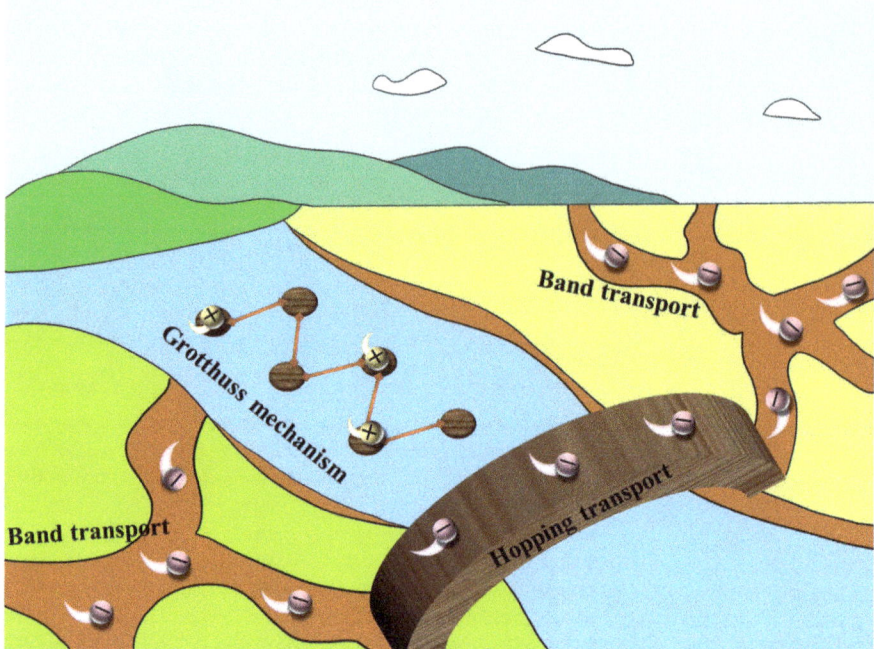

Figure 4. Three conductive mechanisms of c-MOFs.

3. Synthetic Methodologies

Most methods used in crystallography for growing single crystals can be used to prepare c-MOFs materials. The use of different methods or experimental parameters for the synthesis of c-MOFs materials can lead to variations in their properties such as crystallinity and microcrystal size, which can affect their electrical conductivity [55]. Therefore, a search for a suitable method to synthesize c-MOFs materials is vital [39].

3.1. Hydro-/Solvothermal Reactions

Owing to their simple operation and easy control, hydro-/solvothermal reactions are the major scheme for the synthesis of c-MOFs (Figure 5a). The reactants are dissolved in deionized water or other dissolvents, and the solution is heated or pressurized according to the reaction conditions so that the synthesis reaction could proceed normally [39]. The hydro-/solvothermal reactions are characterized by maneuverability and strong adjustability. The conditions of equal pressure and liquid phase reaction are conducive to the growth of perfect crystals with few defects. Through the adjustment of the parameters of the reaction solution, the particle size and morphology of the generated crystals also change [56–58]. In 2017, Li et al. synthesized $Cu_3(HHTP)_2$ (HHTP is 2,3,6,7,10,11–hexahydroxy–triphenylene) as the conductive additive and binder-free electrode for a solid-state supercapacitor [59]. However, the disadvantage of hydro-/solvothermal reactions is also significant. Upon the determination of reaction parameters and conditions, the crystal growth in the process of hydro-/solvothermal reactions could not be monitored. The synthesized polycrystalline films are rough and uneven [60].

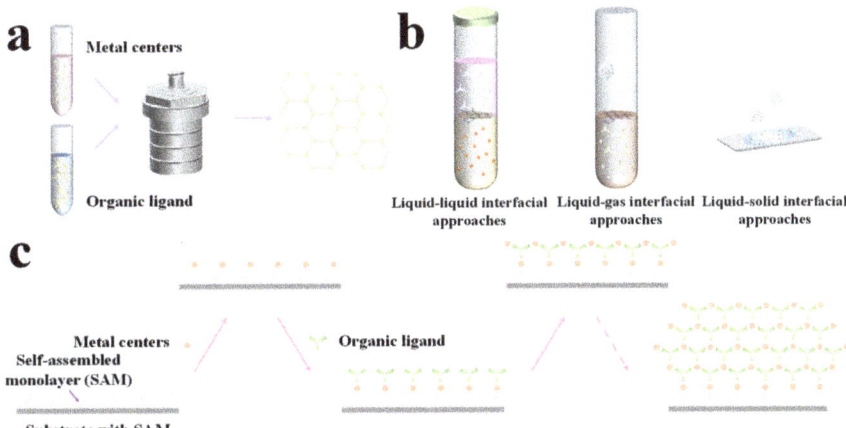

Figure 5. Strategies for the synthesis of c-MOFs. (**a**) Hydro/solvothermal synthesis, (**b**) the interfacial strategies for the synthesis of c-MOFs films, (**c**) the layer-by-layer self-assembly strategies for the synthesis of c-MOFs films.

3.2. Interface-Assisted Synthesis

Figure 5b shows several interfacial synthesis strategies, including liquid–liquid, solid–liquid, and gas–liquid interfaces. Numerous c-MOFs have been synthesized through interfacial synthesis strategies. Compared with a hydro/solvothermal synthesis, interfacial-assisted synthesis is an easier method for preparing two-dimensional (2D) c-MOFs films with high crystallinity and controllable thickness.

In 2015, Pal et al. added BHT in degassed dichloromethane, and then dissolved $K_3[Fe(CN)_6]$ and K_2PdCl_4 in degassed deionized water (dichloromethane and water were mutually insoluble). Subsequently, the two solutions were mixed to form a two-phase system, and a prephenate dehydratase (PdDt) at the interface of the two-phases system after the two solutions contacted at the liquid–liquid interface through self-assembly synthesis [61]. In 2014, Sheberla et al. used the self-assembly synthesis method to mix metal source and organic ligands in an aqueous solution at the gas–liquid interface and then added ammonia to the mixture. Finally, a film was formed between the gas and liquid phases, and $Ni_3(HITP)_2$ was synthesized [62]. In 2019, Song et al. synthesized $Cu_3(HHTP)_2$ on the $La_{0.67}Sr_{0.33}MnO_3$ electrode at the solid–liquid interface via self-assembly synthesis, which was highly crystalline [63].

3.3. Layer-by-Layer (LBL) Self-Assembly

LBL self-assembly as a multifunctional surface modification technology with low operation cost was first proposed by Iler in 1966 and has been developed since the 1990s [64]. Figure 5c illustrates the operation process of LBL self-assembly. First, the substrate is subjected to a functionalization process. Usually, a self-assembled monolayer (SAM) is generated on the surface to intercept metal ions. Then, with the weak interactions between molecules (such as coordination bond, hydrogen bond, and electrostatic attraction), layers spontaneously bond to form films with complete structure, stable performance, and certain specific functions [65,66]. Through the LBL self-assembly method, the gauge of the film can be regulated independently according to the actual needs [67,68]. Stavila et al. proposed a comprehensive mechanism of surface MOFs growth through the generation of $Cu_3(BTC)_2$ films on different substrates [66]. Shekhah et al. obtained $Zn_2(BTC)_3$ with high stability via the LBL self-assembly method, and the polymer did not decompose, regardless of the heating at 100 °C [69]. The LBL approach is relatively flexible in the steps of generating a deposition layer, which can immerse the substrate in a solution or spray the solution on the substrate. However, repeated experimental steps also lead to a longer consumption time for this approach [70].

4. Application of c-MOFs in LIBs

Currently, LIBs are used in most electronic devices in the market. During the discharge of LIBs, the lithium loses electrons and forms lithium ions, which dissolve in the electrolyte and transfer from the anode to lithium cobaltate. During the charging process of LIBs, lithium ions migrate from the cathode to the anode to gain electrons. The principle is shown in Figure 2, and the reaction equation is expressed as follows:

$$LiCoO_2 + C \rightleftharpoons Li_{1-x}CoO_2 + Li_xC \qquad (5)$$

Although LIBs are the most potential batteries, they also have several shortcomings. For example, uncontrollable lithium dendrite growth in the reaction process leads to short life cycle and reduced safety performance [71–75]. Researchers have continuously improved the performance of LIBs through various processes [76]. C-MOFs, as structurally tunable porous materials, can facilitate charge transfer and increase LIBs' reaction rates, owing to their high electrical conductivity. Numerous active sites can be designed according to their tunable structures. The large pore structure of c-MOFs can provide abundant storage sites for lithium ions and inhibit lithium dendrite growth. Therefore, introducing c-MOFs to LIBs is a promising strategy (Figure 6). This section reviews the recent achievements of c-MOFs as anode and cathode materials for LIBs in sequential order.

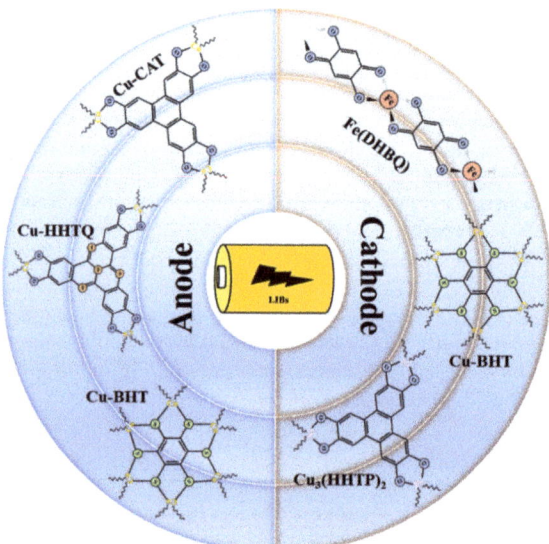

Figure 6. Application of c-MOFs in LIBs.

4.1. LIBs Anode

Designing advanced anode materials has been recognized as an effective approach for constructing suitable LIBs. To find electroactive materials with fast diffusion ability of Li-ions and gratifying capacity, Guo prepared a one-dimensional highly conductive porous Ni–catecholate (Ni–CAT) MOF via a hydrothermal method in 2019 and tested them as anode materials for LIBs (Figure 7a,b) [77]. The layer spacing of Ni–CAT was ~0.37 nm. According to the XRD diagram of the discharge process, the peak of the Ni–CAT electrode significantly changed. As the discharge process proceeded, the main peak slowly weakened until it disappeared. During the de-insertion process of Li$^+$, the structure of Ni–CAT was gradually recovered. However, the insertion of Li$^+$ disrupted the short-range order of Ni–CAT in the C–axis direction. Ni–CAT was characterized by three lithium storage sites, such as benzene rings, pores, and space between layers (Figure 7c). The reversible capacity of Ni–CAT at high current densities was greater than that of the other MOFs anodes owing to its high lithium diffusion capacity and excellent electron conduction performance, and the one-dimensional porous structure provided an effective channel for lithium-ion diffusion. The reversible capacities of Ni–CAT electrodes were 626 mA·h·g^{-1} and 592 mA·h·g^{-1} after 200 cycles at current densities of 0.2 and 0.5 A·g^{-1}. Excluding the first cycle, the CE value of each cycle was ~100% (Figure 7d). Concurrently, the structure of Ni–CAT remained stable after 300 cycles. Ni–CAT electrode exhibited good crystallinity, with a long-range order on the a and b sides and short-range order on the C–axis. This study shows the excellent performance of Ni–CAT MOF as an anode for LIBs. In 2019, Guo et al. used a solvothermal bottom-up strategy to synthesize one-dimensional c-MOF (Cu–CAT) nanowires [78]. This showed good diffusion coefficient of lithium ions, high electronic conductivity, and excellent lithium storage performance. As Cu–CAT was used as an anode for LIBs, the reversible capacities of Cu–CAT were 631 and 381 mA·h·g^{-1} after 500 cycles at current densities of 0.2 and 2 A·g^{-1}, and the CE value of each cycle was 81% at 0.5 A·g^{-1}. The fading rate was low as 0.038% per cycle. The energy density reached up to 275 W·h·kg^{-1} as the full cell was assembled. In 2022, Mao et al. synthesized Co–CAT c-MOF via a liquid-phase method. Its conductivity and one-dimensional structure could promote the rapid transport of ions, resulting in an excellent lithium storage capacity in LIBs [79]. As the half-cell was assembled for an electrochemical test, its reversible capacity was 800 mA·h·g^{-1} after 200 cycles at a current density of 200 mA·g^{-1}. This showed that the Co–CAT electrode exhibited good structural stability. Upon the assemble of the full battery with LiCoO$_2$, the capacity was 404 mA·h·g^{-1} after 100 cycles at current density of 200 mA·g^{-1}. Co–CAT electrodes exhibited excellent lithium storage capacity in LIBs and good potassium storage capacity in potassium-ion batteries (PIBs). Similarly, as Co–CAT was used as the anode for PIBs, its capacity was 230 mA·h·g^{-1} after 700 cycles at 1.0 A·g^{-1}. Meanwhile, an eight-electron transfer occurred in PIBs with excellent potassium storage performance. Owing to the conductivity of Co–CAT, the modified PIBs exhibited excellent electrochemical performance. Therefore, c-MOFs are very promising materials for applications in PIBs.

The anodes in LIBs usually undergo volume expansion during the lithiation process. In 2020, Aqsa Nazir et al. prepared Si/Ni$_3$(HITP)$_2$ (HITP represents 2,3,6,7,10,11–Hexaiminotriphenylene) composites as anodes for LIBs [80]. Ni$_3$(HITP)$_2$ promoted the rapid movement of lithium ions in the electrode, improved the lithium storage capacity and the electrode conductivity, and inhibited the volume expansion of Si owing to its open pore structure, high conductivity, and uniformly dispersed Ni and N heteroatoms. At any rate, the capacity of a silicon nanoparticle anode was lower than that of the Si/Ni$_3$(HITP)$_2$ anode, indicating that the Si/Ni$_3$(HITP)$_2$ anode exhibited excellent high-rate capability. The Si/Ni$_3$(HITP)$_2$ anode featured a higher CE close to 100%, regardless of the high discharge rate at 20 C, which indicates the excellent lithiation/de-lithiation reversibility of this electrode. The reversible capacity of Si/Ni$_3$(HITP)$_2$-assembled batteries was 876 mA·h·g^{-1} after 1000 cycles at 1 C, and CE was ~100%, indicating that Si/Ni$_3$(HITP)$_2$ electrode exhibited good cycling performance. Aqsa Nazir et al. also compared the c-MOFs electrode with the previously reported electrode. Compared with other electrodes, the c-MOF electrode exhibited a high reversible capacity of 2657 mA·h·g^{-1} after 100 cycles at 0.1 C. In 2021, Meng et al. synthesized a composite of Cu–BHT 2D c-MOF and rGO to solve the densification problem of Cu–BHT [81]. Compared with pristine Cu–BHT MOF composite, Cu–BHT and rGO composite exhibited good electrical conductivity and more redox-active sites. The composites were electrochemically tested as anodes for LIBs. The reversible specific capacities of rGO and Cu–BHT as a composite electrode were 1190.4, 1230.8, 1131.4, and 898.7 mA·h·g^{-1} with a ratio of 1:1 at current densities of 100, 200, 500, and 1000 mA·g^{-1}, respectively. The reversible capacities of rGO and Cu–BHT were higher than those of pristine Cu–BHT MOFs.

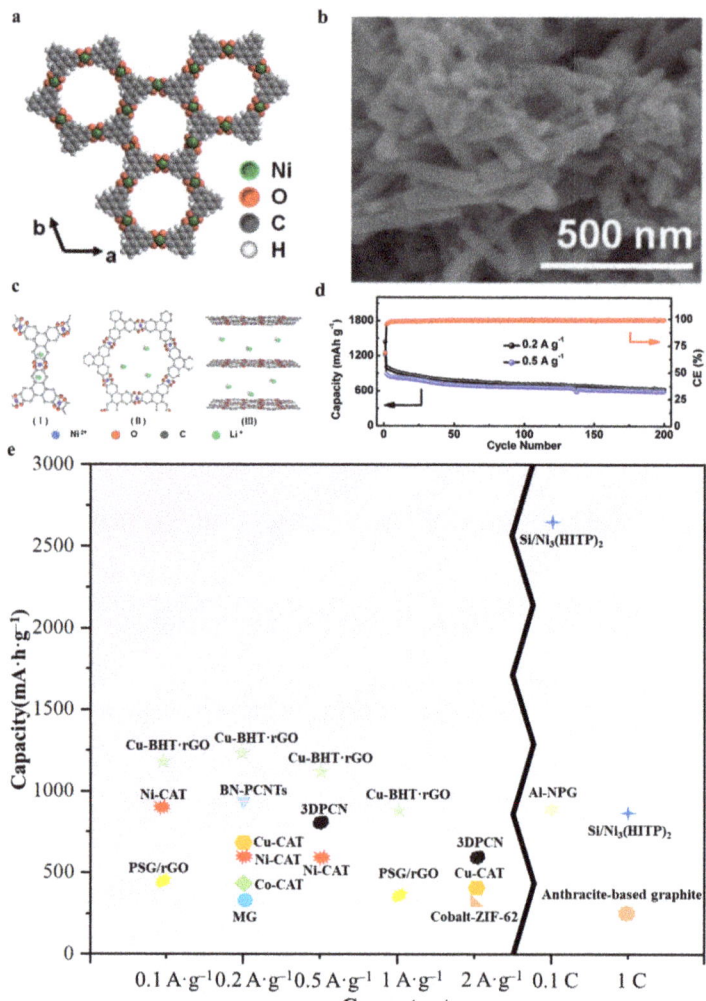

Figure 7. (a) Structural model and (b) FESEM images of Ni–CAT, (c) three lithium storage sites of Ni–CAT located in (I) benzene ring, (II) pores, and (III) space between layers, (d) cycling performance of Ni–CAT at 0.2 and 0.5 A·g^{-1}. Reproduced with permission from Ref. [77]. Copyright 2019 RSC. (e) Comparison of c-MOFs with some conventional carbon-based materials as anodes.

In addition, in 2021, Yan synthesized the tricycloquinazoline (TQ) and two-dimensional c-MOF Cu–HHTQ (HHTQ = 2,3,7,8,12,13–hexahydroxytricycloquinazoline) cooperating with CuO$_4$ [82]. The c-MOF possessed excellent lithium storage ability and high rate capability. The redox activity of TQ was verified by theoretical calculation and experiments for the first time. Electrochemical tests were carried out with Cu–HHTQ as the active material and Li as the electrode. In the first cycle, the discharge capacity was 1716 mA·h·g^{-1}, and the charging capacity was 989 mA·h·g^{-1}. The charging capacity was already one of the highest values in the existing reports. When the current density was 600 mA·g^{-1}, the specific capacity of Cu–HHTQ was 657.6 mA·h·g^{-1}, and the charge–discharge capacity remained at 82% for 200 cycles. Its specific capacity was also one of the highest values, as illustrated in Table 1. It was found that the Cu–HHTQ had partial charge capacitor storage in the scanning rate range of 0.2–1 mV·S^{-1}, and contributed 39% to 58% of the capacitance. This was mainly attributed to the high conductivity, large specific surface area and continuous pores, as well as multiple redox activities.

Table 1. Performance of c-MOFs in rechargeable lithium batteries.

c-MOFs	Dimension	Application	Current Rate	Cycles	Capacity (mA·h·g^{-1})	Reference
Ni–CAT	1	LIBs anode	0.1 A·g^{-1}	/	889	[77]
			0.2 A·g^{-1}	200	626	
			0.5 A·g^{-1}	200	592	
Cu–CAT	1	LIBs anode	0.2 A·g^{-1}	320	646	[78]
			2.0 A·g^{-1}	/	381	
Co–CAT	/	LIBs anode	200 mA·g^{-1}	100	404	[79]
Si/Ni$_3$(HITP)$_2$	2	LIBs anode	0.1 C	100	2657	[80]
			1 C	1000	876	
Cu–BHT·rGO	2	LIBs anode	100 mA·g^{-1}	/	1190.4	[81]
			200 mA·g^{-1}	/	1230.8	
			500 mA·g^{-1}	/	1131.4	
			1000 mA·g^{-1}	/	898.7	
Cu–HHTQ	2	LIBs anode	600 mA·g^{-1}	200	657.6	[82]
Cu$_3$(HHTP)$_2$	/	LIBs cathode	1 C	20	94.9	[83]
Cu–BHT	2	LIBs cathode	300 mA·g^{-1}	500	175	[84]
(NBu$_4$)$_2$Fe$_2$(DHBQ)$_3$	3	LIBs cathode	500 mA·g^{-1}	350	103.1	[85]
Co$_3$(HITP)$_2$	2	Li–S batteries separator	/	/	762	[86]
Ni$_3$(HITP)$_2$	2	Li–S batteries separator	1 C	500	716	[87]
Ni$_3$(HITP)$_2$	2	Li–S batteries separator	0.5 C	300	585.4	[88]
Ni–TABQ	2	Li–S batteries separator	1 C	1000	820	[89]
Ni$_3$(HITP)$_2$	2	Li–S batteries cathode	0.2 C	100	1302.9	[90]
			0.5 C	150	807.4	
			1 C	300	629.6	
Ni–HHTP@CP	/	Li–S batteries cathode	0.2 C	200	910	[91]
NiRu–HTP	/	Li–air batteries cathode	500 mA·g^{-1}	200	/	[92]
Cu–THQ	/	Li–air batteries cathode	1–2 A·g^{-1}	100–300	1000–2000	[93]

Figure 7e shows the comparison of c-MOFs anode (Ni–CAT, Cu–CAT, Co–CAT, Cu–BHT·rGO, Si/Ni$_3$(HITP)$_2$) with some conventional carbon-based materials as anodes, such as porous graphite/rGO (PSG/rGO), cobalt–ZIF–62, modified graphite (MG), anthracite-base graphite, layers of nanoporous graphene (NPG) on the surface of Al thin films (Al–NPG), porous carbon nanotubes webs with high level of boron and nitrogen co-doping (BN–PCNTs), three-dimensional interconnected porous carbon nanoflakes (3DPCNs) [94–100]. As can be seen from the figure, the Cu–BHT·rGO and Si/Ni$_3$(HITP)$_2$ have obvious advantages at current densities of 0.1, 0.2, 0.5 and 1 A·g^{-1} or at rates of 0.1 and 1 C.

4.2. LIBs Cathode

In addition to the construction of c-MOFs anodes, cathodes use c-MOFs to enhance battery performance. As cathode materials, the large aperture structure and high conductivity of c-MOFs could effectively improve the cathode capacity. In 2019, Gu et al. prepared a Cu$_3$(HHTP)$_2$ c-MOF cathode for LIBs (Figure 8a) [83]. The size of Cu$_3$(HHTP)$_2$ nanosheets was ~20–40 nm, showing an irregular shape. Figure 8b shows the high-resolution transmission electron microscope images. According to Brunauer–Emmett–Teller analysis, the specific surface area of Cu$_3$(HHTP)$_2$ reached 506.08 m^2·g^{-1}. The ideal specific surface area and effective pore structure of c-MOF were essential for lithium-ion embedding. During the charge/discharge cycling test with $Cu_3(HHTP)_2$ as the cathode, its discharge capacity increased with the number of cycles, indicating that the $Cu_3(HHTP)_2$ cathode was relatively stable. The capacity of the coin LIBs assembled with c-MOF as the cathode was 95 mA·h·g^{-1} after 60 cycles, and the CE value was 50% after 500 cycles. The capacity decay rate was 0.09% per cycle, and the CE value was 100%. During the discharge process, each unit of Cu^{2+} was reduced to Cu^+ through the insertion of a Li^+ unit. Correspondingly, a unit of Li^+ was released during the charging process, and each unit of Cu^+ was oxidized to Cu^{2+}. Figure 8c,d show the capacity change of coin LIBs assembled with $Cu_3(HHTP)_2$ as the cathode g at a high current rate. The capacity of coin LIBs gradually decreased with enhanced current rates. As a low current rate was restored, the battery capacity was recovered, exhibiting good reversibility. The capacity retention rate of coin LIBs increased with the increasing current rate. After 500 cycles at 20 C, more than 85% of the battery capacity was maintained with a decay rate of 0.023% per cycle. In 2020, Wu et al. synthesized a Cu-BHT 2D c-MOF through the self-assembly reaction of a BHT monomer and Cu(II) salt in ethanol (Figure 9a), exhibiting room temperature conductivity of 231 S·cm^{-1} [84]. The thermogravimetric analysis and acid-base tests revealed that the c-MOF exhibited high thermal and chemical stability. Figure 9b shows the experimental and theoretical calculation results. The c-MOF was used as a cathode material for LIBs with the potential range of 1.5–3.0 V (versus Li^+/Li), and a four-lithium-ion storage reaction at redox-active sulfur atoms occurred on each Cu-BHT unit. The electrochemical test revealed that the reversible capacity of Cu-BHT as the cathode was 232 mA·h·g^{-1}, which was close to its theoretical capacity of 236 mA·h·g^{-1}, and each unit of Cu-BHT stored ~ 3.94 lithium ions. The lifespan and rate capacity of the Cu-BHT cathode were tested. The results show that after 500 cycles at a current density of 300 mA·g^{-1}, the capacity decay rate was 0.048% per cycle. After 10 days of shelving, the capacity decay rate was 98.1% per cycle. Owing to its inherent high conductivity and the strong coordination of Cu(II) and BHT in the complete π-d conjugated system, the Cu-BHT cathode exhibited good cycling performance and longer service life.

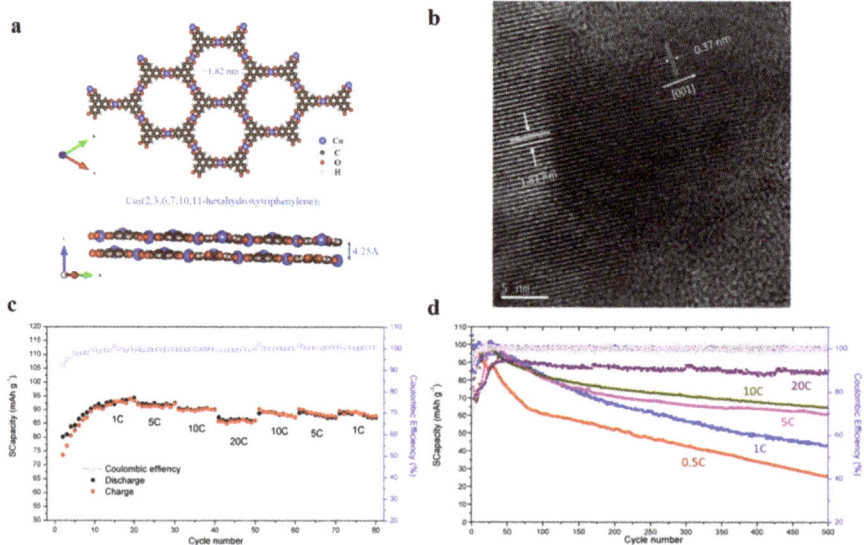

Figure 8. (**a**) Crystal structure of $Cu_3(HHTP)_2$ from top and side views, (**b**) HR−TEM image of $Cu_3(HHTP)_2$, (**c**) the rate capability of $Cu_3(HHTP)_2$ at current rates from 1 C to 20 C, (**d**) the cycling performance of $Cu_3(HHTP)_2$ at different current rates for 500 cycles. Reproduced with permission from Ref. [83]. Copyright 2019 Elsevier B.V.

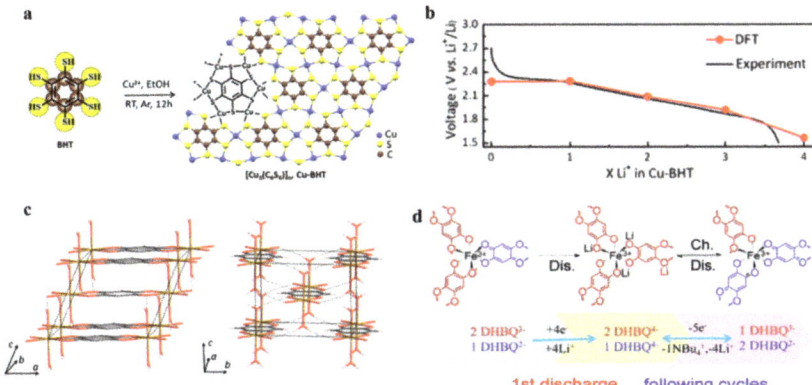

Figure 9. (a) Synthesis process of the Cu–BHT C-MOF, (b) experimental and theoretically predicted DFT voltage curves for the Cu–BHT C-MOF cathode in the voltage range of 1.5 to 3.0 V (versus Li$^+$/Li). Reproduced with permission from Ref. [84]. Copyright 2020 ACS. (c) Perspective view of the crystal structure of Fe(DHBQ)–(H$_2$O)$_2$ along the b- and a-axes. Reproduced with permission from Ref. [101]. Copyright 2021 ACS. (d) (NBu$_4$)$_2$Fe$_2$(DHBQ)$_3$ reaction mechanism during charge and discharge cycles. Reproduced with permission from Ref. [85]. Copyright 2021 ACS.

The performance of LIBs was improved with increasing conductivity of the active cathode. In 2021, Kazuki Kon et al. reported a one-dimensional c-MOF Fe(DHBQ)–(H$_2$O)$_2$ (DHBQ is 2,5–dihydroxy–1,4–benzoquinone) [101]. Figure 9c shows the electron transfer interaction between ligands and metal ions of c-MOF. Fe(DHBQ)–(H$_2$O)$_2$ has no pore structure owing to the removal of water, while Fe(DHBQ) has a permanent porous structure in an anhydrous phase. The conductivity of c-MOF increased to 1.0×10^4 S·cm^{-1} during the desolvation process, and the conductivity at room temperature was 5×10^{-6} S·cm^{-1}. As Fe(DHBQ), acetylene black, and polytetrafluoroethylene were mixed in a certain proportion to design a cathode for LIBs, the initial discharge capacity of LIBs reached 264 mA·h·g^{-1}.

Similarly, in 2021, Dong et al. synthesized a conductive three-dimensional MOF (NBu$_4$)$_2$Fe$_2$(DHBQ)$_3$ via a simple reaction using mixed of valence 2,5–dihydroxyben–zoquinone (DHBQ$^{2-/3-}$) as the linker and Fe^{3+} as the metal center with a conductivity of 1.07 mS·cm^{-1} [85]. The battery based on the c-MOF cathode exhibited a capacity of 91.4% and a high CE of ~100% after 350 cycles at a current density of 500 mA·g^{-1}. With the current density up to 1000 mA·g^{-1}, the reversible capacity reached 94.4 mA·g^{-1}, and the capacity retention rate was 71.5% after 1000 cycles. The outstanding electrochemical properties of c-MOF cathode were attributable to the high conductivity and hollow structure of (NBu$_4$)$_2$Fe$_2$(DHBQ)$_3$, which can promote the migration kinetics of electrons and ions. Figure 9d shows the reaction mechanism of the charging and discharging process. During the discharge process, both DHBQ^{2-} and DHBQ^{3-} underwent a four-electron reaction. In the charging process, DHBQ^{4-} underwent a five-electron reaction with NBu$_4^+$ and four lithium ions. Summarily, the inherent pore structure of c-MOFs enhances the specific surface area of electrodes in LIBs, and increases the number of active sites. This has a positive effect on restraining the volume expansion of the battery during the charging and discharging process. The inherent high conductivity of c-MOFs will promote the speed of ion and electron transfer which is essential for enhancing the performance of LIBs.

5. Application of c-MOFs in Li–S Batteries and Li–Air Batteries

Unlike LIBs that require lithium compounds and graphite as electrodes, Li–S batteries generally use sulfur as the cathode and lithium foil as the anode, and Li–air batteries can directly use lithium and oxygen in the air as electrodes. Li–S and Li–air batteries are ascribed to an electrochemical mechanism different from the ion extraction–insertion mechanism of LIBs. Additionally, Li–S and non-liquid Li–air batteries exhibit high theoretical energy densities of 2600 and 11,200 W·h·kg^{-1}, respectively, [102–105] compared with LIBs (250–300 W·h·kg^{-1}) [106,107]. The following section focuses on the achievements of using c-MOFs to enhance Li–S batteries and Li–air batteries (Figure 10).

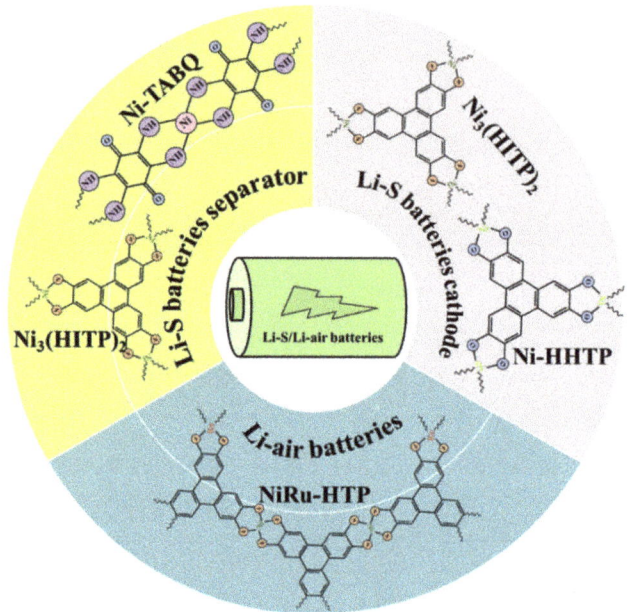

Figure 10. Application of c-MOFs in Li–S and Li–air batteries.

5.1. Li–S Batteries

A Li–S battery consists of a sulfur composite cathode, a lithium metal anode, and an electrolyte between them (Figure 2). Since elemental sulfur is a poor conductor of electrons, sulfur composite positive electrodes consist of the elemental sulfur, a conductive agent, and a polymer binder. Currently, widely used Li–S batteries are organic electrolytes based on organic solvents and lithium salts. According to Equation (6), the metal lithium anode loses electrons in the discharge process to generate lithium ions, and the metal lithium is continuously dissolved in the electrolyte. During the charging process, lithium ions gain electrons from the electrolyte, and metallic lithium is recovered, which continuously forms sediments on the pole.

$$2Li^+ + S + 2e^- \rightleftharpoons Li_2S \qquad (6)$$

Owing to the high theoretical specific energy and capacity (1675 mA·h·g^{-1}) of sulfur as a cathode for lithium batteries, Li–S batteries have received great attention as a next-generation energy storage system [108–110]. Moreover, sulfur is cheap (<USD 150 per metric ton), environmentally friendly, and abundant (17th richest element) [111–113]. However, the poor cyclability and slow charging/discharging rate of Li–S batteries are the bottlenecks restricting their development [114–116]. To address the problems associated with Li–S batteries such as the insulation performance of sulfur [117], the large volume change during the battery discharge process [118], and the reciprocating migration (shuttle effect) of polysulfides (Li$_2$S$_n$, $4 \leq n \leq 8$) formed between positive and negative electrodes [119], the application of MOFs in Li–S batteries has received great attention owing to the high specific surface area, adjustable structures, ideal crystallization, and rich catalytic sites [86,120–124]. However, the inherent electronic conductivity of conventional MOFs materials is usually poor, resulting in unsatisfactory reaction efficiency and low utilization of active species [40]. In contrast, c-MOFs have shown great potential for wide applications in Li–S batteries owing to their excellent electronic conductivity.

In 2018, Li et al. performed first-principles calculations via the projected-augmented wave method according to density functional theory (DFT) [125]. The adsorption of S_8/lithium polysulfides (LiPSs Li_2S_n, n = 1, 2, 4, 6, and 8) on Cu–BHT monolayers was systematically studied. The results revealed that the outstanding conductivity of the Cu–BHT monolayer was the key factor for improving sulfur availability. Owing to the combination of Li–S bonds and S_a–Cu bonds, the Cu–BHT monolayer interacted with LiPSs and guided the uniform diffusion of Li_2S. The above conclusions indicated that the Cu–BHT monolayer played a vital role in inhibiting the shuttling of soluble LiPSs and improving the charging–discharging rate and cyclability.

5.1.1. Li–S Batteries Separators

Experimentally, the surface in situ modification of traditional separators using c-MOFs materials is a promising strategy for separator modification and interfacial stabilization of lithium metal batteries. In 2018, Zang et al. developed a liquid–solid interface via a self-assembly method to design and prepare $Ni_3(HITP)_2$ materials [87]. The material was a large area of the microporous membrane without cracks. Concurrently, the concept of a polysulfide barrier in Li–S batteries was put forward by Zang et al. Crystalline microporous membrane was a beneficial barrier layer that could improve the performance of Li–S batteries. The microporous film exhibited a low density, a large area (more than 75 cm^2), adjustable thickness (90–970 nm), and high conductivity (3720 S·m^{-1}). Additionally, the microporous film exhibited a highly neat pore structure and excellent ability to adsorb polysulfides, which was an ideal barrier. The results showed that the capacity, rate function, and cyclability of the Li–S batteries were significantly improved after the film was used to optimize the Li–S batteries, and the average capacity decay rate was 0.032% per cycle.

In 2019, Chen et al. prepared a membrane modified using $Ni_3(HITP)_2$ layers through a simple filtration method [88]. The battery separators achieved a specific capacity of 1220.1 mA·h·g^{-1} at a discharge rate of 0.1 C, a specific capacity of 800.2 mA·h·g^{-1} at 2 C, and a specific capacity of 1008.0 mA·h·g^{-1} at 0.1 C (Figure 11a). The designed diaphragm exhibited good conductivity, a well-distributed pore structure, and good hydrophilicity, which adsorbed the polysulfide to reduce blockage and improved the cycling stability of the battery. Therefore, the separator can slow down the shuttle effect of Li–S batteries and enhance their capacity rate.

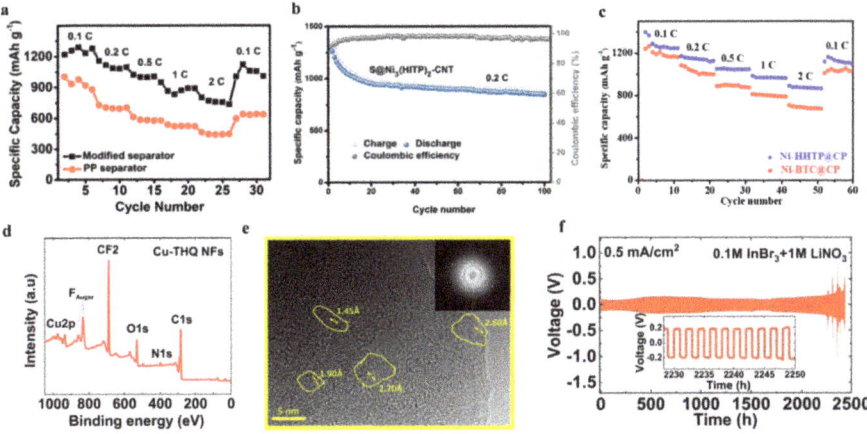

Figure 11. (**a**) Rate performance of Li–S batteries with PP and $Ni_3(HITP)_2$ modified separators at different current densities of 0.1, 0.2, 0.5, 1, 2 and 0.1 C. Reproduced with permission from Ref. [88]. Copyright 2019 ACS. (**b**) Cycling performance of S@$Ni_3(HITP)_2$-CNT cathode after 100 cycles at 0.2 C. Reproduced with permission from Ref. [90]. Copyright 2019 Wiley-VCH. (**c**) Rate performance of Ni–HHTP@CP cathode and Ni-BTC@CP cathode in Li–S batteries. Reproduced with permission from Ref. [91]. Copyright 2021 Elsevier B.V. (**d**) XPS survey spectrum of Cu–THQ NFs coated on GDE,

(**e**) High resolution TEM image of the Li$_2$O$_2$ product after the 10th discharge. The corresponding fast Fourier transform (FFT) pattern is shown in the inset, (**f**) long-term cycling of Li | | Li symmetrical cell with current density of 0.5 mA·cm^{-2} and fixed capacity of capacity of 0.5 mA·h·cm^{-2}. The inset presents the details of the voltage versus time profile towards the end of cycling. Reproduced with permission from Ref. [93]. Copyright 2022 Wiley-VCH Verlag.

In 2022, Xiao et al. proposed a strategy to simultaneously integrate two catalytic centers with different properties (Ni–N$_4$ active site and quinone chemical group) into one c-MOF Ni–tetraaminobenzoquinone (Ni–TABQ) [89]. This strategy could effectively elucidate the concept of efficient multi-step catalytic conversion of LiPSs using MOFs. Additionally, a Ni–TABQ film prepared in situ on the polypropylene separator had a density of 0.075 mg·cm^{-2} and a thickness of 1.8 µm. Systematic electrochemical experiments and detailed DFT simulation calculations show that the c-MOF film can simultaneously realize the sieving, adsorption, and multi-step catalytic conversion of polysulfide ions, which significantly improves the capacity and lifespan of Li–S batteries.

5.1.2. Li–S Batteries Cathode

Regarding the use of c-MOFs to modify the membrane, its modification on the cathode has also received attention. In 2019, Cai et al. synthesized the c-MOF Ni$_3$(HITP)$_2$ via a straightforward hydrothermal approach and studied its electrochemical behavior as a cathode for Li–S batteries [90]. Ni$_3$(HITP)$_2$ is hydrophilic and can adsorb polysulfides, and its shape is similar to the two-dimensional layered structure of graphene. Ni$_3$(HITP)$_2$ was an effective physical barrier that could inhibit the shuttle effect and improve cycling stability. The sulfur body in Li–S batteries required the use of c-MOFs to enhance its performance. The experimental results showed that the carbon nanotube-based S@Ni$_3$(HITP)$_2$ cathode displayed excellent sulfur availability, rate performance, and stable cycling endurance. After 100 cycles at 0.2 C, a high primary capacity of 1302.9 mA·h·g^{-1} and a good capacity maintenance of 848.9 mA·h·g^{-1} occurred (Figure 11b).

In 2021, Wang et al. adjusted the surface chemistry of self-supporting carbon paper (CP) using Ni–HHTP materials to facilitate polysulfide conversion in Li–S batteries [91]. The electronic conductivity of Ni–HHTP was higher than that of traditional MOFs (6 × 10^{-3} S·cm^{-1}). Ni–HHTP exhibited strong chemisorption of polysulfides, which suppressed the shuttle effect, thus improving the utilization rate of active materials in Li–S batteries. The Li–S batteries loaded with Ni–HHTP@CP materials under the conditions of high sulfur load and reduced electrolyte exhibited excellent rate performance, a specific capacity of 892 mA·h·g^{-1} at a discharge rate of 2 C (Figure 11c). The Ni–HHTP@CP material proves that the synergistic effect of strong polysulfide adsorption and excellent electronic conductivity is essential for the design of cathode materials for Li–S batteries.

The aforementioned studies have shown that whether c-MOFs were used for the diaphragm or positive electrode of Li–S batteries, as the c-MOFs adsorbed polysulfides, the shuttle effect was restrained, thus improving the battery performance. This discovery provides a clear direction for the design of Li–S battery modification.

5.2. Li–Air Batteries

As a new generation of large-capacity batteries, Li–air batteries have attracted special interest. The Li–air battery consists of a pure lithium metal sheet, an electrolyte, and an air cathode containing catalysts (Figure 2). The Li–air battery is a simple layered structure. The diaphragm is immersed in the electrolyte and arranged between the cathode and the anode. The positive side of the Li–air battery is covered with oxygen pores. In addition, the positive electrode uses a gas diffusion layer electrode with composite porous carbon as a collector. In the discharge reaction, the metal lithium of the cathode dissolves and reacts with oxygen on the positive electrode, and lithium peroxide is precipitated (Li_2O_2). Charging is the reverse of the discharge reaction. The Li_2O_2 of the positive electrode is decomposed to release oxygen, and the metal *Li* is precipitated on the negative electrode. The overall equation is written as follows:

$$2Li^+ + O_2 + 2e^- \rightleftharpoons Li_2O_2 \tag{7}$$

In theory, since oxygen is not limited as a positive reactant, the battery capacity only depends on the lithium electrode. The theoretical energy density of Li–air batteries is higher than that of LIBs and it is very promising for automotive batteries [126,127]. However, owing to their fatal defects, Li–air batteries have not been popularized. During discharge and charging reactions, solid reaction products (such as Li_2O_2) accumulate on the cathode surface, and the contact between oxygen and electrolyte is blocked; thus, the discharge would be stopped [92,128,129]. Additionally, the decomposition efficiency of Li_2O_2 generated during battery operation is low, and Li_2O_2 would diffuse and accumulate on the cathode surface, which affect the cycling performance of the battery [130]. To optimize battery performance, MOFs have been used as ideal candidates for studying electrode materials in electrochemical energy applications, particularly in secondary batteries such as Li–air batteries, owing to their excellent properties. However, because of their insulation, traditional MOFs are distant from the standard of practical positive electrode materials. Therefore, the synthesis of c-MOFs with high conductivity and structural stability via various design strategies is crucial for their application in Li–air batteries [131,132].

In 2022, Majidi et al. peeled bulk Cu–THQ into 2D nanosheets (NFs) via stripping technology, and then coated the peeled Cu–THQ–NFs on a gas diffusion electrode (GDE) [93]. Figure 11d shows the XPS survey spectrum of Cu–THQ–NFs coated on a GDE. DFT calculation of Li_2O_2 growth on Cu–THQ framework showed that Cu was the growth site, and its formation was thermodynamically favorable. Because the surface of c-MOFs was highly active, it could promote the formation of nanocrystalline Li_2O_2 in the amorphous Li_2O_2 region (Figure 11e). These characteristics, combined with the $InBr_3$ electrolyte additive, enabled the battery to operate at a high charge/discharge current density (Figure 11f).

Overall, c-MOF is a promising material, regardless of the numerous challenges. The progress of c-MOF synthesis will lead to rapid and sustainable development in the fields of electronics and electrochemistry. Moreover, c-MOF has made inspiring progress in numerous aspects when applied to potential rechargeable lithium batteries, although we are a significant distance from commercialization. As the remaining problems in these batteries have been solved by several researchers in the exponentially growing battery market, other batteries are likely to coexist with LIBs and may dominate the rechargeable lithium battery market position over the next decade.

6. Theoretical Calculation

First-principles calculations that rely on DFT are crucial in scientific research. For example, DFT calculations can effectively predict the application of new materials, which is instructive for subsequent experiments and validate the experimental results. In the field of energy storage, DFT calculations can determine the stability of a material structure, calculate the free energy, and then elucidate the electrochemical reactions that occurred in the reaction process. DFT calculations can also calculate the distribution of electrons in new materials. For example, band structure and density of state (DOS) can determine whether a novel electrode material is a metal, a semiconductor or an insulator. Concurrently, DFT calculations can not only simulate the ion diffusion kinetics to elucidate the electrochemical reaction rate but also simulate the ion adsorption kinetics [133,134]. DFT calculations help to guide the experiment and save cost and time. Currently, the application of c-MOFs in the modification of rechargeable lithium batteries has not been sufficiently developed. Therefore, providing theoretical predictions and guidance for the research and development of advanced materials in this field is vital. This section reviews several works conducted via the DFT method, such as the differences between $Cu_3(HHTP)_2$ before and after lithiation, the structural changes of TQ during lithiation, the stable adsorption sites of lithium in Cu–BHT, and the advantages of $Ga_3C_6N_6$ as the anode for LIBs.

Gu et al. performed DFT calculations to elucidate the structural differences between the original and lithiated $Cu_3(HHTP)_2$ in LIBs (Figure 12a) [83]. As lithium ions were intercalated into $Cu_3(HHTP)_2$, each lithium ion was adsorbed on the copper atoms between the layers, which led to the structure relaxation, and a slight decrease in the α and β angles. Other lattice parameters were almost unchanged, indicating the stability of the hexagonal frame. However, the Cu^{2+} of $Cu_3(HHTP)_2$ was reduced to Cu^+ to adsorb a Li^+, resulting in the reduction in the repulsion between the layers in the framework and a decrease in the crystal volume. Gu et al. also calculated the average redox potential of $Cu_3(HHTP)_2$, which was consistent with the experimental results. Similarly, Wu et al. calculated the stable adsorption sites of lithium in Cu–BHT via the DFT method [84]. The benzene ring locates at the center of the structure, and the five-membered ring composed of S, C, Cu and the six-membered ring structure composed of S, Cu are calculated and analyzed (Figure 12b). The results revealed that the C site of the benzene ring was suitable for lithium storage, and the conductivity of lithium atoms increased after the insertion of Cu–BHT. The original Cu–BHT exhibited metallic properties and a small band gap, resulting in high conductivity (Figure 12c). The emergence of new electronic states in the six-membered rings and carbon ring proves that the conductivity is higher compared with that of the original Cu–BHT. To elucidate the lithium storage mechanism of TQ in LIBs, Yan et al. used DFT to calculate the structural change of TQ during the lithiation process [82]. Computational results showed that TQ underwent a nine-electron reaction during the lithiation and delithiation. Figure 12d shows the structural change from TQ to TQ-9Li. 1 Li^+–3 Li^+ combined with three pyrimidine N atoms to form TQ-Li, TQ-2Li, and TQ-3Li. The 4 Li^+–6 Li^+ were distributed around the central N atom, and the C=C bond was adjacent to the C=N bond owing to electrostatic interactions. As 6 Li^+ were inserted, the electrostatic potential of the terminal benzene ring became negative, and the 7 Li^+–9 Li^+ combined with the terminal benzene ring, respectively. Additionally, DFT calculation used in the design of c-MOFs also occupied a vital position in the c-MOFs structure. Wu et al. theoretically designed a 2D MOF of $Ga_3C_6N_6$ based on Cu–BHT via the isoelectronic substitution strategy [135]. The structure of the $Ga_3C_6N_6$ MOF monolayer remained stable at 2400 K, indicating that the MOF exhibited excellent thermal stability. The theoretical calculation results show that $Ga_3C_6N_6$ had a moderate open circuit potential (0.96 V), a low diffusion barrier (1.12 eV), and a high theoretical specific capacity (330 $mA·h·g^{-1}$). According to the calculated band structure and DOS of $Ga_3C_6N_6$ adsorbed by lithium atoms in Figure 12e,f, a single lithium atom was adsorbed, and the lithium atom did not contribute to DOS around the Fermi level, but with increasing adsorption quantity of lithium atoms, the contribution of lithium atoms gradually increased. Therefore, $Ga_3C_6N_6$ exhibited excellent lithium storage capacity and good electronic conductivity. The above calculation results show that $Ga_3C_6N_6$ is a promising anode material for LIBs.

Although the experimental results are important for a new study, DFT calculations can expose some unobservable details, such as the lithium storage sites for c-MOFs, the changes in the structure of c-MOFs during the lithiation and delithiation processes, and the reaction mechanism. These theoretical findings are vital for improving the performance of rechargeable lithium batteries.

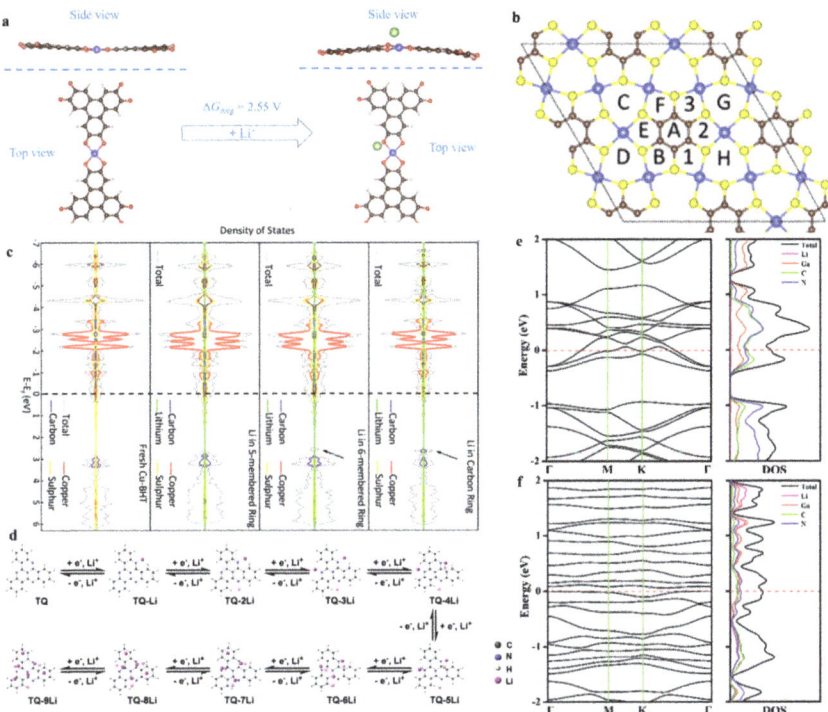

Figure 12. (a) Side view and top view of a single unit of relaxed and lithiated $Cu_3(HHTP)_2$ framework. The average potential for each lithium-ion insertion was 2.55 V, taking the lithium anode as a reference. The green, red, brown, blue, and white spheres are lithium, oxygen, carbon, copper, and hydrogen atoms, respectively. Reproduced with permission from Ref. [83]. Copyright 2019 Elsevier B.V. (b) Fully relaxed 2 × 2 crystallographic structure of the 2D Cu–BHT MOF. The numbers and letters represent different positions for Li^+ adsorption. C atoms in brown, Cu atoms in blue, and S atoms in yellow, (c) total and projected density of states of the fresh Cu–BHT monolayer and 1 Li^+ loaded in the essential three rings of the Cu–BHT structure. Reproduced with permission from Ref. [84]. Copyright 2020 ACS. (d) Structure evolution of TQ lithiation pathway in LIBs calculated by DFT. Reproduced with permission from ref. [82]. Copyright 2021, John Wiley and Sons Ltd. Energy band structures and density of states (DOS) of (e) $Li_{0.25}Ga_3C_6N_6$, (f) $Li_{4.5}Ga_3C_6N_6$. The dashed red lines represent the Fermi levels. Reproduced with permission from Ref. [135]. Copyright 2022 Elsevier B.V.

7. Conclusions and Outlook

As one of the better-developed batteries currently, the development of rechargeable lithium batteries is hindered by problems, such as dendrite generation during operation, volume expansion, polysulfide formation, and loss of active materials. To effectively solve the above-mentioned problems, the specific surface area of the electrode, the conductivity of the electrode surface, and the lithium storage capacity of the electrode have been increased. In recent years, MOFs have emerged as a promising material for several industrial applications owing to their permanent pore structures, large specific surface areas, adjustable structures and simple synthesis methods. However, the insulating nature of most MOFs limits their application in the energy field. C-MOFs have received great attention owing to their high conductivity compared with conventional MOFs. The applications of c-MOFs are outlined as follows. (1) For LIBs, the inherent large specific surface area, porous structure, and high electrical conductivity of c-MOFs provide LIBs with more active sites and lithium storage sites, which promote the migration rate of ions. (2) According to Li–S batteries, c-MOFs can adsorb polysulfide generated during the reaction, either as a cathode or as a diaphragm, which play a vital role in inhibiting the shuttle effect. (3) For Li–air batteries, c-MOFs have been used in these novel energy systems, and they could operate stably at high current densities. (4) DFT calculations play a vital role in predicting and guiding the synthesis of composite electrodes. Presently, c-MOFs are gradually used in rechargeable lithium batteries and energy and electronic devices. This review elucidates the conductive mechanism of c-MOFs, then introduces the preparation methods of c-MOFs, and summarizes the applications of c-MOFs in rechargeable LIBs, Li–S batteries, and Li–air batteries according to the classification of batteries. The performance of c-MOFs in rechargeable lithium batteries is summarized in Table 1.

C-MOFs combine the advantages of metal ions and organic ligands to solve the problem of low conductivity of materials owing to the difficulty in functionalization of metal ions and the inability of organic ligands to achieve a long-range order. However, c-MOFs require several developments. (1) The elucidation of the conduction and energy storage mechanism of electrode materials helps to control the synthesis of c-MOFs electrodes and enhance the battery performance. Different c-MOFs have distinct conduction and energy storage mechanisms. Therefore, exploring the mechanism of c-MOFs is vital. DFT calculations are of great significance in this aspect. (2) Presently, several preparation methods of c-MOFs are relatively complex owing to the harsh conditions and long-synthesis time, which is not conducive for large-scale production. The optimization of the preparation method of c-MOFs through a decrease in the synthesis time and adjusting the synthesis conditions to achieve low-cost, large-scale production still requires continuous improvement. (3) DFT calculations are significant for developing various new organic ligands and guiding the synthesis of new c-MOFs. c-MOFs combined several functional materials with a specific ability to provide a stronger binding force. The heat resistance and acid and alkaline resistance of c-MOFs can be effectively improved. Continuous innovation is required for the improvement of c-MOFs. (4) C-MOFs can be used in the fields of lithium batteries, other alkali metal batteries (sodium-ion batteries and KIBs), and multivalent metal-ion batteries (zinc-ion batteries and magnesium-ion batteries). Although c-MOFs have several applications presently, they still have huge development potential in this field. We believe that the research and application of c-MOFs in the field of electrochemical energy storage and conversion will be more extensive as shown in Figure 13. In this field, opportunities and challenges coexist.

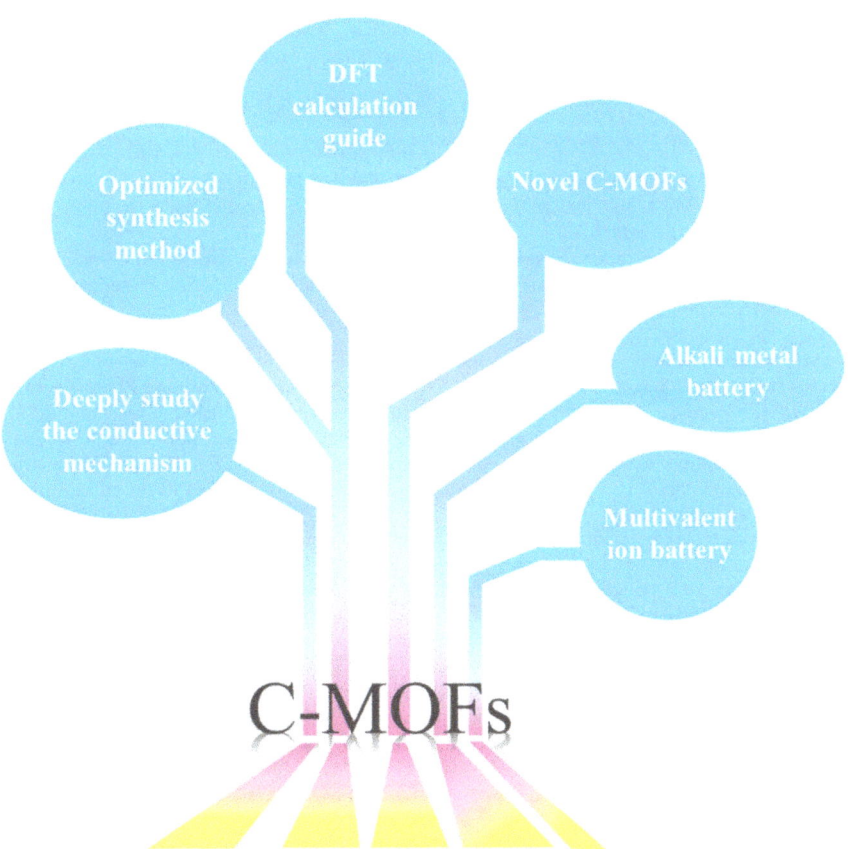

Figure 13. Further research and application of c-MOFs in other batteries.

Author Contributions: F.D. and Y.Z.: Writing—original draft preparation; Y.Y.: Writing—reviewing and editing, supervision, funding acquisition. All authors have read and agreed to the published version of the manuscript.

Funding: This work was financially supported by National Natural Science Foundation of China (No. 61904073), Spring City Plan-Special Program for Young Talents (ZX20210014), Yunnan Talents Support Plan for Yong Talents, Yunnan Local Colleges Applied Basic Research Projects (202101BA070001-138), Scientific Research Fund of Yunnan Education Department (2022Y717), Scientific Research Fund of Yunnan Education Department (2023Y0883).

Institutional Review Board Statement: Not applicable.

Informed Consent Statement: Not applicable.

Data Availability Statement: Not applicable.

Conflicts of Interest: The authors declare no conflict of interest.

References

1. Xu, X.; Liu, J.; Liu, J.; Ouyang, L.; Hu, R.; Wang, H.; Yang, L.; Zhu, M. A General Metal-Organic Framework (MOF)-Derived Selenidation Strategy for In Situ Carbon-Encapsulated Metal Selenides as High-Rate Anodes for Na-Ion Batteries. *Adv. Funct. Mater.* **2018**, *28*, 1870108. [CrossRef]
2. Shen, K.; Xu, X.; Tang, Y. Recent progress of magnetic field application in lithium-based batteries. *Nano Energy* **2022**, *92*, 106703. [CrossRef]

3. Yu, L.; Liu, J.; Xu, X.; Zhang, L.; Hu, R.; Liu, J.; Yang, L.; Zhu, M. Metal–Organic Framework-Derived NiSb Alloy Embedded in Carbon Hollow Spheres as Superior Lithium-Ion Battery Anodes. *ACS Appl. Mater. Interfaces* **2017**, *9*, 2516–2525. [CrossRef] [PubMed]
4. Yu, Y.; Wang, D.; Luo, J.; Xiang, Y. First-principles study of ZIF-8 as anode for Na and K ion batteries. *Colloids Surf. A Physicochem. Eng. Asp.* **2023**, *659*, 130802. [CrossRef]
5. Nasser, O.A.; Petranikova, M. Review of Achieved Purities after Li-Ion Batteries Hydrometallurgical Treatment and Impurities Effects on the Cathode Performance. *Batteries* **2021**, *7*, 60. [CrossRef]
6. Suzanowicz, A.M.; Mei, C.W.; Mandal, B.K. Approaches to Combat the Polysulfide Shuttle Phenomenon in Li–S Battery Technology. *Batteries* **2022**, *8*, 45. [CrossRef]
7. Yu, Y.; Hu, S. The applications of semiconductor materials in air batteries. *Chin. Chem. Lett.* **2021**, *32*, 3277–3287. [CrossRef]
8. Yu, Y.; Hu, S.; Huang, J. Germanium-modified silicon as anodes in Si-Ge air batteries with enhanced properties. *J. Phys. Chem. Solids* **2021**, *157*, e110226. [CrossRef]
9. Wang, K.; Pei, S.; He, Z.; Huang, L.-A.; Zhu, S.; Guo, J.; Shao, H.; Wang, J. Synthesis of a novel porous silicon microsphere@carbon core-shell composite via in situ MOF coating for lithium ion battery anodes. *Chem. Eng. J.* **2019**, *356*, 272–281. [CrossRef]
10. Liu, W.; Cheng, P.; Yan, X.; Gou, H.; Zhang, S.; Shi, S. Facile One-Step Solution-Phase Route to Synthesize Hollow Nanoporous Cu_xO Microcages on 3D Copper Foam for Superior Li Storage. *ACS Sustain. Chem. Eng.* **2021**, *9*, 4363–4370. [CrossRef]
11. Yu, Y.; Hu, S.; Huang, J. Adsorption and diffusion of lithium and sodium on the silicon nanowire with substrate for energy storage application: A first principles study. *Mater. Chem. Phys.* **2020**, *253*, 123243. [CrossRef]
12. Collins, G.A.; Geaney, H.; Ryan, K.M. Alternative anodes for low temperature lithium-ion batteries. *J. Mater. Chem. A* **2021**, *9*, 14172–14213. [CrossRef]
13. Hou, J.; Ma, X.; Fu, J.; Vanaphuti, P.; Yao, Z.; Liu, Y.; Yang, Z.; Wang, Y. A green closed-loop process for selective recycling of lithium from spent lithium-ion batteries. *Green Chem.* **2022**, *24*, 7049–7060. [CrossRef]
14. Dai, X.; Zhang, X.; Wen, J.; Wang, C.; Ma, X.; Yang, Y.; Huang, G.; Ye, H.-M.; Xu, S. Research progress on high-temperature resistant polymer separators for lithium-ion batteries. *Energy Storage Mater.* **2022**, *51*, 638–659. [CrossRef]
15. Yan, C.; Zhang, X.-Q.; Huang, J.-Q.; Liu, Q.; Zhang, Q. Lithium-Anode Protection in Lithium-Sulfur Batteries. *Trends Chem.* **2019**, *1*, 693–704. [CrossRef]
16. Zhou, L.; Danilov, D.L.; Qiao, F.; Wang, J.; Li, H.; Eichel, R.-A.; Notten, P.H.L. Sulfur Reduction Reaction in Lithium-Sulfur Batteries: Mechanisms, Catalysts, and Characterization. *Adv. Energy Mater.* **2022**, 2202094. [CrossRef]
17. Ko, S.; Yoo, Y.; Choi, J.; Lim, H.-D.; Park, C.B.; Lee, M. Discovery of organic catalysts boosting lithium carbonate decomposition toward ambient air operational lithium-air batteries. *J. Mater. Chem. A* **2022**, *10*, 20464–20472. [CrossRef]
18. Wu, Z.; Tian, Y.; Chen, H.; Wang, L.; Qian, S.; Wu, T.; Zhang, S.; Lu, J. Evolving aprotic Li–air batteries. *Chem. Soc. Rev.* **2022**, *51*, 8045–8101. [CrossRef]
19. Wada, K.; Sakaushi, K.; Sasaki, S.; Nishihara, H. Multielectron-Transfer-based Rechargeable Energy Storage of Two-Dimensional Coordination Frameworks with Non-Innocent Ligands. *Angew. Chem. Int. Ed.* **2018**, *57*, 8886–8890. [CrossRef]
20. Liang, Z.; Qu, c.; Guo, w.; Zou, R.; Xu, Q. Pristine Metal-Organic Frameworks and their Composites for Energy Storage and Conversion. *Adv. Mater.* **2018**, *30*, 1702891. [CrossRef]
21. Huang, W.; Huang, S.; Chen, G.; Ouyang, G. Biocatalytic Metal-Organic Frameworks: Promising Materials for Biosensing. *ChemBioChem* **2022**, *23*, 202100567. [CrossRef] [PubMed]
22. Khoshbin, Z.; Davoodian, N.; Taghdisi, S.M.; Abnous, K. Metal organic frameworks as advanced functional materials for aptasensor design. *Spectrochim. Acta A Mol. Biomol. Spectrosc.* **2022**, *276*, 121251. [CrossRef] [PubMed]
23. Ma, Y.; Qu, X.; Liu, C.; Xu, Q.; Tu, K. Metal-Organic Frameworks and Their Composites Towards Biomedical Applications. *Front. Mol. Biosci.* **2021**, *8*, 805228. [CrossRef] [PubMed]
24. Wang, X.; Wang, Y.; Han, M.; Liang, J.; Zhang, M.; Bai, X.; Yue, T.; Gao, Z. Evaluating the changes in phytochemical composition, hypoglycemic effect, and influence on mice intestinal microbiota of fermented apple juice. *Food Res. Int.* **2022**, *155*, 110998. [CrossRef] [PubMed]
25. Jiang, Y.; Fan, R.; Zhang, J.; Fang, X.; Sun, T.; Zhu, K.; Zhou, X.; Xu, Y.; Yang, Y. Sequentially epitaxial growth multi-guest encapsulation strategy in MOF-on-MOF platform: Biogenic amine detection and systematic white light adjustment. *Chem. Eng. J.* **2022**, *436*, 135236. [CrossRef]
26. Tarasi, S.; Ramazani, A.; Morsali, A.; Hu, M.-L.; Ghafghazi, S.; Tarasi, R.; Ahmadi, Y. Drug Delivery Using Hydrophilic Metal-Organic Frameworks (MOFs): Effect of Structure Properties of MOFs on Biological Behavior of Carriers. *Inorg. Chem.* **2022**, *61*, 13125–13132. [CrossRef]
27. Peng, X.; Chen, L.; Li, Y. Ordered macroporous MOF-based materials for catalysis. *Mol. Catal.* **2022**, *529*, 112568. [CrossRef]
28. Wang, T.-S.; Liu, X.; Zhao, X.; He, P.; Nan, C.-W.; Fan, L.-Z. Regulating Uniform Li Plating/Stripping via Dual-Conductive Metal-Organic Frameworks for High-Rate Lithium Metal Batteries. *Adv. Funct. Mater.* **2020**, *30*, 2000786. [CrossRef]
29. Chen, N.; Li, Y.; Dai, Y.; Qu, W.; Xing, Y.; Ye, Y.; Wen, Z.; Guo, C.; Wu, F.; Chen, R. A Li^+ conductive metal organic framework electrolyte boosts the high-temperature performance of dendrite-free lithium batteries. *J. Mater. Chem. A* **2019**, *7*, 9530–9536. [CrossRef]
30. Li, C.; Zhang, L.; Chen, J.; Li, X.; Sun, J.; Zhu, J.; Wang, X.; Fu, Y. Recent development and applications of electrical conductive MOFs. *Nanoscale* **2021**, *13*, 485–509. [CrossRef]

31. Mu, X.; Wang, W.; Sun, C.; Wang, J.; Wang, C.; Knez, M. Recent Progress on Conductive Metal-Organic Framework Films. *Adv. Mater. Interfaces* **2021**, *8*, 2002151. [CrossRef]
32. Zhu, B.; Wen, D.; Liang, Z.; Zou, R. Conductive metal-organic frameworks for electrochemical energy conversion and storage. *Coord. Chem. Rev.* **2021**, *446*, 214119. [CrossRef]
33. Zhang, S.; Li, L.; Lu, Y.; Liu, D.; Zhang, J.; Hao, D.; Zhang, X.; Xiong, L.; Huang, J. Sensitive humidity sensors based on ionically conductive metal-organic frameworks for breath monitoring and non-contact sensing. *Appl. Mater. Today* **2022**, *26*, 101391. [CrossRef]
34. Chen, Y.; Tian, Y.; Zhu, P.; Du, L.; Chen, W.; Wu, C. Electrochemically Activated Conductive Ni-Based MOFs for Non-enzymatic Sensors Toward Long-Term Glucose Monitoring. *Front. Chem.* **2020**, *8*, 602752. [CrossRef]
35. Chen, J.; Huang, X.; Ye, R.; Huang, D.; Wang, Y.; Chen, S. Fabrication of a novel electrochemical sensor using conductive MOF Cu–CAT anchored on reduced graphene oxide for BPA detection. *J. Appl. Electrochem.* **2022**, *52*, 1617–1628. [CrossRef]
36. Niu, L.; Wu, T.; Chen, M.; Yang, L.; Yang, J.; Wang, Z.; Kornyshev, A.A.; Jiang, H.; Bi, S.; Feng, G. Conductive Metal-Organic Frameworks for Supercapacitors. *Adv. Mater.* **2022**, 2200999. [CrossRef]
37. Wang, Y.; Wang, S.; Ma, Z.-L.; Yan, L.-T.; Zhao, X.-B.; Xue, Y.-Y.; Huo, J.-M.; Yuan, X.; Li, S.-N.; Zhai, Q.-G. Competitive Coordination-Oriented Monodispersed Ruthenium Sites in Conductive MOF/LDH Hetero-Nanotree Catalysts for Efficient Overall Water Splitting in Alkaline Media. *Adv. Mater.* **2022**, *34*, 2107488. [CrossRef]
38. Meng, H.; Han, Y.; Zhou, C.; Jiang, Q.; Shi, X.; Zhan, C.; Zhang, R. Conductive Metal–Organic Frameworks: Design, Synthesis, and Applications. *Small Methods* **2020**, *4*, 2000396. [CrossRef]
39. Lin, L.; Zhang, Q.; Ni, Y.; Shang, L.; Zhang, X.; Yan, Z.; Zhao, Q.; Chen, J. Rational design and synthesis of two-dimensional conjugated metal-organic polymers for electrocatalysis applications. *Chem* **2022**, *8*, 1822–1854. [CrossRef]
40. Li, W.-H.; Deng, W.-H.; Wang, G.-E.; Xu, G. Conductive MOFs. *EnergyChem* **2020**, *2*, 100029. [CrossRef]
41. Deng, X.; Hu, J.-Y.; Luo, J.; Liao, W.-M.; He, J. Conductive Metal-Organic Frameworks: Mechanisms, Design Strategies and Recent Advances. *Top. Curr. Chem.* **2020**, *378*, 27. [CrossRef] [PubMed]
42. Zhang, G.; Jin, L.; Zhang, R.; Bai, Y.; Zhu, R.; Pang, H. Recent advances in the development of electronically and ionically conductive metal-organic frameworks. *Coord. Chem. Rev.* **2021**, *439*, 213915. [CrossRef]
43. Zhang, K.; Wen, G.-H.; Yang, X.-J.; Lim, D.-W.; Bao, S.-S.; Donoshita, M.; Wu, L.Q.; Kitagawa, H.; Zheng, L.-M. Anhydrous Superprotonic Conductivity of a Uranyl-Based MOF from Ambient Temperature to 110 °C. *ACS Mater. Lett.* **2021**, *3*, 744–751. [CrossRef]
44. Tang, H.; Lv, X.; Du, J.; Liu, Y.; Liu, J.; Guo, L.; Zheng, X.; Hao, H.; Liu, Z. Improving proton conductivity of metal organic framework materials by reducing crystallinity. *Appl. Organomet. Chem.* **2022**, *36*, e6777. [CrossRef]
45. Ko, M.; Mendecki, L.; Mirica, K.A. Conductive two-dimensional metal-organic frameworks as multifunctional materials. *Chem. Commun.* **2018**, *54*, 7873–7891. [CrossRef]
46. Thanasekaran, P.; Su, C.-H.; Liu, Y.-H.; Lu, K.-L. Weak interactions in conducting metal-organic frameworks. *Coord. Chem. Rev.* **2021**, *442*, 213987. [CrossRef]
47. Nath, A.; Asha, K.S.; Mandal, S. Conductive Metal-Organic Frameworks: Electronic Structure and Electrochemical Applications. *Chem. Eur. J.* **2021**, *27*, 11482–11538. [CrossRef]
48. Takaishi, S.; Hosoda, M.; Kajiwara, T.; Miyasaka, H.; Yamashita, M.; Nakanishi, Y.; Kitagawa, Y.; Yamaguchi, K.; Kobayashi, A.; Kitagawa, H. Electroconductive Porous Coordination Polymer Cu[Cu(pdt)$_2$] Composed of Donor and Acceptor Building Units. *Inorg. Chem.* **2009**, *49*, 9048–9050. [CrossRef]
49. Park, S.S.; Hontz, E.R.; Sun, L.; Hendon, C.H.; Walsh, A.; Voorhis, T.V.; Dinca, M. Cation-Dependent Intrinsic Electrical Conductivity in Isostructural Tetrathiafulvalene-Based Microporous Metal-Organic Frameworks. *J. Am. Chem. Soc.* **2015**, *137*, 1774–1777. [CrossRef]
50. Hua, C.; Doheny, P.W.; Ding, B.; Chan, B.; Yu, M.; Kepert, C.J.; D'Alessandro, D.M. Through-Space Intervalence Charge Transfer as a Mechanism for Charge Delocalization in Metal-Organic Frameworks. *J. Am. Chem. Soc.* **2018**, *140*, 6622–6630. [CrossRef]
51. Talin, A.A.; Centrone, A.; Ford, A.C.; Foster, M.E.; Stavila, V.; Haney, P.; Kinney, R.A.; Szalai, V.; Gabaly, F.E.; Yoon, H.P.; et al. Tunable Electrical Conductivity in Metal-Organic Framework Thin-Film Devices. *Science* **2014**, *343*, 66–69. [CrossRef] [PubMed]
52. Pauliukaitė, R.; Juodkazytė, J.; Ramanauskas, R. Theodor von Grotthuss' Contribution to Electrochemistry. *Electrochim. Acta* **2017**, *236*, 28–32. [CrossRef]
53. Xie, X.-X.; Yang, Y.-C.; Dou, B.-H.; Li, Z.-F.; Li, G. Proton conductive carboxylate-based metal-organic frameworks. *Coord. Chem. Rev.* **2020**, *403*, 213100. [CrossRef]
54. Li, A.-L.; Gao, Q.; Xu, J.; Bu, X.-H. Proton-conductive metal-organic frameworks: Recent advances and perspectives. *Coord. Chem. Rev.* **2017**, *344*, 54–82. [CrossRef]
55. Xie, L.S.; Skorupskii, G.; Dinca, M. Electrically Conductive Metal-Organic Frameworks. *Chem. Rev.* **2020**, *120*, 8536–8580. [CrossRef]
56. Chen, T.; Dou, J.-H.; Yang, L.; Sun, C.; Libretto, N.J.; Skorupskii, G.; Miller, J.T.; Dinca, M. Continuous Electrical Conductivity Variation in M$_3$(Hexaiminotriphenylene)$_2$ (M = Co, Ni, Cu) MOF Alloys. *J. Am. Chem. Soc.* **2020**, *142*, 12367–12373. [CrossRef]
57. Hmadeh, M.; Lu, Z.; Liu, Z.; Gándara, F.; Furukawa, H.; Wan, S.; Augustyn, V.; Chang, R.; Liao, L.; Zhou, F.; et al. New Porous Crystals of Extended Metal-Catecholates. *Chem. Mater.* **2012**, *24*, 3511–3513. [CrossRef]

58. Day, R.W.; Bediako, D.K.; Rezaee, M.; Parent, L.R.; Skorupskii, G.; Arguilla, M.Q.; Hendon, C.H.; Stassen, I.; Gianneschi, N.C.; Kim, P.; et al. Single Crystals of Electrically Conductive Two-Dimensional Metal-Organic Frameworks: Structural and Electrical Transport Properties. *ACS Cent. Sci.* **2019**, *5*, 1959–1964. [CrossRef]
59. Li, W.-H.; Ding, K.; Tian, H.-R.; Yao, M.-S.; Nath, B.; Deng, W.-H.; Wang, Y.; Xu, G. Conductive Metal-Organic Framework Nanowire Array Electrodes for High-Performance Solid-State Supercapacitors. *Adv. Funct. Mater.* **2017**, *27*, 1702067. [CrossRef]
60. Zacher, D.; Baunemann, A.; Hermes, S.; Fischer, R.A. Deposition of microcrystalline [Cu_3(btc)$_2$] and [Zn_2(bdc)$_2$(dabco)] at alumina and silica surfaces modified with patterned self assembled organic monolayers: Evidence of surface selective and oriented growth. *J. Mater. Chem.* **2007**, *17*, 2785–2792. [CrossRef]
61. Pal, T.; Kambe, T.; Kusamoto, T.; Foo, M.L.; Matsuoka, R.; Sakamoto, R.; Nishihara, H. Interfacial Synthesis of Electrically Conducting Palladium Bis(dithiolene) Complex Nanosheet. *ChemPlusChem* **2015**, *80*, 1255–1258. [CrossRef]
62. Sheberla, D.; Sun, L.; Blood-Forsythe, M.A.; Er, S.; Wade, C.R.; Brozek, C.K.; Aspuru-Guzik, A.; Dinca, M. High Electrical Conductivity in Ni_3(2,3,6,7,10,11-hexaiminotriphenylene)$_2$, a Semiconducting Metal-Organic Graphene Analogue. *J. Am. Chem. Soc.* **2014**, *136*, 8859–8862. [CrossRef]
63. Song, X.; Wang, X.; Li, Y.; Zheng, C.; Zhang, B.; Di, C.-A.; Li, F.; Jin, C.; Mi, W.; Chen, L.; et al. 2D Semiconducting Metal-Organic Framework Thin Films for Organic Spin Valves. *Angew. Chem. Int. Ed. Engl.* **2020**, *59*, 1118–1123. [CrossRef]
64. Zacher, D.; Shekhah, O.; Woll, C.; Fischer, R.A. Thin films of metal-organic frameworks. *Chem. Soc. Rev.* **2009**, *38*, 1418–1429. [CrossRef]
65. Summerfield, A.; Cebula, I.; Schroder, M.; Beton, P.H. Nucleation and Early Stages of Layer-by-Layer Growth of Metal Organic Frameworks on Surfaces. *J. Phys. Chem. C* **2015**, *119*, 23544–23551. [CrossRef]
66. Stavila, V.; Volponi, J.; Katzenmeyer, A.M.; Dixon, M.C.; Allendorf, M.D. Kinetics and mechanism of metal–organic framework thin film growth: Systematic investigation of HKUST-1 deposition on QCM electrodes. *Chem. Sci.* **2012**, *3*, 1531–1540. [CrossRef]
67. Ladnorg, T.; Welle, A.; Heissler, S.; Woll, C.; Gliemann, H. Site-selective growth of surface-anchored metal-organic frameworks on self-assembled monolayer patterns prepared by AFM nanografting. *Beilstein J. Nanotechnol.* **2013**, *4*, 638–648. [CrossRef]
68. Arslan, H.K.; Shekhah, O.; Wohlgemuth, J.; Franzreb, M.; Fischer, R.A.; Wöll, C. High-Throughput Fabrication of Uniform and Homogenous MOF Coatings. *Adv. Funct. Mater.* **2011**, *21*, 4228–4231. [CrossRef]
69. Shekhah, O.; Wang, H.; Strunskus, T.; Cyganik, P.; Zacher, D.; Fischer, R.; Wöll, C. Layer-by-Layer Growth of Oriented Metal Organic Polymers on a Functionalized Organic Surface. *Langmuir* **2007**, *23*, 7440–7442. [CrossRef]
70. Contreras-Pereda, N.; Pané, S.; Puigmartí-Luis, J.; Ruiz-Molina, D. Conductive properties of triphenylene MOFs and COFs. *Coord. Chem. Rev.* **2022**, *460*, 214459. [CrossRef]
71. Li, J.; Cai, Y.; Wu, H.; Yu, Z.; Yan, X.; Zhang, Q.; Gao, T.Z.; Liu, K.; Jia, X.; Bao, Z. Polymers in Lithium-Ion and Lithium Metal Batteries. *Adv. Energy Mater.* **2021**, *11*, 2003239. [CrossRef]
72. Zhang, L.; Zhu, C.; Yu, S.; Ge, D.; Zhou, H. Status and challenges facing representative anode materials for rechargeable lithium batteries. *J. Energy Chem.* **2022**, *66*, 260–294. [CrossRef]
73. Su, Y.-S.; Hsiao, K.-C.; Sireesha, P.; Huang, J.-Y. Lithium Silicates in Anode Materials for Li-Ion and Li Metal Batteries. *Batteries* **2022**, *8*, 2. [CrossRef]
74. Liu, W.; Lu, B.; Liu, X.; Gan, Y.; Zhang, S.; Shi, S. In Situ Synthesis of the Peapod-Like Cu–SnO_2@Copper Foam as Anode with Excellent Cycle Stability and High Area Specific Capacity. *Adv. Funct. Mater.* **2021**, *31*, 2101999. [CrossRef]
75. Liu, W.; Chen, X.; Zhang, J.; Zhang, S.; Shi, S. In-Situ synthesis of freestanding porous SnO_x-decorated Ni_3Sn_2 composites with enhanced Li storage properties. *Chem. Eng. J.* **2021**, *412*, 128591. [CrossRef]
76. Liu, W.; Xiang, P.; Dong, X.; Yin, H.; Yu, H.; Cheng, P.; Zhang, S.; Shi, S. Two advantages by a single move: Core-bishell electrode design for ultrahigh-rate capacity and ultralong-life cyclability of lithium ion batteries. *Compos. B Eng.* **2021**, *216*, 108883. [CrossRef]
77. Guo, L.; Sun, J.; Sun, X.; Zhang, J.; Hou, L.; Yuan, C. Construction of 1D conductive Ni-MOF nanorods with fast Li^+ kinetic diffusion and stable high-rate capacities as an anode for lithium ion batteries. *Nanoscale Adv.* **2019**, *1*, 4688–4691. [CrossRef]
78. Guo, L.; Sun, J.; Zhang, W.; Hou, L.; Liang, L.; Liu, Y.; Yuan, C. Bottom-Up Fabrication of 1D Cu-based Conductive Metal-Organic Framework Nanowires as a High-Rate Anode Towards Efficient Lithium Storage. *ChemSusChem* **2019**, *12*, 5051–5058. [CrossRef]
79. Mao, P.; Fan, H.; Liu, C.; Lan, G.; Huang, W.; Li, Z.; Mahmoud, H.; Zheng, R.; Wang, Z.; Sun, H.; et al. Conductive Co-based metal organic framework nanostructures for excellent potassium- and lithium-ion storage: Kinetics and mechanism studies. *Sustain. Energy Fuels* **2022**, *6*, 4075–4084. [CrossRef]
80. Nazir, A.; Le, H.T.T.; Min, C.-W.; Kasbe, A.; Kim, J.; Jin, C.S.; Park, C.J. Coupling of a conductive Ni_3(2,3,6,7,10,11-hexaiminotriphenylene)$_2$ metal-organic framework with silicon nanoparticles for use in high-capacity lithium-ion batteries. *Nanoscale* **2020**, *12*, 1629–1642. [CrossRef]
81. Meng, C.; Hu, P.; Chen, H.; Cai, Y.; Zhou, H.; Jiang, Z.; Zhu, X.; Liu, Z.; Wang, C.; Yuan, A. 2D conductive MOFs with sufficient redox sites: Reduced graphene oxide/Cu-benzenehexathiolate composites as high capacity anode materials for lithium-ion batteries. *Nanoscale* **2021**, *13*, 7751–7760. [CrossRef]
82. Yan, J.; Cui, Y.; Xie, M.; Yang, G.-Z.; Bin, D.S.; Li, D. Immobilizing Redox-Active Tricycloquinazoline into a 2D Conductive Metal-Organic Framework for Lithium Storage. *Angew. Chem. Int. Ed.* **2021**, *60*, 24467–24472. [CrossRef]
83. Gu, S.; Bai, Z.; Majumder, S.; Huang, B.; Chen, G. Conductive metal–organic framework with redox metal center as cathode for high rate performance lithium ion battery. *J. Power Sources* **2019**, *429*, 22–29. [CrossRef]

84. Wu, Z.; Adekoya, D.; Huang, X.; Kiefel, M.J.; Xie, J.; Xu, W.; Zhang, Q.; Zhu, D.; Zhang, S. Highly Conductive Two-Dimensional Metal-Organic Frameworks for Resilient Lithium Storage with Superb Rate Capability. *ACS Nano* **2020**, *14*, 12016–12026. [CrossRef]
85. Dong, H.; Gao, H.; Geng, J.; Hou, X.; Gao, S.; Wang, S.; Chou, S. Quinone-Based Conducting Three-Dimensional Metal–Organic Framework as a Cathode Material for Lithium-Ion Batteries. *J. Phys. Chem. C* **2021**, *125*, 20814–20820. [CrossRef]
86. Gu, S.; Xu, S.; Song, X.; Li, H.; Wang, Y.; Zhou, G.; Wang, N.; Chang, H. Electrostatic Potential-Induced Co-N_4 Active Centers in a 2D Conductive Metal-Organic Framework for High-Performance Lithium-Sulfur Batteries. *ACS Appl. Mater. Interfaces* **2022**, *14*, 50815–50826. [CrossRef]
87. Zang, Y.; Pei, F.; Huang, J.; Fu, Z.; Xu, G.; Fang, X. Large-Area Preparation of Crack-Free Crystalline Microporous Conductive Membrane to Upgrade High Energy Lithium-Sulfur Batteries. *Adv. Energy Mater.* **2018**, *8*, 1802052. [CrossRef]
88. Chen, H.; Xiao, Y.; Chen, C.; Yang, J.; Gao, C.; Chen, Y.; Wu, J.; Shen, Y.; Zhang, W.; Li, S.; et al. Conductive MOF-Modified Separator for Mitigating the Shuttle Effect of Lithium-Sulfur Battery through a Filtration Method. *ACS Appl. Mater. Interfaces* **2019**, *11*, 11459–11465. [CrossRef]
89. Xiao, Y.; Xiang, Y.; Guo, S.; Wang, J.; Ouyang, Y.; Li, D.; Zeng, Q.; Gong, W.; Gan, L.; Zhang, Q.; et al. An ultralight electroconductive metal-organic framework membrane for multistep catalytic conversion and molecular sieving in lithium-sulfur batteries. *Energy Storage Mater.* **2022**, *51*, 882–889. [CrossRef]
90. Cai, D.; Lu, M.; Li, L.; Cao, J.; Chen, D.; Tu, H.; Li, J.; Han, W. A Highly Conductive MOF of Graphene Analogue $Ni_3(HITP)_2$ as a Sulfur Host for High-Performance Lithium-Sulfur Batteries. *Small* **2019**, *15*, 1902605. [CrossRef]
91. Wang, S.; Huang, F.; Zhang, Z.; Cai, W.; Jie, Y.; Wang, S.; Yan, P.; Jiao, S.; Cao, R. R. Conductive metal-organic frameworks promoting polysulfides transformation in lithium-sulfur batteries. *J. Energy Chem.* **2021**, *63*, 336–343. [CrossRef]
92. Lv, Q.; Zhu, Z.; Ni, Y.; Wen, B.; Jiang, Z.; Fang, H.; Li, F. Atomic Ruthenium-Riveted Metal–Organic Framework with Tunable d-Band Modulates Oxygen Redox for Lithium–Oxygen Batteries. *J. Am. Chem. Soc.* **2022**, *144*, 23239–23246. [CrossRef] [PubMed]
93. Majidi, L.; Ahmadiparidari, A.; Shan, N.; Singh, S.K.; Zhang, C.; Huang, Z.; Rastegar, S.; Kumar, K.; Hemmat, Z.; Ngo, A.T.; et al. Nanostructured Conductive Metal Organic Frameworks for Sustainable Low Charge Overpotentials in Li-air Batteries. *Small* **2022**, *18*, 2102902. [CrossRef] [PubMed]
94. Chen, K.; Yang, H.; Liang, F.; Xue, D. Microwave-irradiation-assisted combustion toward modified graphite as lithium ion battery anode. *ACS Appl. Mater. Interfaces* **2018**, *10*, 909–914. [CrossRef] [PubMed]
95. Gao, C.; Jiang, Z.; Qi, S.; Wang, P.; Jensen, L.; Johansen, M.; Christensen, C.; Zhang, Y.; Ravnsbæk, D.; Yue, Y. Metal-organic framework glass anode with an exceptional cycling-induced capacity enhancement for lithium-ion batteries. *Adv. Mater.* **2022**, *34*, 2110048. [CrossRef]
96. Zhong, M.; Yan, J.; Wang, L.; Huang, Y.; Li, L.; Gao, S.; Tian, Y.; Shen, W.; Zhang, J.; Guo, S. Hierarchic porous graphite/reduced graphene oxide composites generated from semi-coke as high-performance anodes for lithium-ion batteries. *Sustain. Mater. Techno.* **2022**, *33*, e00476. [CrossRef]
97. Qiu, T.; Yu, Z.; Xie, W.; He, Y.; Wang, H.; Zhang, T. Preparation of Onion-like Synthetic Graphite with a Hierarchical Pore Structure from Anthracite and Its Electrochemical Properties as the Anode Material of Lithium-Ion Batteries. *Energy Fuels* **2022**, *36*, 8256–8266. [CrossRef]
98. Kwon, G.D.; Moyen, E.; Lee, Y.J.; Joe, J.; Pribat, D. Graphene-coated aluminum thin film anodes for lithium-ion batteries. *ACS Appl. Mater. Interfaces* **2018**, *10*, 29486–29495. [CrossRef]
99. Zhang, L.; Xia, G.; Guo, Z.; Li, X.; Sun, D.; Yu, X. Boron and nitrogen co-doped porous carbon nanotubes webs as a high-performance anode material for lithium ion batteries. *Int. J. Hydrogen Energy* **2016**, *41*, 14252–14260. [CrossRef]
100. Zhang, D.; Su, W.; Li, Z.; Wang, Q.; Yuan, F.; Sun, H.; Li, Y.; Zhang, Y.; Wang, B. Three-dimensional interconnected porous carbon nanoflakes with improved electron transfer and ion storage for lithium-ion batteries. *J. Alloys Compd.* **2022**, *904*, 164122. [CrossRef]
101. Kon, K.; Uchida, K.; Fuku, K.; Yamanaka, S.; Wu, B.; Yamazui, D.; Iguchi, H.; Kobayashi, H.; Gambe, Y.; Honma, I.; et al. Electron-Conductive Metal-Organic Framework, Fe(dhbq)(dhbq = 2,5-Dihydroxy-1,4-benzoquinone): Coexistence of Microporosity and Solid-State Redox Activity. *ACS Appl. Mater. Interfaces* **2021**, *13*, 38188–38193. [CrossRef]
102. Cheng, Q.; Yin, Z.; Pan, Z.; Zhong, X.; Rao, H. Lightweight Free-Standing 3D Nitrogen-Doped Graphene/TiN Aerogels with Ultrahigh Sulfur Loading for High Energy Density Li–S Batteries. *ACS Appl. Energy Mater.* **2021**, *4*, 7599–7610. [CrossRef]
103. McCreary, C.; An, Y.; Kim, S.U.; Hwa, Y. A Perspective on Li/S Battery Design: Modeling and Development Approaches. *Batteries* **2021**, *7*, 82. [CrossRef]
104. Lee, J.-S.; Kim, S.T.; Cao, R.; Choi, N.-S.; Liu, M.; Lee, K.T.; Cho, J. Metal-Air Batteries with High Energy Density: Li–air Versus Zn-Air. *Adv. Energy Mater.* **2011**, *1*, 34–50. [CrossRef]
105. Chen, D.; Li, Y.; Zhang, X.; Hu, S.; Yu, Y. Investigation of the discharging behaviors of different doped silicon nanowires in alkaline Si-air batteries. *J. Ind. Eng. Chem.* **2022**, *112*, 271–278. [CrossRef]
106. Gao, Y.; Pan, Z.; Sun, J.; Liu, Z.; Wang, J. High-Energy Batteries: Beyond Lithium-Ion and Their Long Road to Commercialisation. *Nano-Micro Lett.* **2022**, *14*, 94. [CrossRef]
107. Yu, Y.; Gao, S.; Hu, S. Si modified by Zn and Fe as anodes in Si-air batteries with amelioratve properties. *J. Alloys Compd.* **2021**, *883*, 160902. [CrossRef]

108. Li, F.; Zhao, J. Atomic Sulfur Anchored on Silicene, Phosphorene, and Borophene for Excellent Cycle Performance of Li–S Batteries. *ACS Appl. Mater. Interfaces* **2017**, *9*, 42836–42844. [CrossRef]
109. Li, F.; Liu, Q.; Hu, J.; Feng, Y.; He, P.; Ma, J. Recent advances in cathode materials for rechargeable lithium-sulfur batteries. *Nanoscale* **2019**, *11*, 15418–15439. [CrossRef]
110. Song, Y.; Cai, W.; Kong, L.; Cai, J.; Zhang, Q.; Sun, J. Rationalizing Electrocatalysis of Li–S Chemistry by Mediator Design: Progress and Prospects. *Adv. Energy Mater.* **2019**, *10*, 1901075. [CrossRef]
111. Ye, H.; Li, M.; Liu, T.; Li, Y.; Lu, J. Activating Li_2S as the Lithium-Containing Cathode in Lithium–Sulfur Batteries. *ACS Energy Lett.* **2020**, *5*, 2234–2245. [CrossRef]
112. Li, W.; Li, S.; Bernussi, A.A.; Fan, Z. 3-D Edge-Oriented Electrocatalytic $NiCo_2S_4$ Nanoflakes on Vertical Graphene for Li–S Batteries. *Energy Mater. Adv.* **2021**, *2021*, 2712391. [CrossRef]
113. Song, Y.; Zhao, W.; Kong, L.; Zhang, L.; Zhu, X.; Shao, Y.; Ding, F.; Zhang, Q.; Sun, J.; Liu, Z. Synchronous immobilization and conversion of polysulfides on a VO_2–VN binary host targeting high sulfur load Li–S batteries. *Energy Environ. Sci.* **2018**, *11*, 2620–2630. [CrossRef]
114. Li, S.; Leng, D.; Li, W.; Qie, L.; Dong, Z.; Cheng, Z.; Fan, Z. Recent progress in developing Li_2S cathodes for Li–S batteries. *Energy Storage Mater.* **2020**, *27*, 279–296. [CrossRef]
115. Wu, T.; Yang, T.; Zhang, J.; Zheng, X.; Liu, K.; Wang, C.; Chen, M. CoB and BN composites enabling integrated adsorption/catalysis to polysulfides for inhibiting shuttle-effect in Li–S batteries. *J. Energy Chem.* **2021**, *59*, 220–228. [CrossRef]
116. Kim, Y.; Noh, Y.; Bae, J.; Ahn, H.; Kim, M.; Kim, W.B. N-doped carbon-embedded TiN nanowires as a multifunctional separator for Li–S batteries with enhanced rate capability and cycle stability. *J. Energy Chem.* **2021**, *57*, 10–18. [CrossRef]
117. Zhou, J.; Yang, X.; Zhang, Y.; Jia, J.; He, X.; Yu, L.; Pan, Y.; Liao, J.; Sun, M.; He, J. Interconnected $NiCo_2O_4$ nanosheet arrays grown on carbon cloth as a host, adsorber and catalyst for sulfur species enabling high-performance Li–S batteries. *Nanoscale Adv.* **2021**, *3*, 1690–1698. [CrossRef]
118. Pan, Z.; Brett, D.J.L.; He, G.; Parkin, I.P. Progress and Perspectives of Organosulfur for Lithium–Sulfur Batteries. *Adv. Energy Mater.* **2022**, *12*, 2103483. [CrossRef]
119. Song, Y.; Zhao, W.; Wei, N.; Zhang, L.; Ding, F.; Liu, Z.; Sun, J. In-situ PECVD-enabled graphene-V_2O_3 hybrid host for lithium–sulfur batteries. *Nano Energy* **2018**, *53*, 432–439. [CrossRef]
120. Song, Y.; Zhao, S.; Chen, Y.; Cai, J.; Li, J.; Yang, Q.; Sun, J.; Liu, Z. Enhanced Sulfur Redox and Polysulfide Regulation via Porous VN-Modified Separator for Li–S Batteries. *ACS Appl. Mater. Interfaces* **2019**, *11*, 5687–5694. [CrossRef]
121. Bonnett, B.L.; Smith, E.D.; Garza, M.D.L.; Cai, M.; Haag, J.V.; Serrano, J.M.; Cornell, H.D.; Gibbons, B.; Martin, S.M.; Morris, A.J. PCN-222 Metal-Organic Framework Nanoparticles with Tunable Pore Size for Nanocomposite Reverse Osmosis Membranes. *ACS Appl. Mater. Interfaces* **2020**, *12*, 15765–15773. [CrossRef]
122. Gong, X.; Gnanasekaran, K.; Chen, Z.; Robison, L.; Wasson, M.C.; Bentz, K.C.; Cohen, S.M.; Farha, O.K.; Gianneschi, N.C. Insights into the Structure and Dynamics of Metal-Organic Frameworks Via Transmission Electron Microscopy. *J. Am. Chem. Soc.* **2020**, *142*, 17224–17235. [CrossRef]
123. Saroha, R.; Oh, J.H.; Seon, Y.H.; Kang, Y.C.; Lee, J.S.; Jeong, D.W.; Cho, J.S. Freestanding interlayers for Li–S batteries: Design and synthesis of hierarchically porous N-doped C nanofibers comprising vanadium nitride quantum dots and MOF-derived hollow N-doped C nanocages. *J. Mater. Chem. A* **2021**, *9*, 11651–11664. [CrossRef]
124. Feng, Y.; Wang, G.; Kang, W.; Deng, N.; Cheng, B. Taming polysulfides and facilitating lithium-ion migration: Novel electrospinning MOFs@PVDF-based composite separator with spiderweb-like structure for Li–S batteries. *Electrochim. Acta* **2021**, *365*, 137344. [CrossRef]
125. Li, F.; Zhang, X.; Liu, X.; Zhao, M. Novel Conductive Metal-Organic Framework for a High-Performance Lithium-Sulfur Battery Host: 2D Cu-Benzenehexathial (BHT). *ACS Appl. Mater. Interfaces* **2018**, *10*, 15012–15020. [CrossRef]
126. Asadi, M.; Sayahpour, B.; Abbasi, P.; Ngo, A.T.; Karis, K.; Jokisaari, J.R.; Liu, C.; Narayanan, B.; Gerard, M.; Yasaei, P.; et al. A lithium-oxygen battery with a long cycle life in an air-like atmosphere. *Nature* **2018**, *555*, 502–506. [CrossRef]
127. Balaish, M.; Jung, J.-W.; Kim, I.-D.; Ein-Eli, Y. A Critical Review on Functionalization of Air-Cathodes for Nonaqueous Li–O_2 Batteries. *Adv. Funct. Mater.* **2019**, *30*, 1808303. [CrossRef]
128. Majidi, L.; Yasaei, P.; Warburton, R.E.; Fuladi, S.; Cavin, J.; Hu, X.; Hemmat, Z.; Cho, S.B.; Abbasi, P.; Voros, M.; et al. New Class of Electrocatalysts Based on 2D Transition Metal Dichalcogenides in Ionic Liquid. *Adv. Mater.* **2019**, *31*, 1804453. [CrossRef]
129. Majidi, L.; Hemmat, Z.; Warburton, R.E.; Kumar, K.; Ahmadiparidari, A.; Hong, L.; Guo, J.; Zapol, P.; Klie, R.F.; Cabana, J.; et al. Highly Active Rhenium-, Ruthenium-, and Iridium-Based Dichalcogenide Electrocatalysts for Oxygen Reduction and Oxygen Evolution Reactions in Aprotic Media. *Chem. Mater.* **2020**, *32*, 2764–2773. [CrossRef]
130. Wu, F.; Yu, Y. Toward True Lithium-Air Batteries. *Joule* **2018**, *2*, 815–817. [CrossRef]
131. Li, C.; Sun, X.; Yao, Y.; Hong, G. Recent advances of electrically conductive metal-organic frameworks in electrochemical applications. *Mater. Today Nano* **2021**, *13*, 100105. [CrossRef]
132. Dou, J.-H.; Arguilla, M.Q.; Luo, Y.; Li, J.; Zhang, W.; Sun, L.; Mancuso, J.L.; Yang, L.; Chen, T.; Parent, L.R.; et al. Atomically precise single-crystal structures of electrically conducting 2D metal-organic frameworks. *Nat. Mater.* **2021**, *20*, 222–228. [CrossRef] [PubMed]
133. Liao, X.; Lu, R.; Xia, L.; Liu, Q.; Wang, H.; Zhao, K.; Wang, Z.; Zhao, Y. Density Functional Theory for Electrocatalysis. *Energy Environ. Mater.* **2021**, *5*, 157–185. [CrossRef]

134. Zhao, T.; Zhang, Y.; Wang, D.; Chen, D.; Zhang, X.; Yu, Y. Graphene-coated Ge as anodes in Ge-air batteries with enhanced performance. *Carbon* **2023**, *205*, 86–96. [CrossRef]
135. Wu, Y.; Li, Z.; Hou, J. A novel two-dimensional main group metal organic framework $Ga_3C_6N_6$ as a promising anode material for Li/Na-Ion batteries. *Appl. Surf. Sci.* **2022**, *599*, 153958. [CrossRef]

Disclaimer/Publisher's Note: The statements, opinions and data contained in all publications are solely those of the individual author(s) and contributor(s) and not of MDPI and/or the editor(s). MDPI and/or the editor(s) disclaim responsibility for any injury to people or property resulting from any ideas, methods, instructions or products referred to in the content.

Article

Effect of Initial Structure on Performance of High-Entropy Oxide Anodes for Li-Ion Batteries

Otavio J. B. J. Marques [1,†], Michael D. Walter [2], Elena V. Timofeeva [3,4] and Carlo U. Segre [1,2,*,†]

1. Department of Mechanical, Materials and Aerospace Engineering, Illinois Institute of Technology, Chicago, IL 60616, USA
2. Department of Physics and CSRRI, Illinois Institute of Technology, Chicago, IL 60616, USA
3. Department of Chemistry, Illinois Institute of Technology, Chicago, IL 60616, USA
4. Influit Energy, LLC, Chicago, IL 60612, USA
* Correspondence: segre@iit.edu
† These authors contributed equally to this work.

Abstract: Two different high-entropy oxide materials were synthesized and studied as Li-ion battery anodes. The two materials have the same active metal constituents but different inactive elements which result in different initial crystalline structures: rock salt for (MgFeCoNiZn)O and spinel for (TiFeCoNiZn)$_3$O$_4$. Local structural studies of the metal elements in these two materials over extended electrochemical cycling reveal that the redox processes responsible for the electrode capacity are independent of the initial crystallographic structure and that the capacity is solely dependent on the initial random distribution of the metal atoms and the amount of active metals in the starting material.

Keywords: high-entropy oxides; Li-ion batteries; X-ray absorption spectroscopy; local structure

1. Introduction

Due to the increasing demand on portable electronics, electric vehicles, and green grid solutions, energy storage technologies have received much attention lately. In this scenario, lithium-ion batteries (LIBs) show great advantages compared to other chemistries, namely higher power and energy density, cycle life, and reversibility. The low theoretical capacity of current commercialized negative electrodes for LIBs (i.e., graphite ~372 mAh/g) and sluggish Li-ion diffusivity, limits its utilization for high-energy applications [1–5]. Therefore, the development of new anode materials is crucial for the ever-increasing demand on energy storage technologies, and conversion-type high-entropy oxides (HEOs) have emerged as an alternative for next-generation electrodes [6].

HEOs constitute a new class of materials wherein, during high temperature annealing, the configurational entropy stabilizes a solid solution of five or more transition metals, overcoming the formation enthalpy of competing individual oxide phases, driving the Gibbs free energy to its lowest levels and favoring the formation of a single solid solution. The ability to mix non-soluble components in one entropy-stabilized phase expands the compositional space of elemental mixtures creating new opportunities for advanced energy materials discovery and design [5,7–10].

Previous studies on HEOs as anode in LIBs have focused primarily on materials with a wide variety of elemental combinations forming both the rock salt structure [6,11–13] and the spinel structure [14–20]. These materials show improved capacities, long cycling stability and high ionic conductivities, overcoming some expected drawbacks of conversion-type metal oxide electrodes, such as particle pulverization and fast capacity fading [21]. The key to advanced performance seems to be multiple elements having different roles in the electrochemical reaction, displaying a synergetic storage behavior, and attenuating the volume changes in the electrode. Some of the transition metals (i.e., Zn and Co) present in conversion-type HEO electrodes act as electrochemically active species contributing to the storage capacity, with complete reduction to their metallic state during the lithiation reaction. Others (i.e., Mg, Cu) provide structural stability, contributing to the long cycle life. Adding electrochemically inactive elements to HEO composition (i.e., MgO), is also an effective method to prevent pulverization and agglomeration of active domains.

A still unanswered question is whether the initial structure, rock salt or spinel, provides advantages to a Li-ion HEO electrode in terms of capacity, rate performance and cycle lifetime. In order to probe these effects, in this study, we have prepared two new HEO compounds, (MgFeCoNiZn)O and (TiFeCoNiZn)$_3$O$_4$ and investigated their local structure and conversion mechanism using X-ray Absorption Spectroscopy (XAS).

2. Materials and Methods

Two HEO compositions (MgFeCoNiZn)O and (TiFeCoNiZn)$_3$O$_4$, were prepared from equimolar amounts of the monoxide precursors according to their respective compositions. The purity of the precursors were: MgO (Alfa Aesar, Haverhill, MA, USA, 99.95%), TiO (Sigma Aldrich, St. Louis, MO, USA, 99.9%), Fe$_2$O$_3$ (Nanoamor, Houston, TX, USA, >99%) CoO (Thermo Fisher, Waltham, MA, USA, 99.995%), NiO (Alfa Aesar, 99.998%), and ZnO (Alfa Aesar, 99.999%). Both samples were prepared using a similar route. First, equimolar metal amounts of the oxides were mixed by ball milling (SPEX 8000, Metuchen, NJ, USA) for 1 h using a 4:1 ratio of 1/4″ diameter stainless steel balls to sample. The resulting material was pelletized in 0.5 g batches, sintered for 2 h with an initial heating rate of 10 °C/min, and quenched to room temperature in air. The (MgFeCoNiZn)O sample was sintered at 1300 °C while the (TiFeCoNiZn)$_3$O$_4$ sample at 1600 °C. The difference in temperature is related to the presence of a secondary phase in the (TiFeCoNiZn)$_3$O$_4$ sample, which was eliminated only at 1600 °C. Both high-entropy products were again ball milled for 30 min using a 10:1 ratio of balls to sample and sieved using a 500-mesh sieve (<25 µm) to result in a nanostructured powder. The crystal structures of the sintered and nanostructured samples were analyzed by powder X-ray diffraction (XRD; Bruker AXS, Madison, WI, USA, D2 Phaser, Cu K$_\alpha$ radiation) from 10 to 80 degrees 2θ. The resulting powder patterns were fitted by Pawley refinement using GSAS-II [22]. The particle morphology and elemental composition were analyzed by scanning electron microscopy (SEM; JEOL JSM-IT500HR, Akishima, JP) equipped with an energy dispersive X-ray spectrometer (EDS; Oxford Instruments Ultim-Max, Abingdon, UK).

The nanostructured (MgFeCoNiZn)O and (TiFeCoNiZn)$_3$O$_4$ were tested as anodes in Li-ion batteries using a half-cell configuration. Electrodes were prepared by mixing active materials, carbon black and polyvinylidene fluoride (PVDF) in 70:20:10 wt.% proportions dispersed in N-Methyl-2-pyrrolidone (NMP). The resulting slurry was casted on Cu foil using a doctor blade, dried at 30 °C for 24 h at ambient conditions and cold rolled. CR2032 coin cells were assembled in an Ar-filled glovebox using metallic Li as a counter electrode. Celgard 2325 (25 µm) and LiPF$_6$ (3:7 EC:EMC 1M LIPASTE) were used as separator and electrolyte, respectively. The specific capacity and cycling stability of both compounds were tested by room temperature (~25 °C) galvanostatic cycling of the cells using a BTS400 battery cycler (Neware, Shenzhen, China), at varying current densities (50 mA/g, 100 mA/g, 200 mA/g, 500 mA/g). Cyclic voltammetry was performed using an EZstat-Pro potentiostat (Nuvant Systems, Crown Point, IN, USA) at a series of sweep rates (0.2 mV/sC, 0.5 mV/sC, 0.8 mV/sC, 1 mV/sC, 1.2 mV/sC, 1.5 mV/s, 1.8 mV/sC, 2 mV/sC, 2.5 mV/sC, and 3 mV/sC).

Ex situ XAS measurements were taken on both electrode materials at different charging and cycle states: uncycled, 1st lithiated and delithiated states, and 100th lithiated and delithiated. The electrodes were extracted from the cells in an argon-filled glovebox, cleaned with DMC (dimethyl carbonate), and protected by Kapton tape. In order to prevent oxidation during transport and measurement, the lithiated electrodes were additionally vacuum sealed in polyethylene and removed from the glovebox shortly before measurement. Ti, Fe, Co, Ni, and Zn K-edges were measured in fluorescence mode at the Sector 10, 10-BM beamline of the Materials Research Collaborative Access Team (MRCAT) at the

Advanced Photon Source, Argonne National Laboratory [23]. The resulting X-ray Near Edge Structure (XANES) and Extended X-ray Absorption Fine Structure (EXAFS) data were analyzed using the IFEFFIT-based Athena and Artemis software packages [24,25].

3. Results and Discussion

3.1. Structural Characterization

Figure 1a,c show the XRD results of the sintered and nanostructured (figure inset) high-entropy oxides. The (MgFeCoNiZn)O and (TiFeCoNiZn)$_3$O$_4$, hereafter called HEOR and HEOS, can be indexed as cubic rock salt (Fm-3m) and spinel (Fd-3m) structures, respectively. Both materials remain single-phase after the nanostructuring step. The lattice parameters calculated from the Pawley refinement remain the same after the final nanostructuring, $a_{HEOR} = 4.23 \pm 0.01$ Å and $a_{HEOS} = 8.41 \pm 0.01$ Å. These lattice parameters fall in the middle of those of reference compounds of the individual elements or combinations thereof (Table S2). The average crystallite size extracted from the Pawley refinement is approximately 150 nm for both nanostructured materials. The SEM images (Figure 1b,d) show an average particle of ~1 μm after nanostructuring, indicating that the typical particle contains many crystallites. The elemental mapping confirms that all the metals are distributed uniformly throughout the particles. In the rock salt structure, the metals share the same octahedrally-coordinated lattice site. In the spinel case, there are two available positions for the metals, one tetrahedral and one octahedral. Each metal's choice of lattice position is driven primarily by the crystal field stabilization, which is very difficult to predict in this case [26]. Due to the effect of entropy stabilization, it is likely that the metals are mixed in between both positions.

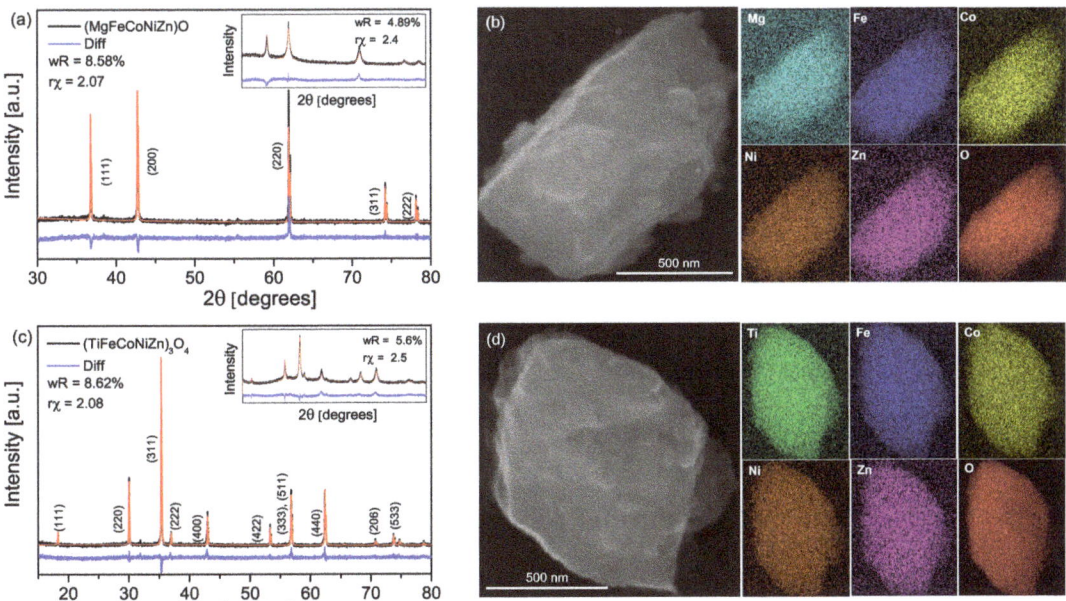

Figure 1. X-ray powder diffraction data (black), Pawley refinement (red) and their difference (blue) for (**a**) HEOR and (**c**) HEOS samples before nanostructuring. The inset images show the nanostructured materials after the final ball milling. SEM images and elemental mapping for typical particles (**b**) HEOR and (**d**) HEOS after nanostructuring.

3.2. Electrochemical Characterization

It is expected that Fe, Co, Ni, and Zn act as electrochemically active species in the reaction, being reduced to their metallic state. For the HEOR, all the metals start in an average state of +2 due to the rock salt structure, and the reactions $Fe^{+2} \to Fe^0$, $Co^{+2} \to Co^0$, $Ni^{+2} \to Ni^0$, $Zn^{+2} \to Zn^0$ can deliver up to 625 mAh/g, in theory. The HEOS, on the other hand, has a mixture of oxidation states, with twice as many octahedral sites as tetrahedral sites. Considering the same active species, and depending on whether Ti is in the +4 or +3 state, the compound can deliver between 636 mAh/g and 704 mAh/g of theoretical capacity. Mg, in the HEOR, and Ti, in the HEOS, are expected to act as structure formers and do not participate directly in the electrochemical conversion reaction. Although the spinel structure has more oxygen and higher molecular weight per mole of metal ions than the rock salt, it can also contribute more electrons to the conversion reaction, contributing to a higher theoretical capacity.

Experimentally, the initial specific capacity of the HEOR (Figure 2a) reaches 575 mAh/g, while the highest capacity for HEOS has a specific capacity of 478 mAh/g at 50 mA/g current density. For both materials, the capacity decreases with increasing current density reaching 304 mAh/g and 246 mAh/g, respectively, at a current density of 500 mA/g. Long-term cycling has been carried out at 100 mA/g current density, where similar patterns were observed on both materials—capacity faded slowly reaching a minimum values of 350 and 275 mAh/g at around 80 cycles followed by an increase in capacity with maximum at 150 cycles, hitting 410 mAh/g and 320 mAh/g, and then slowly decreasing again, reaching 360 mAh/g and 320 mAh/g after 300 cycles. The gradual increase in capacity between cycle 80 and 150 is commonly observed in conversion anodes made with simple metal oxides and the high entropy oxides. The effect is attributed by various authors to different causes. Lin et al. [27] consider the increase as due to an activation process associated with size refinement of the crystallites during conversion while Zhang et al. [28] claim that the electrode undergoes a reorganization that increases the kinetics and capacity of the conversion reaction while reducing the potential. A third possibility is that repeated cycling serves to refine the metal nanoparticle nanostructure and increase the pseudocapacitive portion of the overall capacity suggested by Chen et al. [29]. The HEOS shows remarkable stability after reaching steady value at ~150 cycles and maintaining average capacity of 320 mAh/g until the end of the cycling routine. The difference in specific capacity between the HEOR and HEOS decreases with increasing cycling rate, changing from ~117 mAh/g at 50 mA/g to ~58 mAh/g at 500 mA/g, showing a better response of the HEOS at higher charging/discharging current densities. This could be due to a difference in conductivity of the HEOR and HEOS electrodes due to the inactive metals, MgO and TiO_2, respectively.

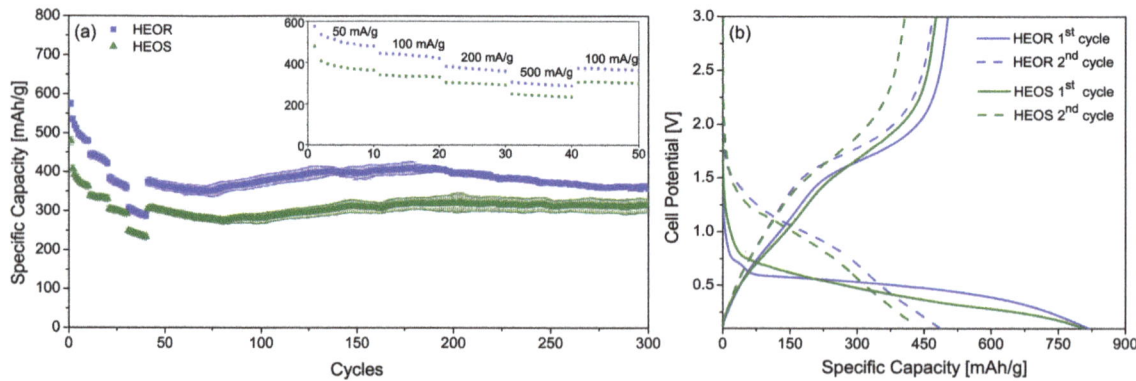

Figure 2. Electrochemical performance of HEOR (blue) and HEOS (green) by (**a**) galvanostatic cycling at different cycling rates averaged over four cells with close-up of cycles 1–50 in the inset and (**b**) the voltage profile of first (solid) and second (dashed) cycles.

Figure 2b shows the voltage profile of HEOR and HEOS for the first two cycles. Both electrodes behave similarly, having a huge drop of specific capacity during the 1st cycle, but maintaining the characteristic features of a reversible conversion reaction subsequently.

During the 1st lithiation, the HEOR reaches a specific capacity of 820 mA/g, while the HEOS 805 mAh/g, both higher than their respective theoretical capacities. This phenomenon is mainly explained due to the complete electrochemical conversion reaction of active species and the SEI layer formation on the anode [30]. The 1st delithiation shows an irreversibility of 61% and 58% for the HEOR and HEOS, reaching a specific capacity of 575 mAh/g and 478 mAh/g, respectively. Both compounds demonstrate similar cell potentials, being 0.7 V and 0.6 V, for the HEOR and HEOS, during the first cycle, and 1.1 V for both electrodes in the 2nd cycle. The difference in specific capacity during the first lithiation and delithiation, and the change in reaction potential between the first two cycles, indicate that some of the active species are not returning to their initial oxidized state upon delithiation. This, along with the likelihood of SEI formation, results in only partial reversibility of the electrochemical reaction in the first cycles.

According to the recent reports on high-entropy oxides as LIB anodes, the presence of metallic species after the conversion reaction can also contribute to a pseudocapacitive behavior of the electrode [15,16,19]. Cyclic-voltammetry (CV) can be used to characterize this behavior by using different voltage sweep rates over the same voltage window, and fitting a logarithmic function to the peak of the sweep currents. This approach can separate the faradaic and non-faradaic contributions to the reaction by calculating the slope, b, of the logarithmic plot of peak current vs. sweep rate. Figure 3 shows the CV for the HEOR (a) and HEOS (b) at different scan rates and the plot of the logarithm of sweep rate vs. the logarithm of the peak current in the respective insets. Both compounds have an intermediate behavior between purely diffusive ($b = 0.5$) and pseudocapacitive ($b = 1$). The b coefficient for the lithiation region is 0.72 and 0.71 for the HEOR and HEOS, respectively, and 0.54 and 0.76 for the delithiation region, respectively. This intermediate behavior indicates a pseudocapacitive contribution to the storage capacity of the electrodes that adds capacity to the electrochemical conversion reaction. Still, it is very difficult to separate the electrochemical contribution of different metals in both compounds, as their CV shows unique and broad redox peaks at each of the charge/discharge steps.

3.3. X-ray Absorption Spectroscopy Analysis

In the uncycled state, the XANES spectrum for Co (Figures S4 and S8) and Zn (Figures S6 and S10) are clearly in a +2 state in both HEOR and HEOS when compared to standard reference compounds. The XANES spectrum for Ni (Figures S5 and S9) are shifted to slightly higher energies than the Ni^{+2} (NiO) reference but show the same fine structure. The Fe edge (Figures S3 and S7) for both HEOR and HEOS is shifted close to the Fe^{+3} reference (Fe_2O_3), with the shape closely resembling that of the Fe_3O_4 XANES.

The Fourier Transform of the EXAFS for Fe, Co, Ni, and Zn K-edges (Figure 4) shows the changes in the local structure between the uncycled electrode and the 1st lithiated and 1st delithiated states for both rock salt (HEOR) and spinel (HEOS) compounds. Initially, all metals in the HEOR show a similar structure (in black), composed of two well-shaped peaks: peak at 1.5 Å is related to the octahedral oxygen shell around the absorbing cation position, while the peak at ~2.6 Å is attributed to the 2nd metallic shell.

The HEOS, on the other hand, has two different sites available for the metallic species, which makes it more difficult to identify their chemical environment. The A site, on the AB_2O_4 spinel structure, is located in a tetrahedral geometry, while the B site is octahedral. All metals in its composition can exist in either geometry, thus their average coordination can be expected to be in between 4 and 6. Due to entropy stabilization, all metals are likely to be randomly and homogeneously distributed on both sites.

During the 1st lithiation, Fe, Co, Ni, and Zn are all reduced to their metallic states as can be seen by the single peak around 2 Å, which is the metallic local structure signature (Figure 4). In the 1st delithiated state, Zn and Fe show the highest degree of reversibility back to an oxidized state as evidenced by re-appearance of the strong metal–oxygen bond peak at 1.5 Å. Co and Ni show very minor re-oxidation (the shoulder on the short distance side of the metallic peak). The XANES data for both HEOR and HEOS presented on the right side of Figures S3–S10 are consistent with these observations and show the spectra for the 100th cycle as well. The Ni (Figures S5 and S9) and Co (Figures S4 and S8) edges shift to the Ni^0 and Co^0 positions on 1st lithiation show little recovery upon delithiation, completely losing the white line by the 100th cycle. The Fe (Figures S3 and S7) and Zn (Figures S6 and S10) show significant changes between lithiated and delithiated states even to the 100th cycle.

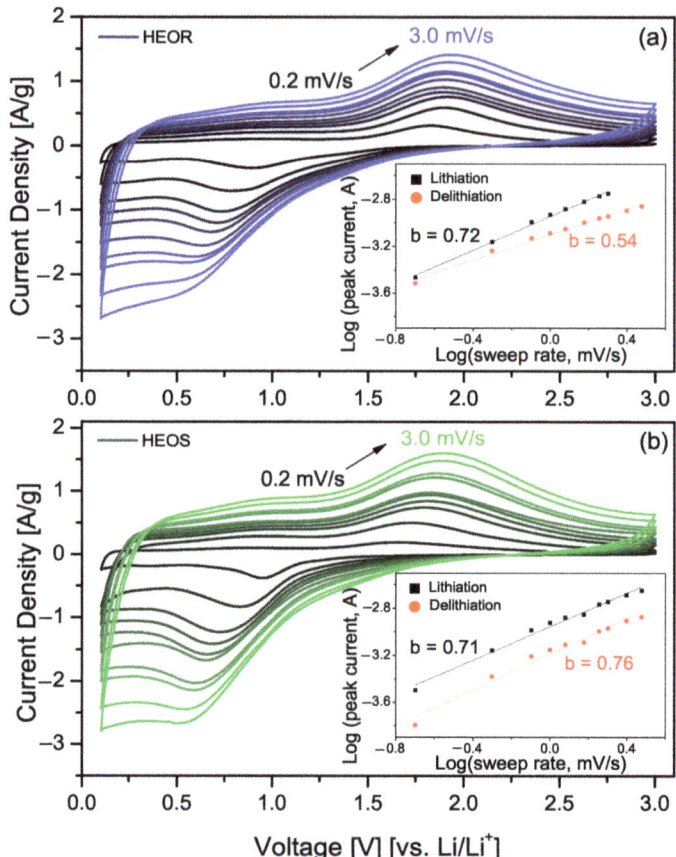

Figure 3. Cyclic voltammetry at different voltage sweep rates: 0.2 darkest shade; 0.5; 0.8; 1.0; 1.2; 1.5; 1.8; 2.0; 2.5, and 3.0 lightest shade mV/sC for (**a**) HEOR and (**b**) HEOS. The insets show the linear fits to the logarithmic plots of peak current and sweep rate for both lithiation (black squares) and delithiation (red circles) currents.

In order to quantify those changes, EXAFS fits of the local structure are performed for the 1st and 100th lithiated and delithiated cycles at the Fe, Co, Ni, and Zn edges. The fits use two paths to model local structure, a TM–O path and a TM–M path, where TM = Fe, Co, Ni, and Zn and M is a general scattering metal. Ni was chosen as M for the models as a representative element with the average scattering properties of the TMs in these compounds. The TM–O paths in the EXAFS fits were calculated using the structural data obtained from the rock salt and spinel Pawley fits. The spinel structure has two possible TM–O paths from the tetrahedral and octahedral metal sites; however, only a single path was needed to fit all the HEOS spectra. The fitting results are presented in Table S3 for HEOR and Table S4 for HEOS.

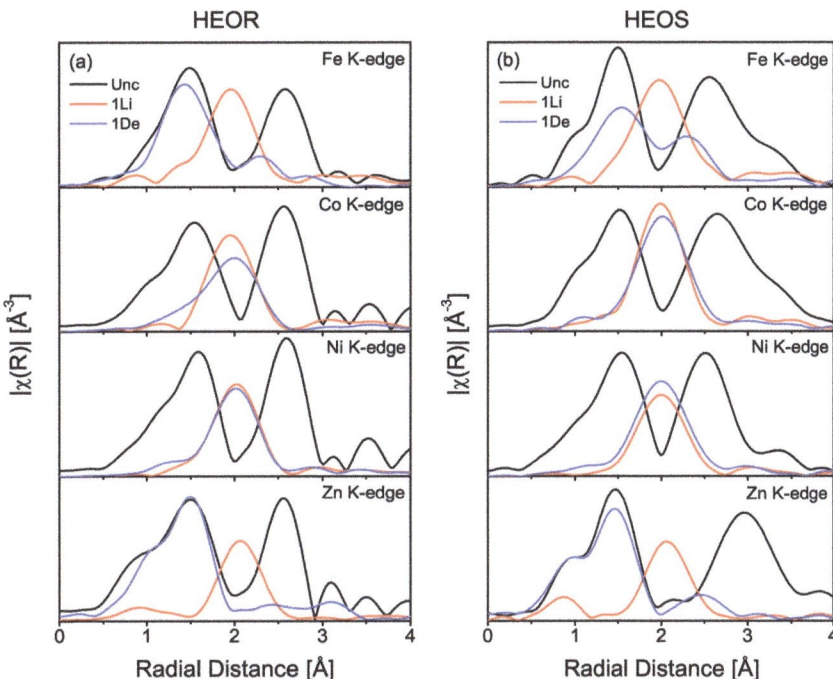

Figure 4. EXAFS spectra in R-space of Fe, Co, Ni, and Zn K-edges for the (**a**) HEOR and (**b**) HEOS electrodes in uncycled (black), 1st lithiated (red), and 1st delithiated states (blue).

3.3.1. TM–Oxygen Model

Figure 5 shows the fit results detailed in Tables S3 and S4 for the TM–oxygen path in uncycled, 1st lithiated, 1st delithiated, 100th lithiated and 100th delithiated states for Fe, Co, Ni, and Zn. In the uncycled state, the number of near neighbors for Co and Ni in the HEOR are close to 6, consistent with an octahedral environment while the Fe and Zn fits give an oxygen coordination number close to 4, indicating that there could be defect structures present in the rock salt structure. The calculated bond distances for Co, Ni, and Zn in the HEOR deviate only slightly from one-half the lattice parameter obtained in the Pawley refinement. The bond distance for Fe, however, is significantly shorter and consistent with its 4-fold coordination. For HEOS, the bond length and coordination numbers for Fe and Zn suggest that they preferentially occupy the tetrahedral sites of the spinel structure, while Co and Ni sit in the octahedral sites.

During the 1st lithiation the number of oxygens around the Co, Ni, and Zn in the HEOR decreases to less than one, and Fe-O neighbors are completely absent, evidence of a reduced environment and destruction of the initial rock salt structure. In the HEOS, TM–oxygen bonds are absent for all four metals in the 1st lithiated state, which means that Fe, Co, Ni, and Zn were completely reduced to their metallic state and the spinel structure is no longer present.

Upon the 1st delithiation, the oxygen neighbors return to all four metals (Fe, Co, Ni, and Zn) in both HEOR and HEOS compounds. As can be seen from the number of near neighbors shown in Figure 5, Zn shows a higher degree of re-oxidation and electrochemical activity followed by Fe, Ni, and Co, in both HEOR and HEOS consistent with the redox conversion activity decreasing in the order of Zn > Fe > Co > Ni. After 100 cycles, Zn, Fe, and Co still show some reversibility with the number of TM–O bonds decreasing in the lithiated state and increasing in the delithiated state. Ni–O bonds completely disappear in both 100th lithiated and delithated states. TM–O distances fluctuate significantly in the initial cycling, consistent with major material restructuring, but remain fairly constant after extended material cycling.

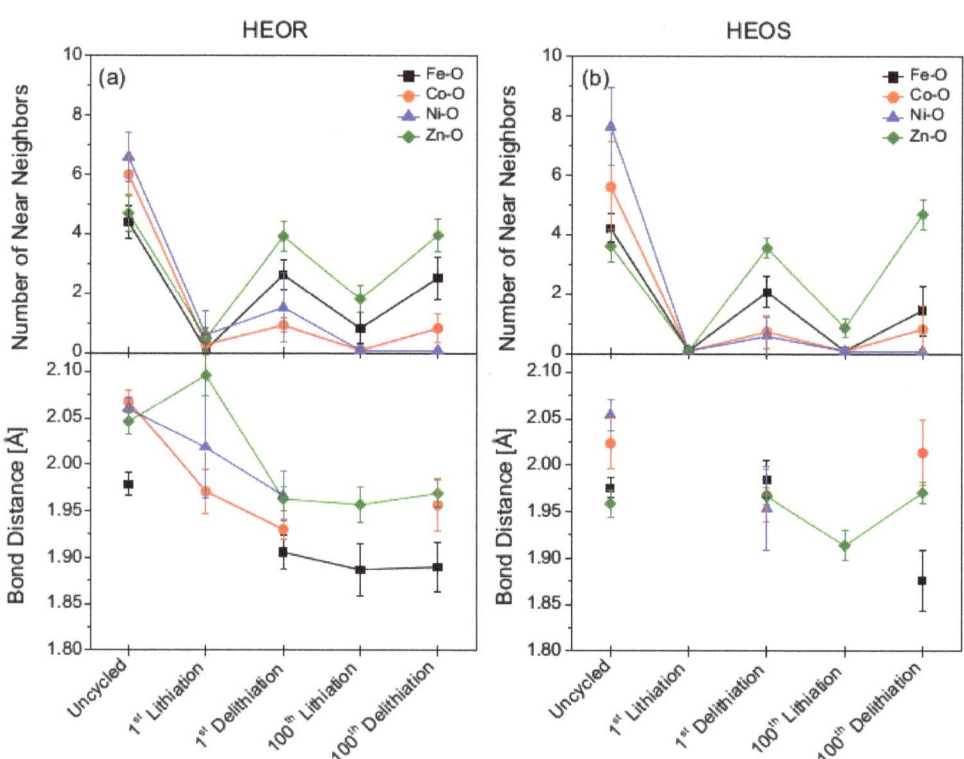

Figure 5. TM–O K-edge EXAFS fit results for (**a**) HEOR and (**b**) HEOS compounds in uncycled, 1st lithiated, 1st delithiated, 100th lithiated and 100th delithiated states for Fe (black squares), Co (red circles), Ni (blue triangles), and Zn (magenta inverted triangles).

3.3.2. TM–Metal Model

The EXAFS fit results of the TM–metal path show that, in the 1st lithiation, conversion of high entropy oxides to metallic Fe, Co, Ni, and Zn occurs to different degrees (Figure 6 and Tables S3 and S4). For both compounds, the respective number of metal near neighbors seems to follow the same trend. Zn has the largest number of near neighbors of ∼9 and ∼8 for HEOR and HEOS, respectively. On the other hand, Fe has the smallest values, being ∼4 for both the HEOR and HEOS. Co has ∼6.5 metal near neighbors for both HEOR and HEOS and Ni has ∼6.9 and ∼6.2 metal near neighbors for HEOR and HEOS, respectively, which are average between the Zn and Fe values. At 1st delithiation, the conversion redox activity of metals decreases in the series Zn > Fe > Co > Ni and the number of metal near neighbors of Ni is unchanged. For both HEOR and HEOS, the Zn–M bond completely disappears even at the 100th delithiation, showing the remarkable reversibility of Zn conversion in these compounds. Fe–M bonds are still present in both 1st and 100th delithiated states, indicating incomplete conversion, but the number of nearest neighbors decreases from ∼4 to ∼1.4 and ∼1.8 for HEOR and HEOS, respectively, with modest growth in near neighbor numbers by the 100th cycle. Co and Ni show only small changes in the number of metal near neighbors on delithiation, but the total number of TM–M near neighbors grows by the 100th cycle to ∼7.5 for Co and ∼10.5 for Ni. It should be noted that this general increase in the number of TM–M near neighbors in both materials after 100 cycles is consistent with the observed decrease in TM–O bonds.

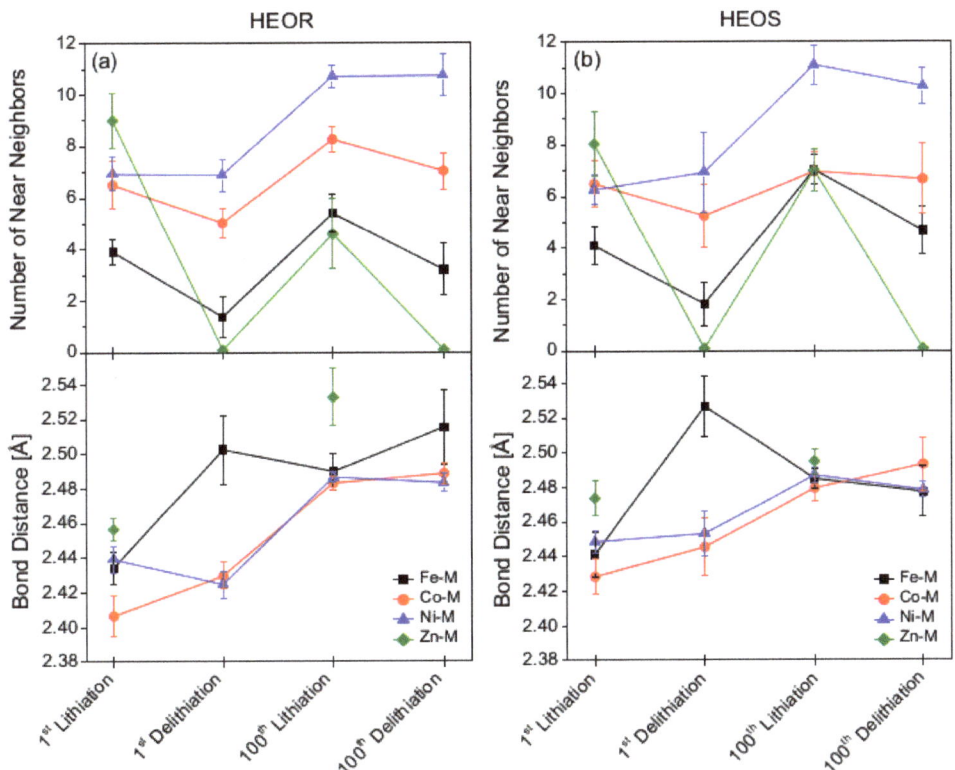

Figure 6. TM–M K-edge EXAFS fit results for (**a**) HEOR and (**b**) HEOS compounds in uncycled, 1st lithiated, 1st delithiated, 100th lithiated and 100th delithiated states for Fe (black squares), Co (red circles), Ni (blue triangles), and Zn (magenta inverted triangles).

These results suggest that the structure of both HEOR and HEOS after the initial lithiation is very similar and consists of slowly growing metal nanoparticles containing primarily Ni and Co but, with some Fe, and a significant amount of the Fe and all of the Zn participating in redox reactions all the way through the 100th cycle. While the Mg in HEOR cannot be probed with XAS, the Ti K-edge in HEOS is accessible and can provide clues as to the role of these two presumably inactive metals.

3.3.3. Titanium K-Edge in the HEOS Electrode

In comparison to Ti reference standards, the position of the Ti K-edge XANES suggests that Ti is initially close to a +4 state, but with a shape similar to that of the TiO rock salt standard (Figure 7). As the electrodes are cycled, the shape and position of the Ti XANES remain substantially unchanged, although some of the fine structure features visible initially are smoothed out by the 100th cycle. The Fourier Transform of the Ti EXAFS (Figure S19) is consistent with the Ti remaining in an oxidized state throughout the cycling, albeit with broadening and attenuation of both the first shell Ti–O and Ti–M peaks. The first shell EXAFS fit to a Ti–O path (Table S5) is consistent, despite large estimated standard deviations, with 6-fold coordination of Ti throughout the cycling [31]. Like the Mg in HEOR, Ti in HEOS clearly has a structure forming role which keeps the metal nanoparticles appearing during the conversion process isolated, slowing their growth, and prolonging the cycle life of the electrode.

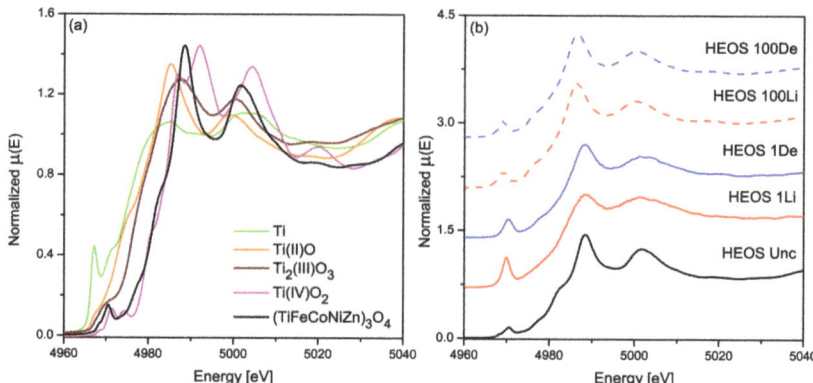

Figure 7. Ti K-edge XANES for uncycled HEOS electrode (black) compared to (**a**) standard reference samples Ti (green), TiO (orange), Ti$_2$O$_3$ (brown) and TiO$_2$ (magenta); and (**b**) electrodes at various states of cycle and charge: 1st lithiation (solid red), 1st delithiation (solid blue), 100th lithiation (dashed red), and 100th delithiation (dashed blue).

4. Conclusions

The XANES and EXAFS local structural analysis reveals that the active elements, Fe, Co, Ni, and Zn behave in substantially identical ways for both HEOR and HEOS. Co and Ni form metallic nanoparticles that grow slowly over cycling time, Fe and Zn continue to be redox active in the conversion process throughout the 100 cycles observed with Zn being fully reduced and re-oxidized each cycle. Mg and Ti do not participate in the electrochemical reaction but fulfill a critical role in maintaining structural integrity of the electrode, ionic conductivity for Li$^+$, and separation of the metal nanoclusters and thus slowing capacity fade. The cycling performance of the HEOR and HEOS electrodes are substantially the same for the 300 cycles investigated with a slight difference in specific capacity (Figure 2a). This difference all but vanishes by noting that HEOR has a single mole oxygen per mole of metal atoms while HEOS has 1.33 moles of oxygen per mole of metal atoms. If the specific capacity is recalculated in terms of molar capacity, the maximum capacities at the 175th cycle are 25.2 ± 0.6 Ah/mol for HEOR and 24.8 ± 1.6 Ah/mol for HEOS (Figure S20). This provides additional evidence that the initial long-range crystallographic structure of an entropy-stabilized oxide conversion anode is not as important for the redox mechanism and ultimate cycling performance of the material. Rather, the critical components for the superior performance compared to single element metal oxides are the amount of active metal ions in the mixture and the initial random spatial distribution of the metal ions resulting from entropy-stabilization.

Supplementary Materials: The following supporting information can be downloaded at: https://www.mdpi.com/article/10.3390/batteries9020115/s1, Figure S1: Rock-salt and spinel structures; Figure S2: Comparison of cyclic voltammetry; Figures S3–S10: Fe, Co, Ni, and Zn XANES spectra of HEOR and HEOS electrodes; Figures S11–S18: Fe, Co, Ni, and Zn EXAFS fits for HEOR and HEOS electrodes; Figure S19: Ti EXAFS spectra of HEOS electrodes; Figure S20: Specific molar capacity for HEOR and HEOS; Table S1: Details of Pawley refinements; Table S2: Comparison of HEOR and HEOS lattice parameters with isostructural compounds including References [32–39]; Table S3: Fe, Co, Ni, and Zn edge EXAFS fit results for HEOR electrodes; Table S4: Fe, Co, Ni, and Zn edge EXAFS fits for HEOS electrodes; Table S5: Ti edge EXAFS fits for HEOS electrodes.

Author Contributions: Conceptualization, C.U.S. and O.J.B.J.M.; methodology, C.U.S. and O.J.B.J.M.; validation, E.V.T.; investigation, O.J.B.J.M. and M.D.W.; resources, C.U.S.; data curation, C.U.S. and O.J.B.J.M.; writing—original draft preparation, O.J.B.J.M.; writing—review and editing, E.V.T. and C.U.S.; visualization, C.U.S. and O.J.B.J.M.; supervision, C.U.S.; project administration, C.U.S. and O.J.B.J.M.; funding acquisition, C.U.S. All authors have read and agreed to the published version of the manuscript.

Funding: This work was supported by the Illinois Institute of Technology Duchossois Leadership Professors Program. MRCAT operations are supported by the Department of Energy and the MRCAT member institutions. This research used resources of the Advanced Photon Source, a U.S. Department of Energy (DOE) Office of Science User Facility operated for the DOE Office of Science by Argonne National Laboratory under Contract No. DE-AC02-06CH11357.

Institutional Review Board Statement: Not applicable.

Informed Consent Statement: Not applicable.

Data Availability Statement: The data presented in this study are available on request from the corresponding author.

Acknowledgments: The authors wish to thank the assistance of High School student teams from Maine South High School under the Exemplary Student Research Program and Stevenson High School under the STEM Professionals As Resource Knowledge program who assisted in sample preparation.

Conflicts of Interest: The authors declare no conflict of interest. The funders had no role in the design of the study; in the collection, analyses, or interpretation of data; in the writing of the manuscript; or in the decision to publish the results.

References

1. Bruce, P.; Scrosati, B.; Tarascon, J.M. Nanomaterials for rechargeable lithium batteries. *Angew. Chem. Int. Ed.* **2008**, *47*, 2930–2946. [CrossRef]
2. Goodenough, J.B.; Kim, Y. Challenges for rechargeable Li batteries. *Chem. Mater.* **2009**, *22*, 587–603. [CrossRef]
3. Li, M.; Lu, J.; Chen, Z.; Amine, K. 30 Years of lithium-ion batteries. *Adv. Mater.* **2018**, *30*, 1800561. [CrossRef] [PubMed]
4. Tarascon, J.M.; Armand, M. Issues and challenges facing rechargeable lithium batteries. *Nature* **2001**, *414*, 359–367. [CrossRef] [PubMed]
5. Lu, J.; Chen, Z.; Pan, F.; Cui, Y.; Amine, K. High-performance anode materials for rechargeable lithium-ion batteries. *Electrochem. Energy Rev.* **2018**, *1*, 35–53. [CrossRef]
6. Sarkar, A.; Velasco, L.; Wang, D.; Wang, Q.; Talasila, G.; de Biasi, L.; Kübel, C.; Brezesinski, T.; Bhattacharya, S.S.; Hahn, H.; et al. High entropy oxides for reversible energy storage. *Nat. Commun.* **2018**, *9*, 3400. [CrossRef]
7. Rost, C.M.; Sachet, E.; Borman, T.; Moballegh, A.; Dickey, E.C.; Hou, D.; Jones, J.L.; Curtarolo, S.; Maria, J.P. Entropy-stabilized oxides. *Nat. Commun.* **2015**, *6*, 8485. [CrossRef]
8. Zhang, R.Z.; Reece, M.J. Review of high entropy ceramics: design, synthesis, structure and properties. *J. Mater. Chem. A* **2019**, *7*, 22148–22162. [CrossRef]
9. Oses, C.; Toher, C.; Curtarolo, S. High-entropy ceramics. *Nat. Rev. Mater.* **2020**, *5*, 295–309. [CrossRef]
10. Musicó, B.L.; Gilbert, D.; Ward, T.Z.; Page, K.; George, E.; Yan, J.; Mandrus, D.; Keppens, V. The emergent field of high entropy oxides: Design, prospects, challenges, and opportunities for tailoring material properties. *APL Mater.* **2020**, *8*, 040912. [CrossRef]
11. Sarkar, A.; Wang, Q.; Schiele, A.; Chellali, M.R.; Bhattacharya, S.S.; Wang, D.; Brezesinski, T.; Hahn, H.; Velasco, L.; Breitung, B. High-entropy oxides: Fundamental aspects and electrochemical properties. *Adv. Mater.* **2019**, *31*, 1806236. [CrossRef] [PubMed]
12. Qiu, N.; Chen, H.; Yang, Z.; Sun, S.; Wang, Y.; Cui, Y. A high entropy oxide ($Mg_{0.2}Co_{0.2}Ni_{0.2}Cu_{0.2}Zn_{0.2}O$) with superior lithium storage performance. *J. Alloys Compd.* **2019**, *777*, 767–774. [CrossRef]
13. Lökçü, E.; Toparli, Ç.; Anik, M. Electrochemical performance of $(MgCoNiZn)_{1-x}Li_xO$ high-entropy oxides in lithium-ion batteries. *ACS Appl. Mater. Interfaces* **2020**, *12*, 23860–23866. [CrossRef]
14. Chen, T.Y.; Wang, S.Y.; Kuo, C.H.; Huang, S.C.; Lin, M.H.; Li, C.H.; Chen, H.Y.T.; Wang, C.C.; Liao, Y.F.; Lin, C.C.; et al. In operando synchrotron X-ray studies of a novel spinel $(Ni_{0.2}Co_{0.2}Mn_{0.2}Fe_{0.2}Ti_{0.2})_3O_4$ high-entropy oxide for energy storage applications. *J. Mater. Chem. A* **2020**, *8*, 21756–21770. [CrossRef]
15. Chen, H.; Qiu, N.; Wu, B.; Yang, Z.; Sun, S.; Wang, Y. A new spinel high-entropy oxide $(Mg_{0.2}Ti_{0.2}Zn_{0.2}Cu_{0.2}Fe_{0.2})_3O_4$ with fast reaction kinetics and excellent stability as an anode material for lithium ion batteries. *RSC Adv.* **2020**, *10*, 9736–9744. [CrossRef] [PubMed]
16. Wang, D.; Jiang, S.; Duan, C.; Mao, J.; Dong, Y.; Dong, K.; Wang, Z.; Luo, S.; Liu, Y.; Qi, X. Spinel-structured high entropy oxide $(FeCoNiCrMn)_3O_4$ as anode towards superior lithium storage performance. *J. Alloys Compd.* **2020**, *844*, 156158. [CrossRef]
17. Nguyen, T.X.; Patra, J.; Chang, J.K.; Ting, J.M. High entropy spinel oxide nanoparticles for superior lithiation–delithiation performance. *J. Mater. Chem. A* **2020**, *8*, 18963–18973. [CrossRef]

18. Xiao, B.; Wu, G.; Wang, T.; Wei, Z.; Sui, Y.; Shen, B.; Qi, J.; Wei, F.; Meng, Q.; Ren, Y.; et al. High entropy oxides (FeNiCrMnX)$_3$O$_4$ (X = Zn, Mg) as anode materials for lithium ion batteries. *Ceram. Int.* **2021**, *47*, 33972–33977. [CrossRef]
19. Tian, K.H.; Duan, C.Q.; Ma, Q.; Li, X.L.; Wang, Z.Y.; Sun, H.Y.; Luo, S.H.; Wang, D.; Liu, Y.G. High-entropy chemistry stabilizing spinel oxide (CoNiZnXMnLi)$_3$O$_4$ (X = Fe, Cr) for high-performance anode of Li-ion batteries. *Rare Met.* **2021**, *41*, 1265–1275. [CrossRef]
20. Zheng, Y.; Wu, X.; Lan, X.; Hu, R. A spinel (FeNiCrMnMgAl)$_3$O$_4$ high entropy oxide as a cycling stable anode material for Li-ion batteries. *Processes* **2021**, *10*, 49. [CrossRef]
21. Poizot, P.; Laruelle, S.; Grugeon, S.; Dupont, L.; Tarascon, J.M. Nano-sized transition-metal oxides as negative-electrode materials for lithium-ion batteries. *Nature* **2000**, *407*, 496–499. [CrossRef]
22. Toby, B.H.; Dreele, R.B.V. *GSAS-II*: The genesis of a modern open-source all purpose crystallography software package. *J. Appl. Crystallogr.* **2013**, *46*, 544–549. [CrossRef]
23. Kropf, A.J.; Katsoudas, J.; Chattopadhyay, S.; Shibata, T.; Lang, E.A.; Zyryanov, V.N.; Ravel, B.; McIvor, K.; Kemner, K.M.; Scheckel, K.G.; et al. The New MRCAT (Sector 10) bending magnet beamline at the Advanced Photon Source. *AIP Conf. Proc.* **2010**, *1234*, 299–302. . [CrossRef]
24. Newville, M. *IFEFFIT*: interactive XAFS analysis and *FEFF* fitting. *J. Synchrotron Radiat.* **2001**, *8*, 322–324. [CrossRef] [PubMed]
25. Ravel, B.; Newville, M. *ATHENA, ARTEMIS, HEPHAESTUS*: Data analysis for X-ray absorption spectroscopy using *IFEFFIT*. *J. Synchrotron Radiat.* **2005**, *12*, 537–541. [CrossRef]
26. Burdett, J.K.; Price, G.D.; Price, S.L. Role of the crystal-field theory in determining the structures of spinels. *J. Am. Chem. Soc.* **1982**, *104*, 92–95. [CrossRef]
27. Lin, Y.M.; Abel, P.R.; Heller, A.; Mullins, C.B. α-Fe$_2$O$_3$ nanorods as anode material for lithium ion batteries. *J. Phys. Chem. Lett.* **2011**, *2*, 2885–2891. [CrossRef]
28. Zhang, J.; Lee, G.H.; Lau, V.W.; Zou, F.; Wang, Y.; Wu, X.; Wang, X.L.; Chen, C.L.; Su, C.J.; Kang, Y.M. Electrochemical grinding-induced metallic assembly exploiting a facile conversion reaction route of metal oxides toward Li ions. *Acta Mater.* **2021**, *211*, 116863. [CrossRef]
29. Chen, H.; Qiu, N.; Wu, B.; Yang, Z.; Sun, S.; Wang, Y. Tunable pseudocapacitive contribution by dimension control in nanocrystalline-constructed (Mg$_{0.2}$Co$_{0.2}$Ni$_{0.2}$Cu$_{0.2}$Zn$_{0.2}$)O solid solutions to achieve superior lithium-storage properties. *RSC Adv.* **2019**, *9*, 28908–28915. [CrossRef]
30. Ng, B.; Faegh, E.; Lateef, S.; Karakalos, S.G.; Mustain, W.E. Structure and chemistry of the solid electrolyte interphase (SEI) on a high capacity conversion-based anode: NiO. *J. Mater. Chem. A* **2021**, *9*, 523–537. [CrossRef]
31. Farges, F.; Brown, G.E.; Rehr, J.J. Ti K-edge XANES studies of Ti coordination and disorder in oxide compounds: Comparison between theory and experiment. *Phys. Rev. B* **1997**, *56*, 1809–1819. [CrossRef]
32. Sasaki, S.; Fujino, K.; Takéuchi, Y. X-ray determination of electron-density distributions in oxides, MgO, MnO, CoO, and NiO, and atomic scattering factors of their constituent atoms. *Proc. Japan Acad. Ser. B* **1979**, *55*, 43–48. [CrossRef]
33. Karzel, H.; Potzel, U.; Potzel, W.; Moser, J.; Schäfer, C.; Steiner, M.; Peter, M.; Kratzer, A.; Kalvius, G. X-Ray diffractometer for high pressure and low temperatures. *Mater. Sci. Forum* **1991**, *79-82*, 419–426. [CrossRef]
34. Fjellvåg, H.; Grønvold, F.; Stølen, S.; Hauback, B. On the crystallographic and magnetic structures of nearly stoichiometric iron monoxide. *J. Solid State Chem.* **1996**, *124*, 52–57. [CrossRef]
35. Will, G.; Masciocchi, N.; Parrish, W.; Hart, M. Refinement of simple crystal structures from synchrotron radiation powder diffraction data. *J. Appl. Crystallogr.* **1987**, *20*, 394–401. [CrossRef]
36. Fleet, M.E. The structure of magnetite: Symmetry of cubic spinels. *J. Solid State Chem.* **1986**, *62*, 75–82. [CrossRef]
37. Pailhé, N.; Wattiaux, A.; Gaudon, M.; Demourgues, A. Correlation between structural features and vis–NIR spectra of α-Fe$_2$O$_3$ hematite and AFe$_2$O$_4$ spinel oxides (A=Mg, Zn). *J. Solid State Chem.* **2008**, *181*, 1040–1047. [CrossRef]
38. Thota, S.; Reehuis, M.; Maljuk, A.; Hoser, A.; Hoffmann, J.U.; Weise, B.; Waske, A.; Krautz, M.; Joshi, D.C.; Nayak, S.; et al. Neutron diffraction study of the inverse spinels Co$_2$TiO$_4$ and Co$_2$SnO$_4$. *Phys. Rev. B* **2017**, *96*, 144104. [CrossRef]
39. da Silva Pedro, S.; López, A.; Sosman, L.P. Zn$_2$TiO$_4$ photoluminescence enhanced by the addition of Cr^{3+}. *SN Appl. Sci.* **2019**, *2*. [CrossRef]

Disclaimer/Publisher's Note: The statements, opinions and data contained in all publications are solely those of the individual author(s) and contributor(s) and not of MDPI and/or the editor(s). MDPI and/or the editor(s) disclaim responsibility for any injury to people or property resulting from any ideas, methods, instructions or products referred to in the content.

Article

Carbon-Coated Si Nanoparticles Anchored on Three-Dimensional Carbon Nanotube Matrix for High-Energy Stable Lithium-Ion Batteries

Hua Fang [1,2,3,*], Qingsong Liu [1], Xiaohua Feng [1], Ji Yan [1,2,3], Lixia Wang [1,2,3], Linghao He [1,*], Linsen Zhang [1,2,3,*] and Guoqing Wang [1,*]

1. College of Material and Chemical Engineering, Zhengzhou University of Light Industry, Zhengzhou 450001, China
2. Ceramic Materials Research Center, Zhengzhou University of Light Industry, Zhengzhou 450001, China
3. Zhengzhou Key Laboratory of Green Batteries, Zhengzhou 450001, China
* Correspondence: fh@zzuli.edu.cn (H.F.); helinghao@zzuli.edu.cn (L.H.); hnlinsenzhang@163.com (L.Z.); gqwang@zzuli.edu.cn (G.W.)

Abstract: An easy and scalable synthetic route was proposed for synthesis of a high-energy stable anode material composed of carbon-coated Si nanoparticles (NPs, 80 nm) confined in a three-dimensional (3D) network-structured conductive carbon nanotube (CNT) matrix (Si/CNT@C). The Si/CNT@C composite was fabricated via in situ polymerization of resorcinol formaldehyde (RF) resin in the co-existence of Si NPs and CNTs, followed by carbonization at 700 °C. The RF resin-derived carbon shell (~10 nm) was wrapped on the Si NPs and CNTs surface, welding the Si NPs to the sidewall of the interconnected CNTs matrix to avoid Si NP agglomeration. The unique 3D architecture provides a highway for Li$^+$ ion diffusion and electron transportation to allow the fast lithiation/delithiation of the Si NPs; buffers the volume fluctuation of Si NPs; and stabilizes solid–electrolyte interphase film. As expected, the obtained Si/CNT@C hybrid exhibited excellent lithium storage performances. An initial discharge capacity of 1925 mAh g^{-1} was achieved at 0.1 A g^{-1} and retained as 1106 mAh g^{-1} after 200 cycles at 0.1 A g^{-1}. The reversible capacity was retained at 827 mAh g^{-1} when the current density was increased to 1 A g^{-1}. The Si/CNT@C possessed a high Si content of 62.8 wt%, facilitating its commercial application. Accordingly, this work provides a promising exploration of Si-based anode materials for high-energy stable lithium-ion batteries.

Keywords: lithium-ion batteries; silicon/carbon anode; carbon nanotubes; in situ polymerization; carbon coating

1. Introduction

Lithium-ion batteries (LIBs) are widely applied to power sources for portable electronic equipment, power tools, electric vehicles, and even grid-scale energy storage systems [1–3]. Advanced LIBs with high energy density, high power density, and long cycle life are urgently needed to cater to the escalating energy density/power demands of the ever-developing portable electronic equipment and electric vehicles [4,5]. As the most successful anode material for current commercial LIBs, graphite materials are approaching their capacity limits and show limited application prospects in next-generation LIBs due to their relatively low theoretical capacity of ~372 mAh g^{-1} [6,7]. To significantly improve the energy density and power density, extensive research has focused on developing next-generation anode materials, including alloy-based anode materials (Si-based [8,9], Sn-based [10,11], Ge-based [12], etc.), transition metal compounds (oxides [13], sulfide [14], etc.) and metal–organic frameworks (MOFs) and their derivatives [15]. Among all the candidates, silicon (Si) is considered a promising alternative for the currently employed graphite, as Si possesses the highest theoretical capacity (4200 mAh g^{-1} for Li$_{22}$Si$_5$ at 415 °C; 3579 mAh g^{-1}

for Li$_{15}$Si$_4$ at room temperature) and relatively low lithium storage potential range (<0.4 V vs. Li/Li$^+$) [5,16,17].

However, extreme volume expansion/contraction (∼300%) in the lithiation/delithiation process can cause Si particle pulverization, electrical contact deterioration, and solid–electrolyte interphase (SEI) instability, which further causes fast capacity decay and low Coulombic efficiency [8,18,19]. In addition, Si shows poor intrinsic electron conductivity, resulting in sluggish electrochemical kinetics and severely limiting the cycle life of the Si-based batteries [9,20]. To date, various strategies have been explored to develop the superior Si-based electrode materials, including the construction of Si nanostructures [21,22], porous Si [23–25], surface coating of protective layers [26–28], Si-based composites [17,29–31], SiO$_x$-based materials [32–34], and new electrode binders [35–39]. Of these, nano-Si materials have been shown to effectively relax the mechanical stress and strain on the electrode during volume fluctuation, avoiding cracking or pulverization of their structures, thereby significantly improving Coulombic efficiency and cyclability [30,40]. In this esteem, Si thin-film [2], Si nanowire [41], Si nanotube [21], and Si with nanoporous structure [42] have been extensively studied to improve the electrochemical performance of Si anodes. However, several challenges remain to be overcome in Si-based materials as anode material of LIBs. Firstly, nano-Si materials show a high specific surface area, leading to excessive formation of unstable SEI film and severe electrolyte decomposition [43]. Secondly, the repeated cracking/regeneration of the SEI films remains difficult to address [44]. Thirdly, Si nanoparticles (NPs) prefer to agglomerate due to their high surface energy, which makes it difficult to disperse Si NPs uniformly in the LIB electrode [45]. These challenges restrict the practical application of nano-Si anode materials.

To tackle the aforementioned problems, one of the most favorable strategies is to construct unique microstructures of Si/carbon (Si/C) composites, due to which carbon materials have the advantages of conductivity, structural stability, low cost, and ease of preparation [6,9]. The introduction of carbon can effectively enhance the rate capability and cycling performance of the Si anode. In this regard, various carbon materials such as porous carbon [46,47], carbon nanotubes (CNTs) [48,49], graphite [50,51], and graphene [52–54], have been composited with nano-Si to improve the cycling stability and rate capability of LIBs. The cross-linked CNT framework with high conductivity and mechanical flexibility can provide an electrically conductive pathway and accommodate the expansion of Si NPs [55]. Additionally, carbon coating, which aims to encapsulate Si NPs with carbon layers on their surface, can facilitate both the formation of stable SEI film and the enhancement of electronic conductivity [56]. Thus, ternary Si-based composites, which integrate Si NPs, carbon coating, and CNTs, have been constructed to produce voids between carbon shells, thereby buffering the volume change and enhancing high electrical conductivity. For example, An et al. [57] reported a three-dimensional (3D) hierarchical nano Si@C/CNT composite prepared via a self-electrostatic route, which exhibited high reversible capacity and outstanding rate performance when it was used as an anode material for LIBs. After enduring 1000 cycles at 0.5 C, it retained a high capacity of 989.5 mAh g^{-1} with high capacity retention of 86.6%. Zhang et al. [55] synthesized a Si/CNTs@(S)-C composite using spray-drying methods. The Si/CNTs@(S)-C composite delivered a high capacity of 943 mAh g^{-1} at a 0.2 C rate after 1000 cycles. Unfortunately, the recent reports on a Si-based anode usually contained a relatively low Si mass content to ensure their cycle stability, which weakens the high capacity superiority in commercial LIBs. Fabricating ternary Si/C/CNT composites with high Si content for next-generation LIBs via a low-cost and scalable route remains a significant challenge.

In this study, a 3D network structured Si/CNT@C composite anode material composed of Si NPs anchored on a conductive CNT matrix assisted by carbon coating, was designed and fabricated via a low-cost and scalable route for high-energy stable lithium-ion batteries. In situ resorcinol formaldehyde (RF) resin polymerization was carried out in the co-existence of Si NPs and CNTs. Through carbonization at 700 °C under N$_2$ atmosphere, the resulting Si/CNT@RF resin composite was transformed into Si/CNT@C composite.

The RF resin-derived carbon can act as an adhesive medium to weld Si NPs on the sidewall of CNTs, resulting in a stable 3D porous structure. Thus, the Si/CNT@C composite showed the advantages of buffering the volume fluctuation, restraining the agglomeration, facilitating fast ion/electron transportation, and stabilizing the SEI film. This unique material design enables the Si/CNT@C composite to deliver a high reversible 1106 mAh g^{-1} after 200 cycles at 0.1 A g^{-1}. The present work provides a facile and scalable strategy to construct the Si-based materials with high lithium storage capacity and excellent cycle stability.

2. Experimental Section

2.1. Chemicals and Reagents

All the chemicals were analytically pure and were used as received. Si powder with a particle size of 50~100 nm was purchased from Shanghai Pantian Powder Material Co., Ltd. (Shanghai, China). CNTs with an outer diameter of ~50 nm were purchased from Chengdu Organic Chemistry Co., Ltd. (Chengdu, China).

2.2. Material Preparation

2.2.1. Pretreatment of Si NPs and CNTs

The Si NPs were first pretreated in a mixed solution of ammonia (25 wt%), hydrogen peroxide (30 wt%), and deionized water with a volume ratio of 1:1:5 at 80 °C for 1 h under magnetic stirring. The pretreated Si NPs were collected via filtration, washed with distilled water, and dried in an oven at 60 °C for 12 h.

The CNTs were pretreated via refluxing in concentrated nitric acid for 6 h. After reflux, the oxidized CNTs were obtained via filtration, washed with deionized water, and finally dried in an oven at 60 °C for 12 h.

2.2.2. Preparation of Si/CNT@RF Resin Composites

For typical preparation, 0.3 g of pretreated Si NPs and 0.92 g of CTAB were ultrasonically dispersed into 28 mL deionized water for 30 min to form a homogeneous suspension. Then, ethanol (11 mL), resorcinol (0.28 g), ammonia (0.1 mL, 25 wt%), and pretreated CNTs (33.3 mg) were added and sonicated at room temperature for 30 min. Thirdly, 0.4 mL formaldehyde (37~40 wt%) was added to the homogeneous suspension. The in situ polymerization of RF resin was achieved through continuous stirring at 35 °C for 6 h and aging at room temperature for 12 h. Finally, the Si/CNT@RF resin composite was obtained via filtration, washed with deionized water and ethanol, and dried in an oven at 60 °C overnight.

2.2.3. Preparation of Si/CNT@C Composites

As for the preparation of the Si/CNT@C composite, the as-prepared Si/CNT@RF resin composite was heated under a N2 atmosphere at 700 °C for 3 h with a heating rate of 2 °C/min. Afterward, the tubular furnace was cooled naturally to room temperature. For comparison, the Si@C composite was also prepared with the same preparation process as the Si/CNT@C composite, but without adding CNTs in the in situ polymerization process.

2.3. Material Characterization

The crystal structures were determined via X-ray diffraction (XRD, Bruker D8 Advance, using Cu Kα) at a scan rate of 5 °C min^{-1}. The morphologies were characterized by field-emission scanning electron microscopy (FE-SEM, JEOL JSM-7001F). Transmission electron microscopy (TEM, JEOL JEM-2100) was employed to characterize the microstructures. The weight contents of Si NPs in composites were determined by thermogravimetric analysis under an air atmosphere (TGA, STA 449) with a heating rate of 10 °C min^{-1}. Raman spectra were performed on a Raman spectrometer system (HORIBA Scientific LabRAM HR Evolution) with an excitation wavelength of 532 nm. X-ray photoelectron spectroscopy (XPS) of the powders was performed using an ESCALAB-250Xi with Al Kα radiation.

2.4. Electrochemical Characterization

The Si electrode, Si@C electrode and Si/CNT@C electrode were prepared via the same process with the bare Si power, Si@C, and Si/CNT@C as active materials, respectively. The electrode slurry was made by mixing the active material, conductive agent (Super P), and binder in a mass ratio of 8:1:1 with deionized water as a solvent. The binder was composed of styrene-butadiene rubber (SBR) and carboxymethyl cellulose (CMC) with a mass ratio of 3:2. The electrode slurry was pasted on copper foil using an automatic coating machine and vacuum dried at 120 °C for 8 h. The coating thickness of the slurry was set as 100 μm for the Si, Si@C and Si/CNT@C electrodes, resulting in comparable mass loading and thickness for all three kinds of electrodes. The obtained electrode film was punched into disc electrodes with a diameter of 14 mm and pressed under 15 MPa. Based on the mass loading of the Si/CNT@C electrode (~0.65 mg cm^{-1}), the mass content of active material (80%), and the bare Si mass ratio in the Si/CNT@C composite (62.8%), the mass loading of bare Si NPs for the Si/CNT@C electrode can be calculated as ~0.65 × 80% × 62.8% = ~0.33 mg cm^{-1}.

Electrochemical measurements were conducted at room temperature on CR2016 coin-type cells, which were fabricated in an argon-filled glove box (MB-10-G, MBRAUN). Lithium metal was used as the counter/reference electrode, Celgard 2400 membrane as the separator, and 1 M LiPF$_6$ in a 1:1:1 (volume ratio) mixture of ethylene carbonate (EC), dimethyl carbonate (DMC), and ethyl methyl carbonate (EMC) as the electrolyte solution. Cyclic voltammetry (CV) tests were carried out on a CHI 660E electrochemical workstation at a scanning rate of 0.1 mV s^{-1} between 1.5 and 0.01 V (vs. Li$^+$/Li). Galvanostatic charge/discharge (GCD) tests were performed at a cell voltage of 1.5–0.01 V (vs. Li$^+$/Li) on a Neware CT-4008T battery tester at different current rates. The GCD test currents was calculated based on the total mass active material. The specific capacity was obtained from $C_m = It/m$, where I represents the charge/discharge current (A); t is the charge/discharge time (s), and m is the total mass of the active material (g). As for the si/CNT@C electrode, the total mass of active material includes the carbon coating shell, CNT matrix, and Si NPs of the composite. Electrochemical impedance spectroscopy (EIS) tests were performed on a CHI 660E electrochemical workstation using newly installed coin-type cells (discharged state). The frequency range and voltage amplitude were 10^5~0.01 Hz and 5 mV, respectively.

3. Results and Discussions

3.1. Morphology, Microstructure and Crystalline Phase Studies

As illustrated in Figure 1a, the Si/CNT@RF resin composite was prepared through in situ polymerization of RF resin in the co-existence of the pretreated Si NPs and CNTs. The hydrophilic treated Si NPs exhibited plenty of Si-O-H surface functional groups after being dissolved in the water–ethanol–ammonia solution [58], while the acid-treated CNTs also possessed rich surface oxygen functional groups [59]. There is a strong hydrogen bonding force between the surface oxygen functional groups, facilitating the fine interface contact among Si NPs, CNTs, and RF resin (shown in Figure 1a). As a result, the RF resin grows directly on the surface of Si NPs and CNT in the in situ polymerization process, thereby forming a uniform RF resin coating shell on the surface of Si NPs and CNTs. During the in situ polymerization procedure, CTAB was used as a surfactant, which can improve the dispersion of Si NPs in the reaction solution and act as the soft template for the formation of a mesostructured RF resin coating shell [60]. After carbonization, the brown Si/CNT@RF resin composite was converted to black Si/CNT@C (Figure S1).

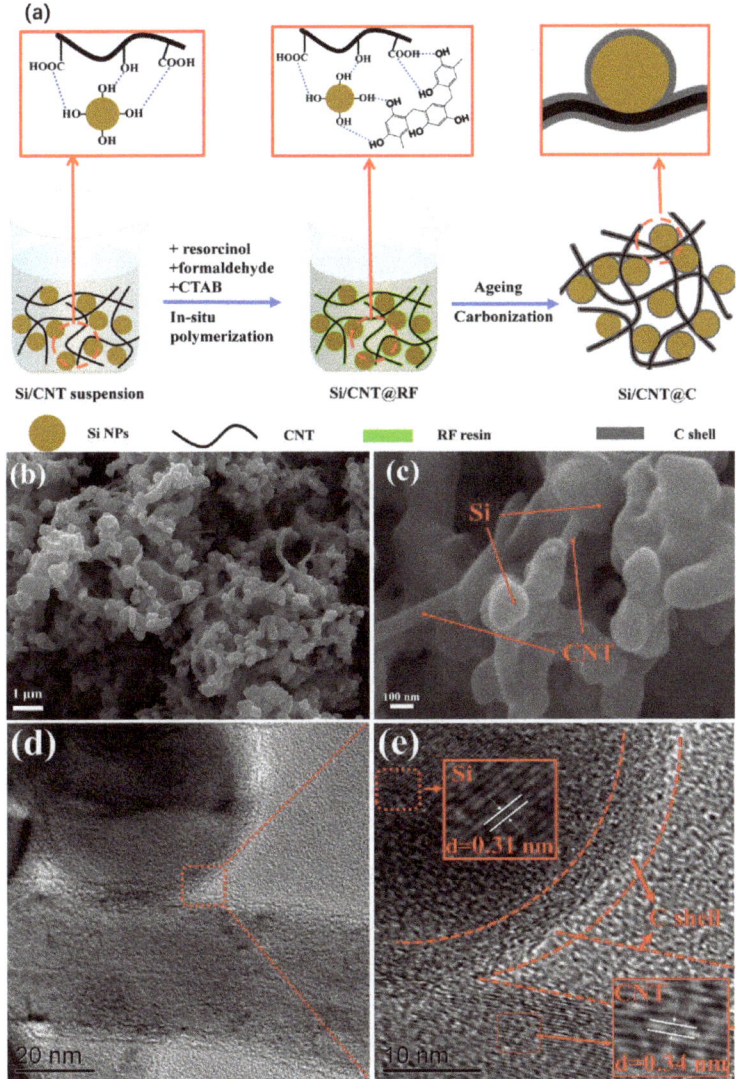

Figure 1. (**a**) Schematic illustration of the fabrication process, (**b**,**c**) SEM images, (**d**) TEM image, and (**e**) HR-TEM image of Si/CNT/C composite.

Figure 1b,c show the FE-SEM of Si/CNT@C composite, which exhibits a 3D nanoporous morphology. It can be clearly observed that Si NPs were well dispersed in the 3D interconnected CNT matrix without aggregation. The microstructure of Si/CNT@C composites was further characterized by TEM and HR-TEM images. As shown in Figure 1d, the RF resin-derived carbon shell, about ~10 nm thick, was uniformly coated on the surface of Si NPs and CNTs. The clear lattice fringe with spacing of 0.31 and 0.34 nm indexed to the (111) plane of the Si NPs and (002) plane of the CNT matrix appears in the high-resolution TEM (HR-TEM) image (Figure 1e) [61]. Here, the carbon coating shell used an adhesive medium, which attached Si NPs to the CNT sidewall and inhibited the aggregation of the Si NPs. Meanwhile, CNTs acted as "bridges" to string the isolated Si NPs together and provide a fast electron transport highway. Different from the Si NPs in the Si/CNT@C composite anchored on the CNT sidewall without agglomeration, the Si NPs both in the bare Si sample and Si@C composite showed a certain level aggregate phenomenon due to their high surface energy (Figure S2). This result reveals that the presence of CNTs in the Si/CNT@C composite effectively avoids the agglomeration of Si NPs.

The XRD patterns of all the samples exhibit six diffraction peaks at 28.4°, 47.2°, 56.2°, 69.2°, 75.5°, and 88.1°, corresponding to (111), (220), (311), (400), (331), and (442) plane of the Si crystal (JCPDS no. 27-1402) [56,62]. In addition, a wide and weak peak centered at 24.5° can be observed in the Si@C and Si/CNT@C samples, which can be indexed to the (002) plane of the amorphous carbon coating layer and CNT matrix. Compared with the pattern of Si@C, the (002) peak of Si/CNT@C is much clearer due to the existence of CNTs. The existence of amorphous carbon and CNTs can be further confirmed by the Raman spectrum. As shown in Figure 2b, peaks at the wavenumbers of 136, 289, 509, and 924 cm^{-1} can be attributed to elemental Si [1,57,63]. The peaks at 1350 and 1590 cm^{-1} are indexed to the characteristic D and G peaks of carbon material, respectively. The D band is associated with the disordered carbon structure (sp^3-hybridized carbon), while the G band is attributed to an ordered graphitic lattice (E$_{2g}$ phonon of sp^2-hybridized carbon) [62,64]. The relative intensity (I_D/I_G) of the D and G band is commonly used to detect the graphitization degree of carbon materials. The increase in I_D/I_G means the materials are more amorphous [65]. The I_D/I_G of the Si@C sample is 0.98, much larger than that of the Si/CNT@C sample (0.91), revealing that the RF resin-derived carbon coating is amorphous. This is consistent with the XRD results mentioned above.

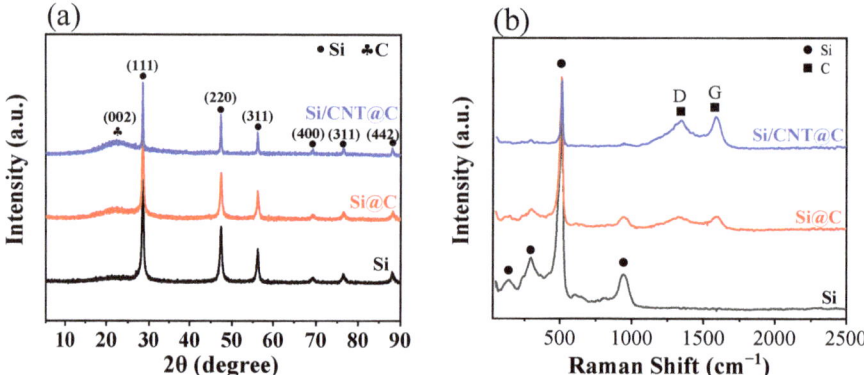

Figure 2. (**a**) XRD patterns and (**b**) Raman spectra of the samples.

The chemical state and molecular environment of the Si/CNT@C sample are further characterized by XPS spectra. As shown in Figure 3a, the signals of Si 2p (102.9 eV), C 1s (285.5), and O 1s (533.0) can be observed in XPS survey spectrum of the Si/CNT@C. The high-resolution XPS spectra of Si, C, and O elements contained in the Si/CNT@C composite were deconvoluted using software named XPSPEAK41, a baseline type of Shirley, and a Lorentzian–Gaussian distribution. The Si 2p XPS spectrum (Figure 3b) can be divided into five peaks, which are attributed to the Si–Si (99.1 and 99.8 eV) and Si–O (100.9, 103.1 and 103.9 eV) bonds, respectively [66]. The Si–O peak originated from the SiO$_2$ and SiO layer on the surface of Si NPs [67]. The surface of Si NPs is usually oxidized to form SiO$_x$, including SiO and SiO$_2$. The existence of SiO$_x$ in the Si-based anode materials has been reported by some recent studies [1,43,66]. The C 1s spectrum in Figure 3c can be fitted to three peaks at 284.8, 286.5, and 289.0 eV, which are assigned to the C–C, C–O, and O–C=O bonds [66,68]. The O1s peak can be deconvoluted into Si–O (532.4 eV) and O–C=O (533.6 eV) bonds [69]. To acquire high specific capacity, it is vital to raise the Si mass content in the composite electrodes [1]. The mass contents of Si NPs were estimated according to the TG plots of Si, Si@C, and Si/CNT@C samples (Figure S3). As exhibited in Table S2, the Si/CNT@C composite contains a very high Si content of 62.8 wt%, which greatly exceeds that of most recently reported Si-based anode materials, such as SMPS−1 (12.37 wt%) [70], GSCC (39.75 wt%) [66], Si/CNTs/C (40.4 wt%) [71], C/Si/CNT (42.17 wt%) [72], and Si@C@v@CNTs (14.1 wt%) [73]. Due to the high Si content of the as-prepared Si/CNT@C composite, high energy density can be expected to facilitate its commercial application.

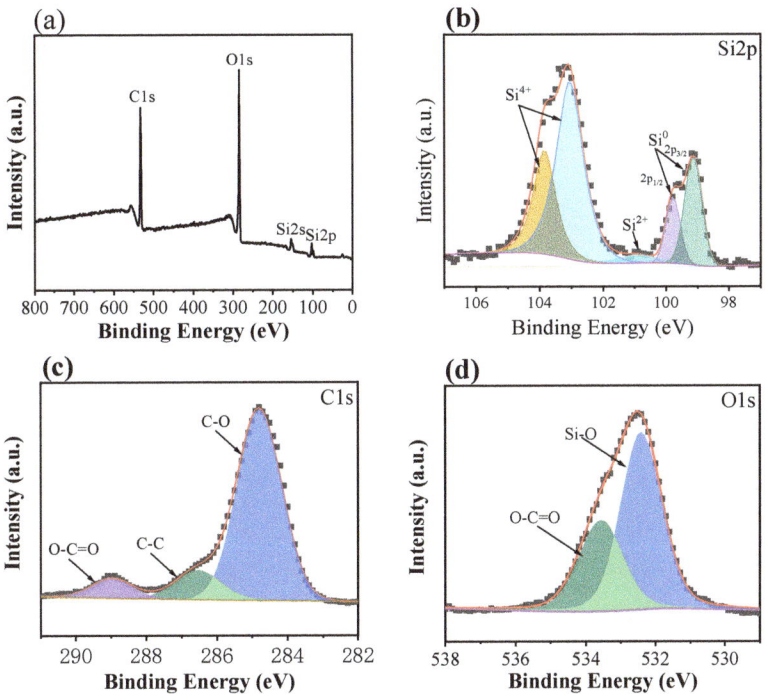

Figure 3. (**a**) XPS spectra, (**b**) Si2p, (**c**) C1s, and (**d**) O1s of Si/CNT@C sample.

3.2. Electrochemical Studies

To characterize the electrochemical behaviors of the Si/CNT@C composite, CV tests were first performed at a scan rate of 0.1 mV s^{-1} with a potential range of 0.01~1.5 V vs. Li/Li$^+$. As shown in Figure 4a, a weak and broad peak centered at 0.8 V in the first cathodic scan is ascribed to the decomposition of electrolyte and the formation of SEI film [43,55]. Such an irreversible peak did not show up in the subsequent cathodic scans, indicating that stable SEI film was achieved in the first cathodic CV scan [74]. The sharp peak below 0.1 V in the first cathodic scan corresponded to the lithiation of silicon to Li$_x$Si [75]. In the first anodic scan, the oxidation peaks at 0.32 and 0.47 V are recognized as the delithiation process of the Li$_x$Si phase to amorphous Si. The following CV scans show two reduction peaks at 0.22 and below 0.1 V, and two oxidation peaks at 0.32 and 0.47 V, corresponding to the reversible lithiation/delithiation behavior of amorphous silicon [76]. As the number of cycles increases, the peak intensity of the CV curves gradually increases, revealing an activation process [69,77]. The bare Si and Si@C electrodes show similar CV behaviors (Figure S4a,b). In contrast, the Si/CNT@C electrode showed the largest response current among the three kinds of electrodes, demonstrating the best lithium storage performance (Figure S4c).

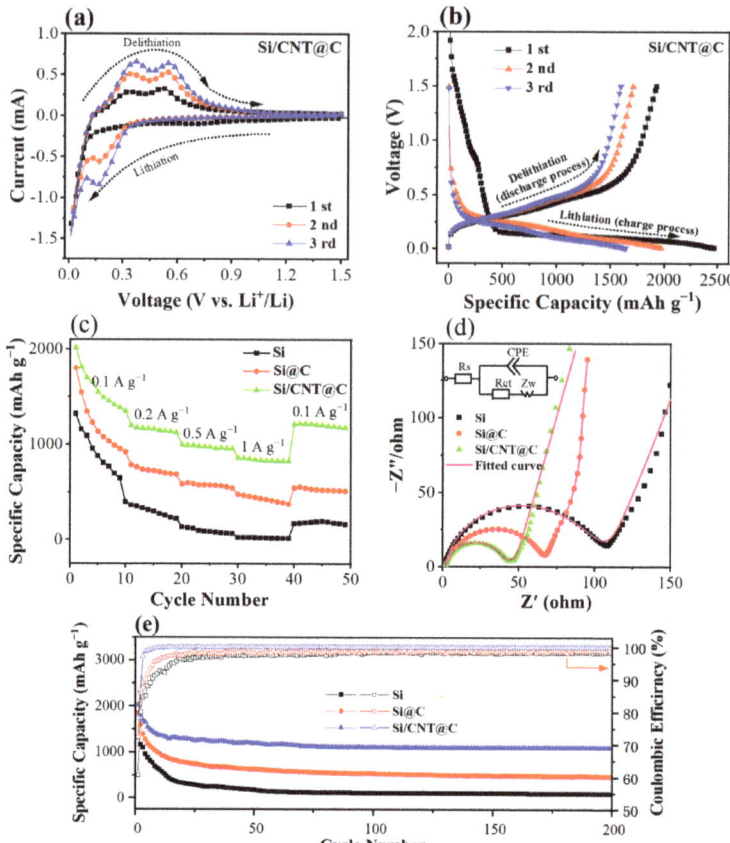

Figure 4. (**a**) CV curves of Si/CNT/C electrode at a scan rate of 0.1 mV s^{-1}. (**b**) Charge/discharge curves of Si/CNT/C electrode. (**c**) Rate performances of the bare Si, Si/C, and Si/CNT/C electrodes. (**d**) EIS curves (inset shows equivalent circuit) of the bare Si, Si/C, and Si/CNT/C electrodes. (**e**) Cycle performances of the bare Si, Si/C and Si/CNT/C electrodes.

The first three GCD curves of the Si/CNT@C, Si and Si@C electrodes at 0.1 A g^{-1} are exhibited in Figure 4b, Figure S5a and Figure S5b, respectively. In the charge curves of the first GCD cycles, all the electrodes show slope charge plateaus between 1.5 V and 0.2 V and voltage plateaus at 0.1 V, which correspond to the irreversible SEI formation and lithiation of silicon to Li$_x$Si [43]. The slope charge plateau between 1.5 V and 0.2 V disappears in the charge curves of the following GCD cycles, indicating that SEI formation is completed in the first cycle. As for the discharge curves, the discharging plateaus are observed at 0.1 V, which corresponds to the delithiation process of the Li$_x$Si alloy. These results are consistent with the CV scanning of the Si/CNT@C electrode (Figure 4a). The initial discharge capacities were 1322, 1802, and 1925 mAh g^{-1}, and the initial Coulomb efficiencies were 61 %, 72%, and 78% for the Si, Si@C, and Si/CNT@C electrodes, respectively. The low initial Coulomb efficiencies are caused by irreversible formation of the SEI layer [78]. The Si/CNT@C electrode showed the largest discharge capacity and the highest Coulomb efficiency, which may be attributed to the synergistic effects of the carbon coating and CNT matrix.

Furthermore, the rate performances were investigated by GCD tests at different current densities. As shown in Figure 4c, the Si/CNT@C exhibits the best rate capability at all current rates compared with those of the Si and Si@C. The Si/CNT@C showed a high charge capacity of 1387 mAh g^{-1} after 10 cycles at 0.1 A g^{-1}, which is higher than those of the Si (648 mAh g^{-1}) and Si@C (956 mAh g^{-1}). When the current rate increased to 1 A g^{-1}, the Si/CNT@C exhibited a high capacity of 827 mAh g^{-1}, which is obviously much higher than those of the Si (15 mAh g^{-1}) and Si@C (378 mAh g^{-1}). The great improved rate performance of the Si/CNT@C can be attributed to the high conductivity of the CNT matrix and the fast Li$^+$ ion transmission in the 3D porous structure. When the current density is recovered from 1 to 0.1 A g^{-1}, the capacity of the Si/CNT@C electrode recovers to 852 mAh g^{-1}, which is much higher than the respective capacities of Si (172 mAh g^{-1}) and Si/C (520 mAh g^{-1}), indicating the good reversibility of the Si/CNT@C electrode.

The remarkable discharge capacitance of Si/CNT/C can be ascribed to the synergetic effects of the CNTs and RF resin-derived carbon coating. Firstly, Si NPs were uniformly dispersed in the 3D interconnected CNT matrix and were attached to the CNT sidewall by the amorphous carbon coating shell, forming a 3D nanoporous structured Si/CNT@C composite. The as-developed 3D nanoporous structure facilitates the inhibition of agglomeration of Si NPs, accelerates Li$^+$ ion transportation, and buffers the volume expansion effect of Si NPs during the delithiation/lithiation process. Secondly, isolated Si NPs were stringed by the CNT matrix, providing a fast electron transport route. Thirdly, the uniform carbon coating layer on the surface of Si NPs and CNTs can restrain the volume expansion effect of Si NPs and thus facilitate the formation of stable SEI film. Based on the above analysis, the Si/CNT@C electrode deserves the greatly enhanced lithium storage capacity and rate capability.

EIS Nquist plots were carried out to eluate the conductivity of the samples. As shown in Figure 4d, all the Nyquist curves reveal a semicircle in the mid-frequency region and a straight line in the low-frequency region, which can be simulated by the inserted equivalent circuit. The R_s, R_{ct}, CPE, and W represent the ohmic resistance of electrolyte and electrode, charge transfer resistance, the constant phase element related to electric double-layer capacitance, and the Warburg impedance, respectively [1,56,62]. As revealed by the fitting results shown in Table S1, the Si/CNT@C electrode shows the smallest R_s (1.6 Ω) and R_{ct} (38.5 Ω), which are much smaller than those of the Si (R_s 2.0 Ω and R_{ct} 92.5 Ω) and Si@C (R_s 1.7 Ω and R_{ct} 62.4 Ω), indicating the significantly enhanced Li$^+$ ion and electron transportation of the Si/CNT@C electrode. Such significant improvement arises from the synergetic effects of the CNTs and RF resin-derived carbon coating, which enhance the electron transportation between Si NPs and CNTs. The 3D nanoporous structure derived from the CNT matrix also facilitates fast Li$^+$ ion transportation, which contributes to a rapid de-alloying reaction between Li and Si [75].

Figure 4e shows the cycling stability of the bare Si, Si@C, and Si/CNT@C electrodes at a current rate of 0.1 A g^{-1}. The Si/CNT@C electrode can retain a high specific capacity of 1106 mAh g^{-1} after 200 cycles and its Coulomb efficiency rapidly increased above 99% after the third cycle. In contrast, the Si@C electrode can only retain a relatively low specific capacity of 485 mAh g^{-1} after 200 cycles. As for the bare Si electrode, the specific capacity dramatically decays to almost zero after 200 cycles. Figure S6 shows that the Si/CNT@C electrode can retain a high specific capacity of 770 mAh g^{-1} after cycling at 0.5 A g^{-1} for 200 cycles, further proving its application prospects. As exhibited by Table S2, the Si/CNT@C electrode shows competitively high specific capacity, long cycle stability, and rate capability compared with the recently reported Si-based anode materials, such as SMPS−1 (501 mAh g^{-1} at 1 A g^{-1} after 500 cycles) [70], GSCC (837.3 mAh g^{-1} at 0.2 A g^{-1} after 100 cycles) [66], Si/CNTs/C (702 mAh g^{-1} at 0.2 A g^{-1} after 300 cycles) [71], C/Si/CNT (696.8 mAh g^{-1} at 0.1A g^{-1} after 50 cycles) [72], Si@C@v@CNTs (912.8 mAh g^{-1} at 0.1 A g^{-1} after 100 cycles) [73], and Si/CNTs@PMMA−C (1024.8 mAh g^{-1} at 0.2 A g^{-1} after 200 cycles) [78]. The significantly enhanced cycle performance can be attributed to the unique structure advantages of the Si/CNT@C electrode, which can effectively buffer the expansion effect of Si NPs during the delithiation/lithiation process.

4. Conclusions

In this study, a 3D network structured Si/CNT@C composite anode material was synthesized via simple in situ polymerization and the following carbonization strategy. In the as-prepared Si/CNT@C composite, Si NPs were anchored on the surface of CNT, with uniform carbon shells coated on both surfaces of Si NPs and CNT. This unique 3D nanoporous structure, with interconnected CNTs as a supporting matrix, alleviated the agglomeration/volume fluctuation of Si NPs and thereby facilitated fast ion/electron transportation during charge/discharge cycles. As a result, the Si/CNT@C sample retained a high specific capacity of 1106 mAh g^{-1} and 770 mAh g^{-1} after 200 charge/discharge cycles at 0.1 A g^{-1} and 0.5 A g^{-1}, respectively. When the current density increased to 1 A g^{-1}, the Si/CNT@C electrode exhibited a relatively high capacity of 827 mAh g^{-1}. Compared with the bare Si NPs and Si@C samples, the Si/CNT@C exhibited significantly improved cycle stability and rate capability. Considering the easy synthetic route and high Si mass content, the Si/CNT@C composite anode material reported here paves a new way for the industrial application of Si-based anode materials in high-energy LIBs.

Supplementary Materials: The following supporting information can be downloaded at: https://www.mdpi.com/article/10.3390/batteries9020118/s1. Figure S1: Photos of the as-prepared Si/C@RF and Si/CNT@C powder samples. Figure S2: (a,b) SEM images of Si NPs, (c,d) SEM images, and (e,f) TEM images of Si@C composite. Figure S3: Thermogravimetric curves of Si, Si@C and Si/CNT@C samples. Figure S4: (a) CV curves of Si, (b) CV curves of Si@C, and (c) comparison of the third cycle CV curves of the Si, Si@C, and Si/CNT@C. Figure S5: GCD curves of Si (a) and Si@C (b). Figure S6: Cycle performances of the Si/CNT/C electrode at 0.5 A g−1. Table S1: Fitting results of the EIS curves. Table S2: Comparison among the recently reported Si-based anode materials [79,80].

Author Contributions: Conceptualization, H.F. and X.F.; methodology, H.F. and X.F.; software, X.F. and Q.L.; validation, H.F., X.F. and Q.L.; formal analysis, X.F. and Q.L.; investigation, X.F. and Q.L.; resources, X.F. and Q.L.; data curation, X.F. and Q.L.; writing—original draft preparation, H.F. and X.F.; writing—review and editing, L.H., J.Y. and L.W.; visualization, H.F., X.F. and Q.L.; supervision, L.Z. and G.W.; project administration, H.F., L.Z. and G.W.; funding acquisition, H.F., L.Z. and G.W. All authors have read and agreed to the published version of the manuscript.

Funding: This research was funded by Key Scientific Research Project in Colleges and Universities of Henan Province of China (grant number 20B530006) sponsored by Henan Provincial Department of Education, China, and the Science and Technology Project of Henan Province (grant number 222102240122) sponsored by Henan Provincial Department of Science and Technology, China.

Institutional Review Board Statement: Not applicable.

Informed Consent Statement: Not applicable.

Data Availability Statement: Not applicable.

Conflicts of Interest: The authors declare no conflict of interest.

References

1. Ma, T.; Xu, H.; Yu, X.; Li, H.; Zhang, W.; Cheng, X.; Zhu, W.; Qui, X. Lithiation Behavior of Coaxial Hollow Nanocables of Carbon-Silicon Composite. *ACS Nano* **2019**, *13*, 2274–2280. [CrossRef] [PubMed]
2. Salah, M.; Murphy, P.; Hall, C.; Francis, C.; Kerr, R.; Fabretto, M. Pure Silicon Thin-Film Anodes for Lithium-Ion Batteries: A Review. *J. Power Sources* **2019**, *414*, 48–67. [CrossRef]
3. Wang, F.; Lin, S.; Lu, X.; Hong, R.; Liu, H. Poly-Dopamine Carbon-Coated Stable Silicon/Graphene/CNT Composite as Anode for Lithium Ion Batteries. *Electrochim. Acta* **2022**, *404*, 139708. [CrossRef]
4. Wang, H.W.; Fu, J.Z.; Wang, C.; Wang, J.Y.; Yang, A.K.; Li, C.C.; Sun, Q.F.; Cui, Y.; Li, H.Q. A Binder-Free High Silicon Content Flexible Anode for Li-Ion Batteries. *Energy Environ. Sci.* **2020**, *13*, 848–858. [CrossRef]
5. Ko, M.; Chae, S.; Ma, J.; Kim, N.; Lee, H.-W.; Cui, Y.; Cho, J. Scalable Synthesis of Silicon-Nanolayer-Embedded Graphite for High-Energy Lithium-Ion Batteries. *Nat. Energy* **2016**, *1*, 16113. [CrossRef]
6. Lv, X.; Wei, W.; Huang, B.; Dai, Y. Achieving High Energy Density for Lithium-Ion Battery Anodes by Si/C Nanostructure Design. *J. Mater. Chem. A* **2019**, *7*, 2165–2171. [CrossRef]

7. Zhang, L.; Wang, C.; Dou, Y.; Cheng, N.; Cui, D.; Du, Y.; Liu, P.; Al-Mamun, M.; Zhang, S.; Zhao, H. A Yolk-Shell Structured Silicon Anode with Superior Conductivity and High Tap Density for Full Lithium-Ion Batteries. *Angew. Chem. Int. Ed.* **2019**, *58*, 8824–8828. [CrossRef]
8. Zhang, C.; Wang, F.; Han, J.; Bai, S.; Tan, J.; Liu, J.; Li, F. Challenges and Recent Progress on Silicon-Based Anode Materials for Next-Generation Lithium-Ion Batteries. *Small Struct.* **2021**, *2*, 2100009. [CrossRef]
9. Wang, Z.; Jing, L.; Zheng, X.; Xu, Y.; Yuan, Y.; Liu, X.; Fu, A.; Guo, Y.-G.; Li, H. Microspheres of Si@Carbon-CNTs Composites with a Stable 3D Interpenetrating Structure Applied in High-Performance Lithium-Ion Battery. *J. Colloid Interface Sci.* **2022**, *629*, 511–521. [CrossRef]
10. Wei, R.; Liu, X.Q.; Tian, Y.; Huang, F.F.; Xu, S.Q.; Zhang, J.J. Sn-Based Glass-Graphite-Composite as a High Capacity Anode for Lithium-Ion Batteries. *J. Am. Ceram. Soc.* **2023**, *106*, 330–338. [CrossRef]
11. Nurpeissova, A.; Adi, A.; Aishova, A.; Mukanova, A.; Kim, S.S.; Bakenov, Z. Synergistic Effect of 3D Current Collector Structure and Ni Inactive Matrix on the Electrochemical Performances of Sn-Based Anodes for Lithium-Ion Batteries. *Mater. Today Energy* **2020**, *16*, 100397. [CrossRef]
12. Chen, Y.; Zou, Y.M.; Shen, X.P.; Qiu, J.X.; Lian, J.B.; Pu, J.R.; Li, S.; Du, F.H.; Li, S.Q.; Ji, Z.Y.; et al. Ge Nanoparticles Uniformly Immobilized on 3D Interconnected Porous Graphene Frameworks as Anodes for High-Performance Lithium-Ion Batteries. *J. Energy Chem.* **2022**, *69*, 161–173. [CrossRef]
13. Wang, Y.R.; Zhuang, Q.F.; Li, Y.; Hu, Y.L.; Liu, Y.Y.; Zhang, Q.B.; Shi, L.; He, C.X.; Zheng, X.; Yu, S.H. Bio-Inspired Synthesis of Transition-Metal Oxide Hybrid Ultrathin Nanosheets for Enhancing the Cycling Stability in Lithium-Ion Batteries. *Nano Res.* **2022**, *15*, 5064–5071. [CrossRef]
14. Hou, T.; Liu, B.; Sun, X.; Fan, A.; Xu, Z.; Cai, S.; Zheng, C.; Yu, G.; Tricoli, A. Covalent Coupling-Stabilized Transition-Metal Sulfide/Carbon Nanotube Composites for Lithium/Sodium-Ion Batteries. *ACS Nano* **2021**, *15*, 6735–6746. [CrossRef]
15. Shen, M.; Ma, H. Metal-Organic Frameworks (MOFs) and Their Derivative as Electrode Materials for Lithium-Ion Batteries. *Coord. Chem. Rev.* **2022**, *470*, 214715. [CrossRef]
16. Li, P.; Hwang, J.-Y.; Sun, Y.-K. Nano/Microstructured Silicon-Graphite Composite Anode for High-Energy-Density Li-Ion Battery. *ACS Nano* **2019**, *13*, 2624–2633. [CrossRef]
17. Fang, S.; Shen, L.; Li, S.; Kim, G.-T.; Bresser, D.; Zhang, H.; Zhang, X.; Maier, J.; Passerini, S. Alloying Reaction Confinement Enables High-Capacity and Stable Anodes for Lithium-Ion Batteries. *ACS Nano* **2019**, *13*, 9511–9519. [CrossRef]
18. Zhang, Y.; Li, B.; Tang, B.; Yao, Z.; Zhang, X.; Liu, Z.; Gong, R.; Zhao, P. Mechanical Constraining Double-Shell Protected Si-Based Anode Material for Lithium-Ion Batteries with Long-Term Cycling Stability. *J. Alloys Compd.* **2020**, *846*, 156437. [CrossRef]
19. Zhao, Z.; Cai, M.; Zhao, H.; Ma, Q.; Xie, H.; Xing, P.; Zhuang, Y.X.; Yin, H. Harvesting Si Nanostructures and C-Si Composites by Paired Electrolysis in Molten Salt: Implications for Lithium Storage. *ACS Appl. Nano Mater.* **2022**, *5*, 3781–3789. [CrossRef]
20. Chae, S.; Choi, S.-H.; Kim, N.; Sung, J.; Cho, J. Integration of Graphite and Silicon Anodes for the Commercialization of High-Energy Lithium-Ion Batteries. *Angew. Chem. Int. Ed.* **2020**, *59*, 110–135. [CrossRef]
21. Zhao, J.; Wei, W.; Xu, N.; Wang, X.; Chang, L.; Wang, L.; Fang, L.; Le, Z.; Nie, P. Dealloying Synthesis of Silicon Nanotubes for High-Performance Lithium Ion Batteries. *ChemPhysChem* **2022**, *23*, e202100832. [CrossRef] [PubMed]
22. Wang, F.; Ma, Y.; Li, P.; Peng, C.; Yin, H.; Li, W.; Wang, D. Electrochemical Conversion of Silica Nanoparticles to Silicon Nanotubes in Molten Salts: Implications for High-Performance Lithium-Ion Battery Anode. *ACS Appl. Nano Mater.* **2021**, *4*, 7028–7036. [CrossRef]
23. Yan, Z.; Jiang, J.; Zhang, Y.; Yang, D.; Du, N. Scalable and Low-Cost Synthesis of Porous Silicon Nanoparticles as High-Performance Lithium-Ion Battery Anode. *Mater. Today Nano* **2022**, *18*, 100175. [CrossRef]
24. Chen, Y.; Liu, L.F.; Xiong, J.; Yang, T.Z.; Qin, Y.; Yan, C.L. Porous Si Nanowires from Cheap Metallurgical Silicon Stabilized by a Surface Oxide Layer for Lithium Ion Batteries. *Adv. Funct. Mater.* **2015**, *25*, 6701–6709. [CrossRef]
25. Pathak, A.D.; Chanda, U.K.; Samanta, K.; Mandal, A.; Sahu, K.K.; Pati, S. Selective Leaching of Al from Hypereutectic Al-Si Alloy to Produce Nano-Porous Silicon (NPs) Anodes for Lithium Ion Batteries. *Electrochim. Acta* **2019**, *317*, 654–662. [CrossRef]
26. Huang, X.; Ding, Y.C.; Li, K.L.; Guo, X.Y.; Zhu, Y.; Zhang, Y.X.; Bao, Z.H. Spontaneous Formation of the Conformal Carbon Nanolayer Coated Si Nanostructures as the Stable Anode for Lithium-Ion Batteries from Silica Nanomaterials. *J. Power Sources* **2021**, *496*, 229833. [CrossRef]
27. Hailu, A.G.; Ramar, A.; Wang, F.M.; Wu, N.L.; Yeh, N.H.; Hsu, C.C.; Chang, Y.J.; Tiong, P.W.L.; Yuwono, R.A.; Khotimah, C.; et al. Investigations of Intramolecular Hydrogen Bonding Effect of a Polymer Brush Modified Silicon in Lithium-Ion Batteries. *Adv. Mater. Interfaces* **2022**, *9*, 2102007. [CrossRef]
28. Yu, C.H.; Chen, X.; Xiao, Z.X.; Lei, C.; Zhang, C.X.; Lin, X.Q.; Shen, B.Y.; Zhang, R.F.; Wei, F. Silicon Carbide as a Protective Layer to Stabilize Si-Based Anodes by Inhibiting Chemical Reactions. *Nano Lett.* **2019**, *19*, 5124–5132. [CrossRef]
29. Gao, X.; Lu, W.Q.; Xu, J. Unlocking Multiphysics Design Guidelines on Si/C Composite Nanostructures for High-Energy-Density and Robust Lithium-Ion Battery Anode. *Nano Energy* **2021**, *81*, 105591. [CrossRef]
30. Bitew, Z.; Tesemma, M.; Beyene, Y.; Amare, M. Nano-Structured Silicon and Silicon Based Composites as Anode Materials for Lithium Ion Batteries: Recent Progress and Perspectives. *Sustain. Energy Fuels* **2022**, *6*, 1014–1050. [CrossRef]
31. She, Z.M.; Uceda, M.; Pope, M.A. Encapsulating a Responsive Hydrogel Core for Void Space Modulation in High-Stability Graphene-Wrapped Silicon Anodes. *ACS Appl. Mater. Interfaces* **2022**, *14*, 10363–10372. [CrossRef] [PubMed]

32. Yang, Z.X.; Du, Y.; Yang, Y.J.; Jin, H.C.; Shi, H.B.; Bai, L.Y.; Ouyang, Y.G.; Ding, F.; Hou, G.L.; Yuan, F.L. Large-Scale Production of Highly Stable Silicon Monoxide Nanowires by Radio-Frequency Thermal Plasma as Anodes for High-Performance Li-Ion Batteries. *J. Power Sources* **2021**, *497*, 229906. [CrossRef]
33. Chen, S.Y.; Xu, Y.N.; Du, H.B. One-Step Synthesis of Uniformly Distributed SiO_x-C Composites as Stable Anodes for Lithium-Ion Batteries. *Dalton Trans.* **2022**, *51*, 11909–11915. [CrossRef] [PubMed]
34. Gu, H.T.; Wang, Y.; Zeng, Y.; Yu, M.; Liu, T.; Chen, J.; Wang, K.; Xie, J.Y.; Li, L.S. Boosting Cyclability and Rate Capability of SiO_x via Dopamine Polymerization-Assisted Hybrid Graphene Coating for Advanced Lithium-Ion Batteries. *ACS Appl. Mater. Interfaces* **2022**, *14*, 17388–17395. [CrossRef] [PubMed]
35. Di, S.H.; Zhang, D.X.; Weng, Z.; Chen, L.; Zhang, Y.; Zhang, N.; Ma, R.Z.; Chen, G.; Liu, X.H. Crosslinked Polymer Binder via Phthalic Acid for Stabilizing SiO_x Anodes. *Macromol. Chem. Phys.* **2022**, *223*, 2200068. [CrossRef]
36. Zhao, J.K.; Wei, D.A.; Wang, J.J.; Yang, K.M.; Wang, Z.L.; Chen, Z.J.; Zhang, S.G.; Zhang, C.; Yang, X.J. Inorganic Crosslinked Supramolecular Binder with Fast Self-Healing for High Performance Silicon Based Anodes in Lithium-Ion Batteries. *J. Colloid Interface Sci.* **2022**, *625*, 373–382. [CrossRef]
37. Kim, J.; Park, Y.K.; Kim, H.; Jung, I.H. Ambidextrous Polymeric Binder for Silicon Anodes in Lithium-Ion Batteries. *Chem. Mater.* **2022**, *34*, 5791–5798. [CrossRef]
38. Rajeev, K.K.; Jang, W.; Kim, S.; Kim, T.H. Chitosan-Grafted-Gallic Acid as a Nature-Inspired Multifunctional Binder for High-Performance Silicon Anodes in Lithium-Ion Batteries. *ACS Appl. Energy Mater.* **2022**, *5*, 3166–3178.
39. Deng, L.; Zheng, Y.; Zheng, X.M.; Or, T.; Ma, Q.Y.; Qian, L.T.; Deng, Y.P.; Yu, A.P.; Li, J.T.; Chen, Z.W. Design Criteria for Silicon-Based Anode Binders in Half and Full Cells. *Adv. Energy Mater.* **2022**, *12*, 2200850. [CrossRef]
40. Li, H.; Li, H.; Lai, Y.; Yang, Z.; Yang, Q.; Liu, Y.; Zheng, Z.; Liu, Y.; Sun, Y.; Zhong, B.; et al. Revisiting the Preparation Progress of Nano-Structured Si Anodes toward Industrial Application from the Perspective of Cost and Scalability. *Adv. Energy Mater.* **2022**, *12*, 2102181. [CrossRef]
41. Wu, Z.; Kong, D. Comparative Life Cycle Assessment of Lithium-Ion Batteries with Lithium Metal, Silicon Nanowire, and Graphite Anodes. *Clean Technol. Environ. Policy* **2018**, *20*, 1233–1244. [CrossRef]
42. Maxwell, T.L.; Balk, T.J. The Fabrication and Characterization of Bimodal Nanoporous Si with Retained Mg through Dealloying. *Adv. Eng. Mater.* **2018**, *20*, 1700519. [CrossRef]
43. Wang, X.; Wen, K.; Chen, T.; Chen, S.; Zhang, S. Supercritical Fluid-Assisted Preparation of Si/CNTs@FG Composites with Hierarchical Conductive Networks as a High-Performance Anode Material. *Appl. Surf. Sci.* **2020**, *522*, 146507. [CrossRef]
44. Yang, Y.; Yuan, W.; Kang, W.; Ye, Y.; Yuan, Y.; Qiu, Z.; Wang, C.; Zhang, X.; Ke, Y.; Tang, Y. Silicon-Nanoparticle-Based Composites for Advanced Lithium-Ion Battery Anodes. *Nanoscale* **2020**, *12*, 7461–7484. [CrossRef]
45. Liu, L.; Lyu, J.; Li, T.; Zhao, T. Well-Constructed Silicon-Based Materials as High-Performance Lithium-Ion Battery Anodes. *Nanoscale* **2016**, *8*, 701–722. [CrossRef]
46. Li, Y.; Liu, X.; Zhang, J.; Yub, H.; Zhang, J. Carbon-Coated Si/N-Doped Porous Carbon Nanofibre Derived from Metal-Organic Frameworks for Li-Ion Battery Anodes. *J. Alloys Compd.* **2022**, *902*, 163635. [CrossRef]
47. Zeng, Y.; Huang, Y.; Liu, N.; Wang, X.; Zhang, Y.; Guo, Y.; Wu, H.-H.; Chen, H.; Tang, X.; Zhang, Q. N-Doped Porous Carbon Nanofibers Sheathed Pumpkin-Like Si/C Composites as Free-Standing Anodes for Lithium-Ion Batteries. *J. Energy Chem.* **2021**, *54*, 727–735. [CrossRef]
48. Park, B.H.; Lee, G.-W.; Choi, S.B.; Kim, Y.-H.; Kim, K.B. Triethoxysilane-Derived SiO_x-Assisted Structural Reinforcement of Si/Carbon Nanotube Composite for Lithium-Ion Battery. *Nanoscale* **2020**, *12*, 22140–22149. [CrossRef]
49. Yi, Z.; Lin, N.; Zhao, Y.; Wang, W.; Qian, Y.; Zhu, Y.; Qian, Y. A Flexible Micro/Nanostructured Si Microsphere Cross-Linked by Highly-Elastic Carbon Nanotubes toward Enhanced Lithium Ion Battery Anodes. *Energy Storage Mater.* **2019**, *17*, 93–100. [CrossRef]
50. Liu, X.; Sun, X.; Shi, X.; Song, D.; Zhang, H.; Li, C.; Wang, K.-Y.; Xiao, C.; Liu, X.; Zhang, L. Low-Temperature and High-Performance Si/Graphite Composite Anodes Enabled by Sulfite Additive. *Chem. Eng. J.* **2021**, *421*, 127782. [CrossRef]
51. Zhao, E.; Luo, G.; Gu, Y.; Yang, L.; Hirano, S.-I. Preactivation Strategy for a Wide Temperature Range in Situ Gel Electrolyte-Based $LiNi_{0.5}Co_{0.2}Mn_{0.3}O_2$ Parallel to Si-Graphite Battery. *ACS Appl. Mater. Interfaces* **2021**, *13*, 59843–59854. [CrossRef] [PubMed]
52. Liu, P.; Sun, X.; Xu, Y.; Wei, C.; Liang, G. By Self-Assembly of Electrostatic Attraction to Encapsulate Protective Carbon-Coated Nano-Si into Graphene for Lithium-Ion Batteries. *Ionics* **2022**, *28*, 1099–1108. [CrossRef]
53. Ren, Y.; Xiang, L.; Yin, X.; Xiao, R.; Zuo, P.; Gao, Y.; Yin, G.; Du, C. Ultrathin Si Nanosheets Dispersed in Graphene Matrix Enable Stable Interface and High Rate Capability of Anode for Lithium-Ion Batteries. *Adv. Funct. Mater.* **2022**, *32*, 2110046. [CrossRef]
54. Shao, F.; Li, H.; Yao, L.; Xu, S.; Li, G.; Li, B.; Zou, C.; Yang, Z.; Su, Y.; Hu, N.; et al. Binder-Free, Flexible, and Self-Standing Non-Woven Fabric Anodes Based on Graphene/Si Hybrid Fibers for High-Performance Li-Ion Batteries. *ACS Appl. Mater. Interfaces* **2021**, *13*, 27270–27277. [CrossRef] [PubMed]
55. Zhang, H.; Zhang, X.; Jin, H.; Zong, P.; Bai, Y.; Lian, K.; Xu, H.; Ma, F. A Robust Hierarchical 3D Si/CNTs Composite with Void and Carbon Shell as Li-Ion Battery Anodes. *Chem. Eng. J.* **2019**, *360*, 974–981. [CrossRef]
56. Yang, Y.; Li, J.; Chen, D.; Fu, T.; Sun, D.; Zhao, J. Binder-Free Carbon-Coated Silicon–Reduced Graphene Oxide Nanocomposite Electrode Prepared by Electrophoretic Deposition as a High-Performance Anode for Lithium-Ion Batteries. *ChemElectroChem* **2016**, *3*, 757–763. [CrossRef]

57. An, W.; Xiang, B.; Fu, J.; Mei, S.; Guo, S.; Huo, K.; Zhang, X.; Gao, B.; Chu, P.K. Three-Dimensional Carbon-Coating Silicon Nanoparticles Welded on Carbon Nanotubes Composites for High-Stability Lithium-Ion Battery Anodes. *Appl. Surf. Sci.* **2019**, *479*, 896–902. [CrossRef]
58. Luo, W.; Wang, Y.; Chou, S.; Xu, Y.; Li, W.; Kong, B.; Dou, S.X.; Liu, H.K.; Yang, J. Critical Thickness of Phenolic Resin-Based Carbon Interfacial Layer for Improving Long Cycling Stability of Silicon Nanoparticle Anodes. *Nano Energy* **2016**, *27*, 255–264. [CrossRef]
59. Fang, H.; Zou, W.; Yan, J.; Xing, Y.; Zhang, S. Facile Fabrication of Fe_2O_3 Nanoparticles Anchored on Carbon Nanotubes as High-Performance Anode for Lithium-Ion Batteries. *ChemElectroChem* **2018**, *5*, 2458–2463. [CrossRef]
60. Guan, B.; Wang, X.; Xiao, Y.; Liu, Y.; Huo, Q. A Versatile Cooperative Template-Directed Coating Method to Construct Uniform Microporous Carbon Shells for Multifunctional Core-Shell Nanocomposites. *Nanoscale* **2013**, *5*, 2469–2475. [CrossRef]
61. Ma, X.Y.; Yin, Z.L.; Tong, H.; Yu, S.; Li, Y.; Ding, Z.Y. 3D Graphene-Like Nanosheets/Silicon Wrapped by Catalytic Graphite as a Superior Lithium Storage Anode. *J. Electroanal. Chem.* **2020**, *873*, 114350. [CrossRef]
62. Wang, M.-S.; Wang, Z.-Q.; Jia, R.; Yang, Y.; Zhu, F.-Y.; Yang, Z.-L.; Huang, Y.; Li, X.; Xu, W. Facile Electrostatic Self-Assembly of Silicon/Reduced Graphene Oxide Porous Composite by Silica Assist as High Performance Anode for Li-Ion Battery. *Appl. Surf. Sci.* **2018**, *456*, 379–389. [CrossRef]
63. Wang, B.; Li, X.L.; Zhang, X.F.; Luo, B.; Jin, M.H.; Liang, M.H.; Dayeh, S.A.; Picraux, S.T.; Zhi, L.J. Adaptable Silicon-Carbon Nanocables Sandwiched between Reduced Graphene Oxide Sheets as Lithium Ion Battery Anodes. *ACS Nano* **2013**, *7*, 1437–1445. [CrossRef]
64. Guo, C.Z.; Liao, W.L.; Li, Z.B.; Sun, L.T.; Chen, C.G. Easy Conversion of Protein-Rich Enoki Mushroom Biomass to a Nitrogen-Doped Carbon Nanomaterial as a Promising Metal-Free Catalyst for Oxygen Reduction Reaction. *Nanoscale* **2015**, *7*, 15990–15998. [CrossRef]
65. Ma, C.; Wang, Z.; Zhao, Y.; Li, Y.; Shi, J. A Novel Raspberry-Like Yolk-Shell Structured Si/C Micro/Nano-Spheres as High-Performance Anode Materials for Lithium-Ion Batteries. *J. Alloys Compd.* **2020**, *844*, 156201. [CrossRef]
66. Huang, Y.; Li, W.; Peng, J.; Wu, Z.; Li, X.; Wang, X. Structure Design and Performance of the Graphite/Silicon/Carbon Nanotubes/Carbon (GSCC) Composite as the Anode of a Li-Ion Battery. *Energy Fuels* **2021**, *35*, 13491–13498. [CrossRef]
67. Wang, S.; Liao, J.; Wu, M.; Xu, Z.; Gong, F.; Chen, C.; Wang, Y.; Yan, X. High Rate and Long Cycle Life of a CNT/RGO/Si Nanoparticle Composite Anode for Lithium-Ion Batteries. *Part. Part. Syst. Charact.* **2017**, *34*, 1700141. [CrossRef]
68. Ramachandran, A.; Sarojiniamma, S.; Varatharajan, P.; Appusamy, I.S.; Yesodha, S.K. Nano Graphene Shell for Silicon Nanoparticles: A Novel Strategy for a High Stability Rechargeable Battery Anode. *ChemistrySelect* **2018**, *3*, 11190–11199. [CrossRef]
69. Zhu, X.; Choi, S.H.; Tao, R.; Jia, X.; Lu, Y. Building High-Rate Silicon Anodes Based on Hierarchical Si@C@CNT Nanocomposite. *J. Alloys Compd.* **2019**, *791*, 1105–1113. [CrossRef]
70. Xu, R.; Wei, R.; Hu, X.; Li, Y.; Wang, L.; Zhang, K.; Wang, Y.; Zhang, H.; Liang, F.; Yao, Y. A Strategy and Detailed Explanations to the Composites of Si/MWCNTs for Lithium Storage. *Carbon* **2021**, *171*, 265–275. [CrossRef]
71. Fu, Z.; Bian, F.; Ma, J.; Zhang, W.; Gan, Y.; Xia, Y.; Zhang, J.; He, X.; Huang, H. In Situ Synthesis of a Si/CNTs/C Composite by Directly Reacting Magnesium Silicide with Lithium Carbonate for Enhanced Lithium Storage Capability. *Energy Fuels* **2021**, *35*, 20386–20393. [CrossRef]
72. Kong, Y.; Luo, S.; Rong, L.; Xie, X.; Zhou, S.; Chen, Z.; Pan, A. Enveloping a Si/N-Doped Carbon Composite in a CNT-Reinforced Fibrous Network as Flexible Anodes for High Performance Lithium-Ion Batteries. *Inorg. Chem. Front.* **2021**, *8*, 4386–4394. [CrossRef]
73. Han, N.; Li, J.; Wang, X.; Zhang, C.; Liu, G.; Li, X.; Qu, J.; Peng, Z.; Zhu, X.; Zhang, L. Flexible Carbon Nanotubes Confined Yolk-Shelled Silicon-Based Anode with Superior Conductivity for Lithium Storage. *Nanomaterials* **2021**, *11*, 699. [CrossRef] [PubMed]
74. Liu, W.; Hu, Y.; Qiao, Y.; Jiang, J.; Huang, M.; Qu, M.; Peng, G.; Xie, Z. 1-Aminopyrene-Modified Functionalized Carbon Nanotubes Wrapped with Silicon as a High-Performance Lithium-Ion Battery Anode. *Solid State Ion.* **2021**, *369*, 115724. [CrossRef]
75. Park, G.D.; Choi, J.H.; Jung, D.S.; Park, J.-S.; Kang, Y.C. Three-Dimensional Porous Pitch-Derived Carbon Coated Si Nanoparticles-CNT Composite Microsphere with Superior Electrochemical Performance for Lithium Ion Batteries. *J. Alloys Compd.* **2020**, *821*, 153224. [CrossRef]
76. Liang, J.; Li, X.; Hou, Z.; Zhang, W.; Zhu, Y.; Qian, Y. A Deep Reduction and Partial Oxidation Strategy for Fabrication of Mesoporous Si Anode for Lithium Ion Batteries. *ACS Nano* **2016**, *10*, 2295–2304. [CrossRef]
77. Luo, Z.; Xiao, Q.; Lei, G.; Li, Z.; Tang, C. Si Nanoparticles/Graphene Composite Membrane for High Performance Silicon Anode in Lithium Ion Batteries. *Carbon* **2016**, *98*, 373–380. [CrossRef]
78. Shi, J.; Jiang, X.; Ban, B.; Li, J.; Chen, J. Carbon Nanotubes-Enhanced Lithium Storage Capacity of Recovered Silicon/Carbon Anodes Produced from Solar-Grade Silicon Kerf Scrap. *Electrochim. Acta* **2021**, *381*, 138269. [CrossRef]
79. Liang, A.-H.; Xu, T.-H.; Liou, S.; Li, Y.-Y. Silicon Single Walled Carbon Nanotube-Embedded Pitch-Based Carbon Spheres Prepared by a Spray Process with Modified Antisolvent Precipitation for Lithium Ion Batteries. *Energy Fuels* **2021**, *35*, 9705–9713. [CrossRef]
80. Saha, S.; Jana, M.; Khanra, P.; Samanta, P.; Koo, H.; Chandra Murmu, N.; Kuila, T. Band gap modified boron doped NiO/Fe_3O_4 nanostructure as the positive electrode for high energy asymmetric supercapacitors. *RSC Adv.* **2016**, *6*, 1380–1387. [CrossRef]

Disclaimer/Publisher's Note: The statements, opinions and data contained in all publications are solely those of the individual author(s) and contributor(s) and not of MDPI and/or the editor(s). MDPI and/or the editor(s) disclaim responsibility for any injury to people or property resulting from any ideas, methods, instructions or products referred to in the content.

Article

Lithiophilic Quinone Lithium Salt Formed by Tetrafluoro-1,4-Benzoquinone Guides Uniform Lithium Deposition to Stabilize the Interface of Anode and PVDF-Based Solid Electrolytes

Yinglu Hu [1], Li Liu [1], Jingwei Zhao [2,*], Dechao Zhang [1], Jiadong Shen [1], Fangkun Li [1], Yan Yang [1], Zhengbo Liu [1], Weixin He [1], Weiming Zhao [1] and Jun Liu [1,*]

[1] Guangdong Provincial Key Laboratory of Advanced Energy Storage Materials, School of Materials Science and Engineering, South China University of Technology, Guangzhou 510641, China
[2] Research and Development Center, Guangzhou Tinci Materials Technology Co., Ltd., Guangzhou 510765, China
* Correspondence: zhaojingwei@tinci.com (J.Z.); msjliu@scut.edu.cn (J.L.)

Abstract: Poly(vinylidene fluoride) (PVDF)-based composite solid electrolytes (CSEs) are attracting widespread attention due to their superior electrochemical and mechanical properties. However, the PVDF has a strong polar group $-CF_2-$, which easily continuously reacts with lithium metal, resulting in the instability of the solid electrolyte interface (SEI), which intensifies the formation of lithium dendrites. Herein, Tetrafluoro-1,4-benzoquinone (TFBQ) was selected as an additive in trace amounts to the PVDF/Li-based electrolytes. TFBQ uniformly formed lithophilic quinone lithium salt (Li_2TFBQ) in the SEI. Li_2TFBQ has high lithium-ion affinity and low potential barrier and can be used as the dominant agent to guide uniform lithium deposition. The results showed that PVDF/Li-TFBQ 0.05 with a mass ratio of PVDF to TFBQ of 1:0.05 had the highest ionic conductivity of 2.39×10^{-4} S cm^{-1}, and the electrochemical stability window reached 5.0 V. Moreover, PVDF/Li-TFBQ CSE demonstrated superior lithium dendrite suppression, which was confirmed by long-term lithium stripping/sedimentation tests over 2000 and 650 h at a current of 0.1 and 0.2 mA cm^{-2}, respectively. The assembled solid-state $LiNi_{0.6}Co_{0.2}Mn_{0.2}O_2$ ||Li cell showed an excellent performance rate and cycle stability at 30 °C. This study greatly promotes the practical research of solid-state electrolytes.

Keywords: solid-state lithium battery; polymer electrolytes; solid electrolytes interface; poly(vinylidene fluoride)

1. Introduction

Lithium-ion batteries (LIBs) are indispensable energy storage devices in daily life and are widely used in many fields, and the increasing demand in applications has promoted research on LIBs with high power/energy density and good safety [1,2]. Almost all LIBs usually use organic electrolytes as ion-conducting media in practical applications; their flammability brings great challenges to the safety of LIBs. Meanwhile, uneven lithium deposition and huge volume effect cause the rupture of the SEI layer and the growth of lithium dendrites, which will reduce the coulomb efficiency and battery life, limiting the increase in energy density [3–5]. The development and use of solid-state electrolytes can effectively alleviate these problems and offer possibilities for the development of solid-state lithium metal batteries [6].

Solid electrolytes are generally classified into three categories [7]: inorganic solid electrolytes, solid polymer electrolytes (SPEs), and composite solid electrolytes (CSEs). For inorganic solid electrolytes, studies of $Li_7La_3Zr_2O_{12}$ (LLZO) [8], $Li_{1-x}Al_xTi_{2-x}(PO_4)_3$

(LATP) [8], $Li_{10}GeP_2S_{12}$ (LGPS) [9], e.g., have been reported. For SPEs, polyethylene oxide (PEO) [6], poly(vinylidene fluoride) (PVDF) [10], polyvinylidene fluoride hexafluoropropylene (PVDF-HFP) [11], polyacrylonitrile (PAN) [12] and polymethylmethacrylate (PMMA) [13], e.g., can be used as polymer matrices. Through component control and structural design, composite solid electrolytes (CSEs) that combine the advantages of different types of electrolytes can improve the overall ionic conductivity and mechanical strength. Common composite solid electrolytes include polymers and inorganic fillers [14], polymers and polymers [15], and other synergistic approaches. Polymer electrolytes are commonly used as matrices, and inorganic fillers play the role of effectively reducing the crystallinity of polymer matrices, expanding amorphous regions, and increasing the content of free Li^+ to improve ionic conductivity [16,17].

PVDF has the advantages of good mechanical properties, non-flammability, strong solubility of lithium salts and low price, so it is very suitable for use as a polymer matrix for composite electrolytes [18]. However, PVDF polymer also has obvious shortcomings in lithium battery research applications; first of all, high crystallinity at room temperature brings excellent mechanical properties but also hinders the improvement in the ion conductivity. Secondly, the incompatibility of PVDF polymer and lithium metal, PVDF has a strong polar group $-CF_2-$, lithium metal reacts easily with PVDF and causes the instability of the contact interface [19,20]. Through extensive attempts and in-depth exploration, it is scientific and feasible to modify PVDF-based solid electrolytes with different design strategies. Modification using inorganic fillers can significantly enhance ion conductivity, mechanical properties, and thermal stability. Zhang et al. [21] used LLZTO as the active filler in the PVDF-HFP polymer matrix, making lithium salt LiFSI more easily dissociated and providing more migration Li^+; LLZTO can provide additional Li^+ migration paths, effectively improving the electrochemical stability of solid-state electrolyte. Li et al. [22] prepared a solid polymer electrolyte with PVDF, using CA (cellulose acetate) as the matrix and modified MMT (montmorillonite) as a filler; CA promoted the sufficient spatial conformation of the polymer molecular segment so Li^+ could be transferred conveniently, improved the mechanical strength of the membrane, and restrained the growth of lithium dendrites. Zeng et al. [23] enhanced the polarization of polymers by copolymerizing trifluoro-ethylene (TrFE) monomer with PVDF monomer and exhibiting the pure all-trans (TTTT) planar zigzag conformation at the TrFE content of 20 to 50 mol%, thereby achieving an improvement in ionic conductivity. Liu et al. [24] used PVDF-HFP as the matrix, through ingenious structural design, to create a novel type of polymer-in-salt solid electrolyte (PISSE), which can penetrate and grow directly in the three-dimensional (3D) electrode on the current collector spacing to achieve the most adequate contact between the electrode and the solid polymer electrolyte.

Like most solid electrolytes, the interface problem of PVDF-based solid electrolytes remains unresolved, and uneven lithium ion conductivity leads to uneven nucleation of lithium deposition, exacerbating the formation of lithium dendrites [25]. Higher current densities result in more susceptible breakage of the SEI, and the fresh lithium metal exposed under the crack has a lower lithium ion transport barrier, which in turn leads to more uneven lithium deposition [26], which may lead to low coulomb efficiency, short long-term operation cycle, and other safety hazards. It greatly affects the improvement in battery energy density and battery safety [27,28]. Therefore, the SEI layer is crucial to the overall performance of lithium metal anodes and lithium batteries; it regulates the current distribution during lithium deposition by changing the physicochemical properties, thereby fundamentally inhibiting the production of lithium dendrites.

The original SEI layer usually exhibits fragile characteristics, and the method of in situ formation can be regarded as the most feasible, practical and constructive, showing outstanding lithium dendrite inhibition ability and promoting the circulation of lithium metal electrodes [29]. The lithiophilic matrix of the SEI layer reduces the potential barrier for lithium nucleation, balances the current distribution, and enhances the binding of lithium ions to other substances. Qian et al. reported an electrolyte additive in lithium metal batteries: tetrachloro-1,4-benzoquinone (TCBQ) and found through experiments that its decomposition product (Li_2TCBQ) in SEI is more lithiophilic [30]. Specifically, TCBQ can be neatly stacked and arranged between molecules because of its unique chemical bonds, and lithium ions are neatly riveted on the surface of TCBQ, and the formed Li_2TCBQ, LiF, and Li_2CO_3 together form a stable SEI layer [31]. It can be seen that benzoquinone based on conjugated carbonyl compounds has excellent performance in reaction kinetics, combined with stable amorphous structure, which is very valuable in the selection of electrolyte additives. As a similar substance, tetrafluoro-1,4-benzoquinone (TFBQ) contains F-groups that are more competitive than TCBQ, and the effect of micro-addition plays a significant role and can be utilized as the high-potential additive for composite electrolyte modification.

Therefore, in this study, PVDF was selected as the polymer matrix, a high-concentration of lithium salt (LiTFSI) was added, and TFBQ was selected as the additive, in the hopes that a trace amount of TFBQ added to the PVDF/Li-based electrolyte could uniformly form a new lithiophilic quinone lithium salt (Li_2TFBQ) component in the SEI layer. Due to its strong lithium-ion affinity and relatively low potential barrier, Li_2TFBQ can be used as a guiding agent of SEI layer to promote uniform lithium deposition. This has been proven in this experiment through related physicochemical performance characterization and electrochemical testing. PVDF-based composite solid electrolytes with trace TFBQ have excellent stability coupled to symmetric lithium electrodes, and under the current density gradient of 0.1–0.4 mA cm^{-2}, lithium symmetric cells can stabilize the cycle with stable polarization voltage and without short circuits. In addition, the PVDF/Li-TFBQ CSE confirmed the significant suppression of lithium dendrite growth through long-term lithium stripping/deposition tests for more than 200 h. It was worth emphasizing that the assembled solid-state $LiNi_{0.6}Co_{0.2}Mn_{0.2}O_2$ | | Li cells exhibited excellent rate performance and cycle stability at 30 °C. In this experiment, the effect of trace benzoquinone additives on the anode interface of composite solid electrolytes was explored, opening up a new idea for the modification of solid electrolyte interfaces.

2. Materials and Methods

2.1. Material Synthesis

In this work, through the solution casting method of synthesized PVDF/Li-TFBQ CSEs membranes: First, polyvinylidene fluoride (PVDF, HF-Kejing, Hefei, China), lithium bis(trifluoromethane sulfonyl)imide (LiTFSI, 99.99%, Sigma-Aldrich, St. Louis, MI, USA), and a certain amount of tetrafluoro-1,4-benzoquinone (TFBQ, Sigma-Aldrich) were dissolved in N, N-dimethylformamide (DMF, 99.8%, Sigma-Aldrich) solution. Next, we stirred the obtained solution continuously at 60 °C for 12 h to form a homogeneous solution. Then, we evenly coated the slurry on the clean polytetrafluoroethylene(PTFE) mold and placed it horizontally to be evaporated and shaped and vacuum-dried at 80 °C for 12 h to obtain a membranes thickness of about 100 μm. In subsequent characterization and testing, the electrolyte membranes were cut to the appropriate size as required (the diameter of all electrolyte membranes assembled into the button cell was 16 mm).

2.2. Preparation of $LiNi_{0.6}Co_{0.2}Mn_{0.2}O_2$ Cathodes

The components of the cathode are $LiNi_{0.6}Co_{0.2}Mn_{0.2}O_2$ powder, poly(vinylidene fluoride) (PVDF), and carbon black (super P); the three components were added to the N-methyl pyrrolidone (NMP) solution in the mass ratio of 8:1:1 and stirred vigorously, the obtained slurry was coated on aluminum foil, and vacuum-heated at 80 °C for 12 h, and then cut into small discs with the diameter of 12 mm, and the average mass loading of the active material on the cathode is approximately 2.0–2.3 mg cm^{-2}.

2.3. Material Characterization

The purity of the samples was identified by XRD, and the crystallinity changes of the composite electrolyte membranes were analyzed with a test angle of 10–70% and a test step size of 0.010 (Cu K_α radiation). The micro-morphology and thickness measurements of the composite electrolyte membranes were analyzed by SEM and EDS, and the electrode materials before and after cycling were also micro-characterized. Electrolyte membranes stability was analyzed by thermogravimetric analysis (TGA). The change of melting temperature point (T_m) of the composite electrolyte was analyzed by DSC, and then the change of crystallinity was reflected. The heating rate was 10 °C min^{-1}, the test temperature range was 70–200 °C, and the protective atmosphere was argon atmosphere. The transformation of the chemical bond of the samples were characterized by Fourier transform infrared spectroscopy (FTIR), and the test wavelength range was 4000–600 cm^{-1}. The interface between the electrolyte and the lithium metal anode was tested by XPS on a Thermo K-Alpha XPS spectrometer equipped with a monochromatic Al-K$_\alpha$ X-ray source to analyze the composition and content change of the main phase substances in the interface layer.

2.4. Electrochemical Measurements

The ionic conductivity of electrolytes in the stainless steel (SS)||CSE||SScell was measured by Electrochemical impedance spectroscopy (EIS) in the temperature range of 10–80 °C with alternating current (AC) amplitude of 0.01 V and frequency range of 1–106 Hz. The linear sweep voltammetry (LSV) of the electrolyte was tested by Li||CSE Licell with a voltage range of 2–6 V and a scan rate of 0.001 V s^{-1}. The lithium-ion transference number (t_{Li+}) of the electrolyte was calculated from direct current polarization and alternating current impedance measurements using Li||CSE||SS cells at 25 °C. At 30 °C, the electrochemical properties of all lithium symmetric cells Li||CSE||SS and LiNi$_{0.6}$Co$_{0.2}$Mn$_{0.2}$O$_2$||Li solid-state cell were measured by the LAND charge–discharge test system. All cells were assembled into CR 2025 coin cells in a glove box filled with argon gas with lithium metal foil as the anode, using only CSE as the electrolyte and no liquid electrolyte added.

3. Results

In this work, PVDF/Li-TFBQ CSE membranes were synthesized via solution casting method, and a series of PVDF/Li-TFBQ samples with different additive contents were prepared; the mass ratio of PVDF to LiTFSI was 1:1.25, and the mass ratio of PVDF to additive TFBQ were 1:0.1, 1:0.05, and 1:0.01, respectively, named PVDF/LiTFSI, PVDF/Li-TFBQ 0.1, PVDF/Li-TFBQ 0.05, and PVDF/Li-TFBQ 0.01, respectively, in which the PVDF/LiTFSI was the control test group without any additives.

The X-ray diffraction (XRD) pattern of Figure 1a shows the phase and crystallinity of LiTFSI, TFBQ, and PVDF/Li-TFBQ CSE membranes. After adding LiTFSI and TFBQ, the characteristic peaks of PVDF at 20° and 40° showed a decrease in diffraction peak intensity, and the position of the diffraction peak did not shift, indicating that the degree of the amorphous state of PVDF was strengthened after mixing with lithium salt and additives. This shows that the addition of TFBQ additive and LiTFSI can increase the amorphous region of the PVDF matrix and reduce crystallinity. At the same time, in the composite electrolyte, neither lithium salt nor additive shows their corresponding characteristic peaks, which indicates that the lithium salt and additive added to the system were homogeneously mixed with the PVDF matrix rather than in a dispersed form. As shown in Figure 1b, the crystallinity of PVDF decreases with the addition of LiTFSI and TFBQ by analyzing the differential scanning calorimetry (DSC) curves of pure PVDF, PVDF/LiTFSI, and PVDF/Li-TFBQ membranes. The endothermic peak of the pure PVDF membranes appeared at 154 °C, that is, the crystalline-amorphous transition temperature (T_m). Due to ensuring the appropriate usage of LiTFSI, the endothermic peak of the electrolyte obviously decreased to 148 °C and decreased more significantly with the addition of TFBQ, which also proved that the addition of LiTFSI and TFBQ could effectively reduce the crystallinity of PVDF polymer, among which PVDF/Li TFBQ 0.05 showed the lowest endothermic peak temperature of 130 °C, which means that there were more amorphous regions in this sample, which is more conducive to the formation of lithium ions. In order to further explore the effect of additives and lithium salts on the composite electrolyte, the chemical bond transition of these samples was investigated by FTIR spectroscopy. In the FTIR spectra shown in Figure 1c, the characteristic peaks of 761 and 1402 cm^{-1} correspond to the α-phase crystals, and the characteristic peaks of 835 and 876 cm^{-1} correspond to the amorphous phases. It is worth noting that the absorption peaks of -CF$_2$-, located at 1072 and 1171 cm^{-1}, were shifted after adding LiTFSI, and a new absorption peak was also generated at 1660 cm^{-1} due to the interaction between the PVDF matrix and LiTFSI. The amorphous phase peak at 876 cm^{-1} also weakened in intensity after adding TFBQ, which corresponds well to the results of DSC and XRD. At the same time, a new characteristic peak appeared at 957 cm^{-1} in the experimental group samples, which strengthens with the increase in additive addition, which is the characteristic peak of the C-F bond in TFBQ, which indicates that the TFBQ additive is successfully mixed inside the composite electrolyte.

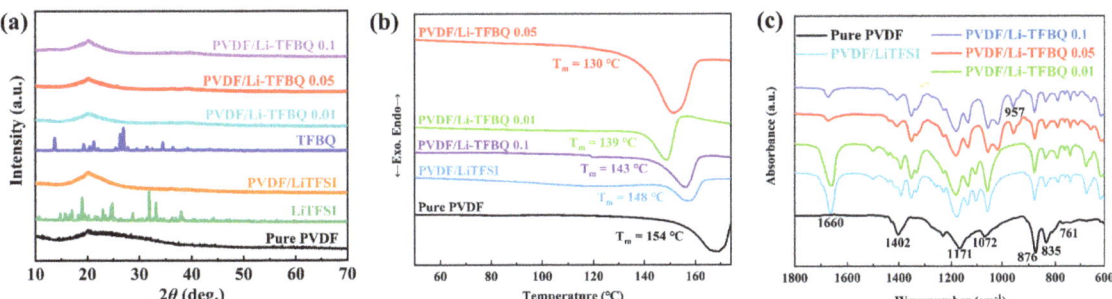

Figure 1. (a) X-ray diffraction (XRD) patterns of TFBQ, LiTFSI, pure PVDF, PVDF/Li, and PVDF/Li-TFBQ CSE membranes, DSC profiles (b) and FTIR spectra (c) of pure PVDF, PVDF/Li, and PVDF/Li-TFBQ CSE membranes.

To explore the appropriate amount of TFBQ, we measured the ionic conductivity of different samples at room temperature, as shown in Figure 2a,b, PVDF/Li-TFBQ 0.05 showed the highest ionic conductivity (2.39 × 10^{-4} S cm^{-1}) at 25 °C after multiple tests and calculations with the same repeatability. To further verify the scientific conclusion, we also tested and summarized the ionic conductivity performance at different temperatures; as shown in Figure 2c, it can be clearly found that PVDF/Li-TFBQ 0.05 has the highest ionic conductivity at different temperature points in the entire temperature range. According to the approximate slope of its curve, it can also be seen that the sample has a low ionic migration barrier. Therefore, PVDF/Li-TFBQ 0.05 was the optimal sample according to the performance of ionic conductivity; that is, when the mass ratio of PVDF to TFBQ was 1:0.05, it was the optimal micro-addition amount of additive, and subsequent characterization and testing were carried out on this sample.

To characterize the thermal stability of the composite electrolyte, the thermo-gravimetric loss of the electrolyte membranes was detected in the temperature range of 40–700 °C by TG, as shown in Figure 2d, after adding lithium salt and TFBQ, the thermal stability of CSE is not destroyed, it still has excellent thermal stability in a wide temperature range as the pure phase PVDF membranes, indicating that the PVDF/Li-TFBQ 0.05 CSE membranes also have better safety performance.

In addition to being expected to play a positive role at the electrolyte interface, the additive TFBQ can also facilitate the dissociation of LiTFSI and improve the ionic conductivity, as demonstrated in additive exploration experiments. At the same time, whether the dissociation of lithium salt is further facilitated can also be reflected by the number of electrolyte ions. Therefore, the results of the lithium-ion migration number test and calculation of the sample are shown in Figure 2e, the lithium-ion migration number of PVDF/Li-TFBQ 0.05 at room temperature can reach 0.42, which is much higher than the ion migration number of ordinary PVDF-based composite solid electrolytes (about 0.3), which further confirms that the addition of TFBQ can facilitate the dissociation of LiTFSI and upgrade the overall electrochemical performance of the composite solid electrolyte. To characterize the adaptability of the composite solid electrolyte to the cathode material, the electrolyte samples were tested for LSV to characterize the electrochemical stability window, and the results are shown in Figure 2f; the overall electrochemical stability of the composite solid electrolyte also improved after TFBQ addition, and the electrochemical window of the PVDF/Li-TFBQ 0.05 reached 5.0 V, possibly because TFBQ not only promoted the dissociation of lithium salt but also made the distribution of lithium salt more uniform and also absorbed some impurities in the electrolyte as a cross-linking center, thus improving overall antioxidant resistance.

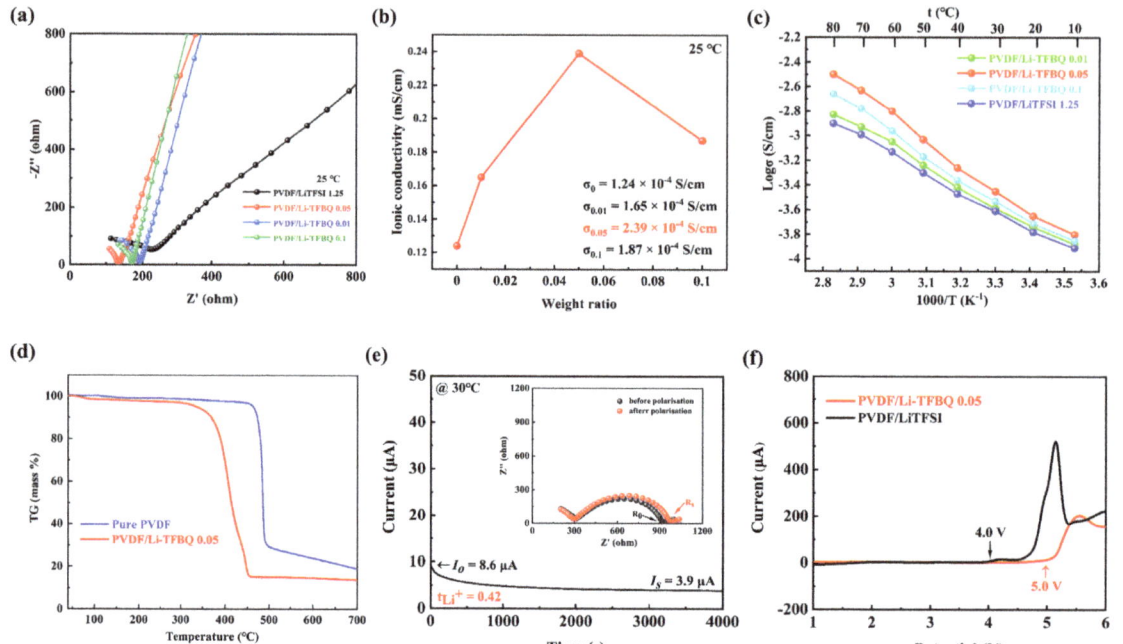

Figure 2. Impendence spectra (**a**) and ionic conductivities (**b**) of PVDF/Li-TFBQ CSE membranes with different contents of TFBQ at room temperature, (**c**) Arrhenius plots of ionic conductivities of PVDF/Li and PVDF/Li-TFBQ CSE membranes. (**d**) TG curves of pure PVDF and PVDF/Li-TFBQ 0.05 CSE membranes. (**e**) Direct current polarization result for the Li | |PVDF/Li-TFBQ 0.05| |Li symmetrical cell and its EIS variation before and after polarization (inset), (**f**) linear sweep voltammetry plot of PVDF/Li and PVDF/Li-TFBQ 0.05 CSE membranes at 25 °C.

The interfacial compatibility and stability of electrolyte and lithium metal have great influence on cell performance. To directly demonstrate the contact stability between lithium metal and electrolyte, the EIS experiment of the lithium-symmetric cell was designed, and then the interface compatibility and stability of PVDF/Li-TFBQ 0.05 CSE and lithium electrode after long-term contact between the electrolyte membranes and lithium metal at room temperature was analyzed. As shown in Figure 3a, the interface between the CES membranes and lithium metal tends to balance and stabilize after three days of contact, with no significant increase in interface resistance, indicating good interface stability. The electrochemical stability of the CSE under high voltage was further evaluated by applying a constant voltage of 4.5 V to the cell for 150 h and performing impedance measurements. Figure 3b shows Li | |PVDF/Li-TFBQ 0.05| |SS asymmetric cell impedance spectrum over time. Through the analysis of experimental results, the almost constant resistance at constant voltage proves the significant electrochemical stability of CSE. The stability of long-term battery cycling is significantly affected by the interface state between the electrolyte and the lithium metal electrode, so the deposition/stripping process of the lithium symmetric battery was analyzed to evaluate the stability between the electrolyte and the lithium metal anode. Figure 3c shows the rate performance of lithium-symmetric cells for different current magnitudes at room temperature. Under the current density gradient of 0.1–0.4 mA cm^{-2}, lithium-symmetric cells can cycle stably with stable polarization voltage and no short circuit. In addition, as shown in Figure 3d,e, PVDF/Li-TFBQ CSE shows outstanding lithium dendrite suppression ability, which can stably cycle for more than 2000 and 650 cycles, respectively, under the current of 0.1 and 0.2 mA cm^{-2}, and the voltage polarization change is small before, during, and after the cycle, which indicates that the interface condition between the electrolyte and lithium metal remains stable throughout the cycle. It can be seen that the addition of TFBQ can improve the interfacial stability of electrolyte and lithium metal.

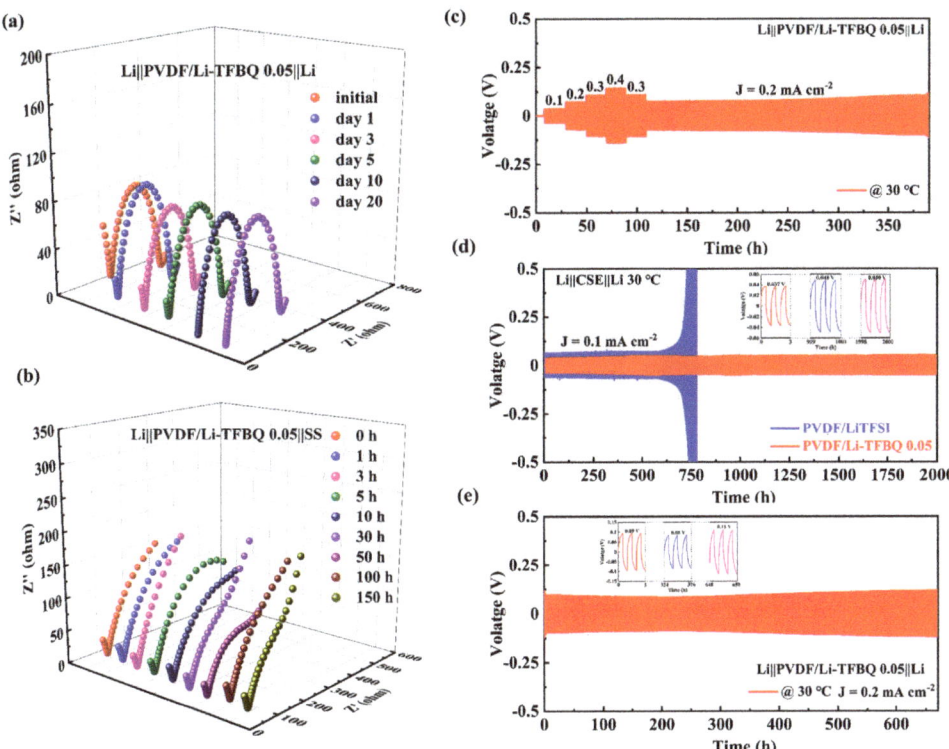

Figure 3. (a) Impedance spectra of Li||PVDF/Li-TFBQ 0.05||Li symmetrical cell for different aging times. (b) Impedance spectra of Li||PVDF/Li-TFBQ 0.05||SS unsymmetrical cell at bias voltage 4.5 V. (c) Galvanostatic rate performance of the Li symmetrical cell with a current density of 0.1–0.3 mA cm^{-2} at 30 °C. Galvanostatic long-term cycling of Li||PVDF/Li-TFBQ 0.05||Li symmetrical cell with a current density of 0.1 mA cm^{-2} (d) and 0.2 mA cm^{-2} (e).

The microscopic morphology of the electrolyte can be intuitively analyzed through SEM images. Due to the absence of inorganic fillers, the PVDF/Li TFBQ 0.05 CSE membranes have a smoother and more uniform surface, as shown in Figure 4a,b, which can bring positive effects on the ion transport inside the electrolyte and the stability of the interface between the electrolyte and the electrode. At the same time, as shown in Figure 4c, the cross-section shows that the thickness of the electrolyte is about 100 μm. The thinner electrolyte shows that the migration path of lithium ions is shortened during cell operation, which has a positive significance for reducing cell polarization and improving the ionic conductivity of the electrolyte. The thinner thickness of the electrolyte is an encouraging sign for the application of all solid-state batteries. The EDS analysis provides insight into the distribution of various elements inside the electrolyte. As shown in Figure 4d–g, N, F, S, and O are characteristic elements of LiTFSI, and it can be intuitively found that various elements are uniformly distributed in the electrolyte section, which indicates that although the composite electrolyte is added with high concentrations of lithium salts, it is still completely dissolved and evenly distributed inside the electrolyte matrix. It can be seen that the simple solution casting method can also obtain high-quality electrolyte membranes with uniform composition and smooth morphology.

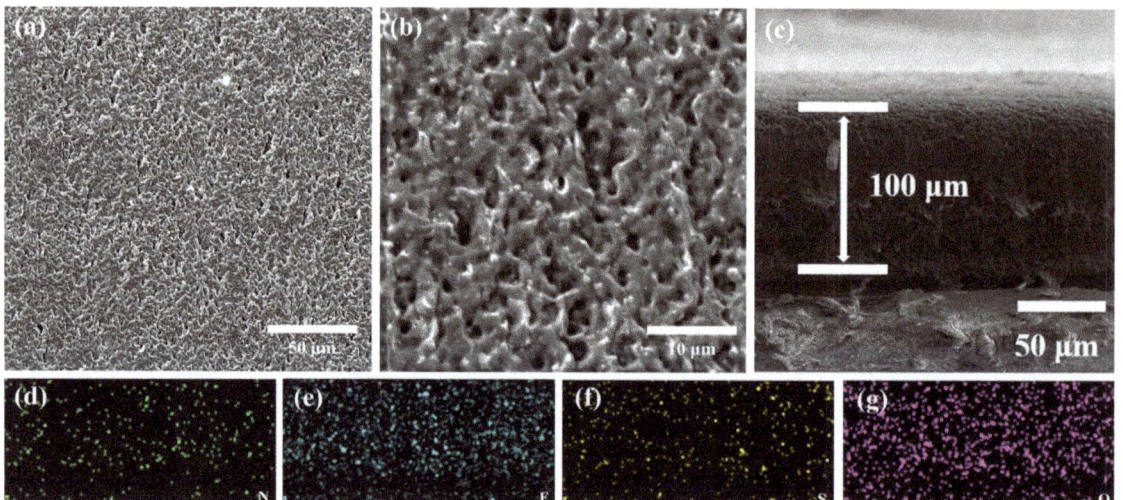

Figure 4. Surface (**a**,**b**) and cross-sectional (**c**) SEM images of PVDF/Li-TFBQ 0.05 CSE membranes, corresponding EDS mapping of nitrogen (**d**), fluorine (**e**), sulfur (**f**), and oxygen (**g**).

The lithium metal interface of Li||PVDF/LiTFSI||Li and Li||PVDF/Li-TFBQ 0.05||Li symmetrical cell after 100 cycles at 0.1 mA cm^{-2} current were analyzed by SEM images. As shown in Figure 5c,d, after contact with PVDF/LiTFSI electrolyte membranes and deposition/stripping 100 times, the surface of the lithium metal electrode plate obviously shows uneven lithium deposition, which is characterized by rough and loose interface and irregular lithium dendrites. In contrast, the surface of the lithium metal electrode after PVDF/Li-TFBQ 0.05 CSE contact, shown in Figure 5a,b, is flatter and smoother, and no lithium dendrite is observed. This indicates that lithium-ions are deposited more uniformly on the surface, which also confirms our assumption that the TFBQ additive can spontaneously form a stable SEI layer on the surface of lithium metal and guide the uniform deposition of lithium ions. To further verify the influence of TFBQ additive on the interface layer of electrolyte and lithium metal, we performed in situ XPS characterization on the lithium metal surface of the lithium symmetric cell after cycling to determine the chemical composition information in the electrochemical process. Figure 5e shows the XPS spectra of F, O, and C elements. After the addition of TFBQ, a more stable interface layer mainly composed of LiF, Li$_2$CO$_3$, and Li$_2$TFBQ was formed on the lithium anode due to the interface reaction between TFSI$^-$ and TFBQ and lithium metal, providing good ion conductivity. At the same time, electronic insulation and mechanical properties were improved, thereby achieving the suppression of side reactions and uniform lithium deposition/stripping.

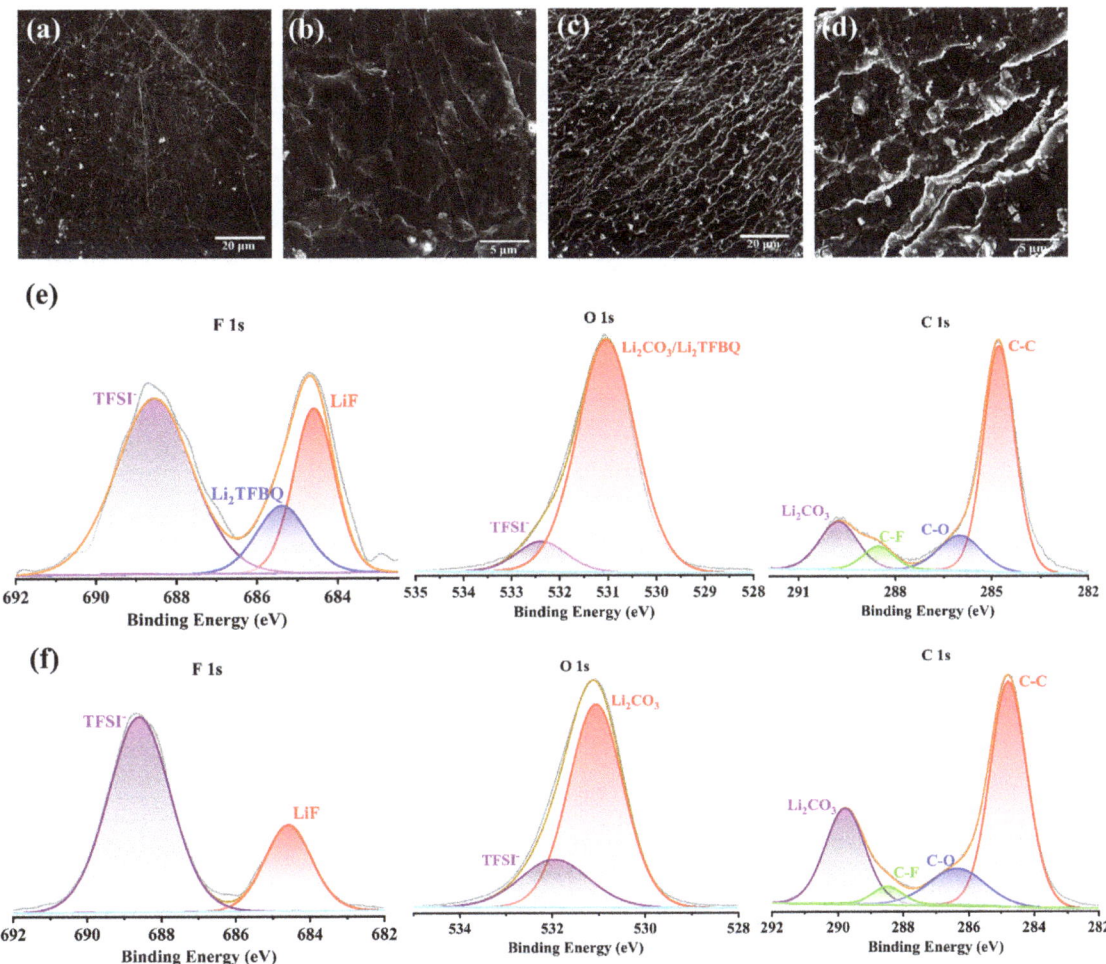

Figure 5. SEM images and XPS spectra of cycled Li metal collected from symmetrical Li cells utilizing PVDF/Li-TFBQ 0.05 (**a,b,e**) and PVDF/Li (**c,d,f**).

To verify the performance of the PVDF/Li-TFBQ 0.05 composite electrolyte cell, we assembled solid LiNi$_{0.6}$Co$_{0.2}$Mn$_{0.2}$O$_2$ ∥ Li button cells and tested the rate and cycle performance of the cells at 30 °C. As shown in Figure 6a, at currents of 0.2, 0.3, 0.5, 1, and 1.5 C, the cells show good discharge specific capacities of 160, 152, 140, 110, and 90 mA h g^{-1}, respectively, which is comparable to the rate performance of most conventional electrolyte cells. The charge–discharge curve of these cells is shown in Figure 6d. Under the current condition of low rate, the voltage polarization is small, while under the condition of high rate, although the polarization is increased, it can still maintain a high-capacity performance. As shown in Figure 6b, in the long-term cycle of the cell, the capacity of nearly 150 mA h g^{-1} can be kept stable for more than 180 cycles under a current of 0.2 C, the capacity retention rate is more than 80%, and the coulomb efficiency also remains at a high level during the cycle. According to the charge–discharge curve in Figure 6e, the capacity of the cell gradually increased to the highest level within the first 50 cycles. The reason is that there is a homogenization process in the initial charge–discharge cycle of the cell, and the internal ion conduction and interface contact conditions of the cell gradually stabilize with the progression of the charge–discharge cycle. As shown in Figure 6e,f, LiNi$_{0.6}$Co$_{0.2}$Mn$_{0.2}$O$_2$ | |Li cell can stably cycle the capacity of nearly 125 mA h g^{-1} for about 100 cycles under a current of 0.5 C, and the capacity retention rate can also reach 65% after 300 cycles and maintain a high coulomb efficiency of nearly 100% in the whole process. It can also be seen from the charge–discharge curve of the cell that voltage polarization is large during the high-current cycle, but it is worth affirming that the cell can still guarantee high-quality capacity output in the first 100 cycles, which is of positive significance to the research of solid-state high-voltage high-capacity lithium batteries. Through the test of LiNi$_{0.6}$Co$_{0.2}$Mn$_{0.2}$O$_2$ | |Li cell assembled by PVDF/Li-TFBQ 0.05 composite electrolyte, it can be seen that the electrolyte ensures the overall excellent multiple and cycle performances of the cell with its good electrochemical performance. In a word, the cell performance of PVDF-based composite electrolyte can be significantly improved by adding the appropriate amount of TFBQ.

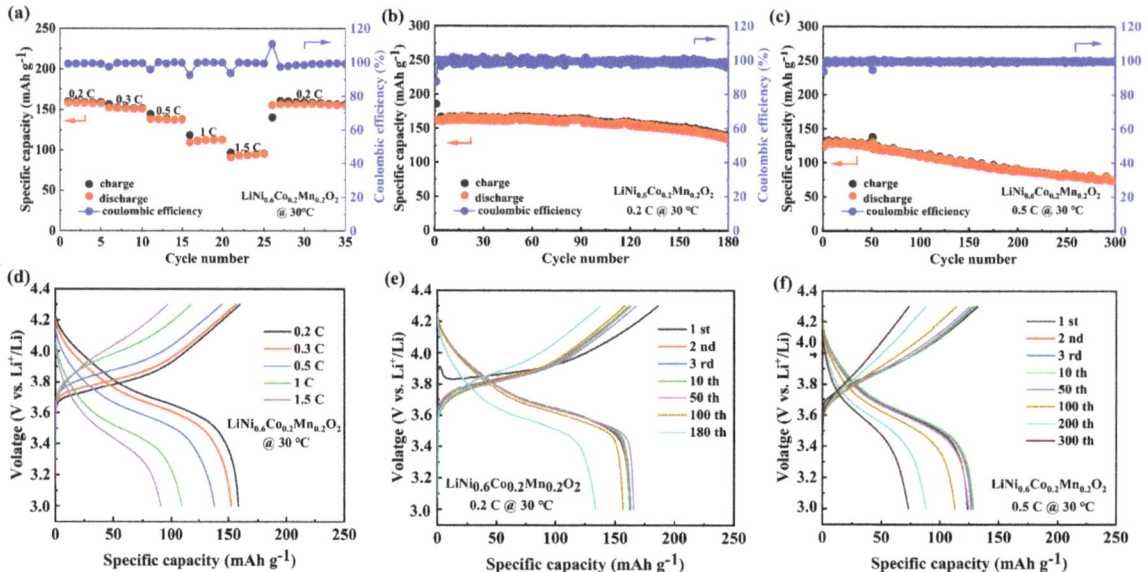

Figure 6. (a) Rate performance at rates of 0.2–1.5 C of LiNi$_{0.6}$Co$_{0.2}$Mn$_{0.2}$O$_2$ cells. (b,c) Cycling performance of cells at different rates. (d–f) Selective charge–discharge curves after different cycles of cells at different rates, cycled at 30 °C.

4. Conclusions

In summary, the addition of benzoquinone additive TFBQ to PVDF-based composite electrolytes can promote the uniform deposition of lithium. In the electrochemical process, due to its unique chemical bonds, the molecules can be neatly stacked and arranged, and it has strong lithiophilicity, and then lithium-ions are neatly riveted on the surface of TFBQ. That is, a small amount of TFBQ in the electrolyte can be sacrificed and decomposed to form a uniform Li$_2$TFBQ layer in the SEI layer on the surface of the Li anode. This component exhibits a relatively low energy barrier, thus effectively improving the stability of the lithium metal anode surface by regulating the deposition of lithium metal, thereby further reducing the side reaction resistance between the electrolyte and lithium metal. The results show that the lithium symmetrical cell assembled with the electrolyte containing an appropriate amount of TFBQ can cycle stably under the condition of stable polarization voltage and no short circuit under the current density gradient of 0.1–0.4 mA cm^{-2}. In addition, PVDF/Li TFBQ CSE exhibits outstanding lithium dendrite suppression ability, which has been demonstrated by long-term lithium stripping/deposition tests conducted at currents of 0.1 and 0.2 mA cm^{-2} for over 2000 and 650 h, respectively. The solid LiNi$_{0.6}$Co$_{0.2}$Mn$_{0.2}$O$_2$ | |Li cell assembled with the CSE also shows outstanding rate and cycling performance at room temperature. Therefore, TFBQ is a promising additive in the composite solid electrolyte system.

Author Contributions: Methodology, L.L., D.Z., F.L. and W.Z.; software, J.S.; investigation, Z.L.; resources, W.H.; data curation, Y.Y.; writing—original draft preparation, Y.H.; supervision, J.Z. and J.L. All authors have read and agreed to the published version of the manuscript.

Funding: This work was financially supported by the National Natural Science Foundation of China (No. U21A2033251771076), R&D Program in Key Areas of Guangdong Province (No. 2020B0101030005), Guangdong Basic and Applied Basic Research Foundation (Nos. 2020B1515120049, 2021A1515010332), and Natural Science Foundation of Guangdong Province (2022A1515010076).

Data Availability Statement: All collected data are presented in the manuscript.

Conflicts of Interest: The authors declare no conflict of interest.

References

1. Li, M.; Lu, J.; Chen, Z.; Amine, K. 30 Years of Lithium-Ion Batteries. *Adv. Mater.* **2018**, *30*, 1800561. [CrossRef] [PubMed]
2. Chen, S.; Wen, K.; Fan, J.; Bando, Y.; Golberg, D. Progress and future prospects of high-voltage and high-safety electrolytes in advanced lithium batteries: From liquid to solid electrolytes. *J. Mater. Chem. A* **2018**, *6*, 11631–11663. [CrossRef]
3. Zheng, L.; Zhang, H.; Cheng, P.; Ma, Q.; Liu, J.; Nie, J.; Feng, W.; Zhou, Z. Li[(FSO$_2$)(n-C$_4$F$_9$SO$_2$)N] versus LiPF$_6$ for graphite/LiCoO$_2$ lithium-ion cells at both room and elevated temperatures: A comprehensive understanding with chemical, electrochemical and XPS analysis. *Electrochim. Acta* **2016**, *196*, 169–188. [CrossRef]
4. Kalhoff, J.; Eshetu, G.G.; Bresser, D.; Passerini, S. Safer Electrolytes for Lithium-Ion Batteries: State of the Art and Perspectives. *ChemSusChem* **2015**, *8*, 2154–2175. [CrossRef]
5. Yang, C.; Fu, K.; Zhang, Y.; Hitz, E.; Hu, L. Protected Lithium-Metal Anodes in Batteries: From Liquid to Solid. *Adv. Mater.* **2017**, *29*, 1701169. [CrossRef]
6. Wu, F.; Maier, J.; Yu, Y. Guidelines and trends for next-generation rechargeable lithium and lithium-ion batteries. *Chem. Soc. Rev.* **2020**, *49*, 1569–1614. [CrossRef]
7. Murugan, R.; Thangadurai, V.; Weppner, W. Fast Lithium Ion Conduction in Garnet-Type Li$_7$La$_3$Zr$_2$O$_{12}$. *Angew. Chem. Int. Ed.* **2007**, *46*, 7778–7781. [CrossRef] [PubMed]
8. Epp, V.; Ma, Q.; Hammer, E.-M.; Tietz, F.; Wilkening, M. Very fast bulk Li ion diffusivity in crystalline Li$_{1.5}$Al$_{0.5}$Ti$_{1.5}$(PO$_4$)$_3$ as seen using NMR relaxometry. *Phys. Chem. Chem. Phys.* **2015**, *17*, 32115–32121. [CrossRef]
9. Kanno, R.; Hata, T.; Kawamoto, Y.; Irie, M. Synthesis of a new lithium ionic conductor, thio-LISICON-lithium germanium sulfide system. *Solid State Ion.* **2000**, *130*, 97–104. [CrossRef]
10. Yao, P.; Zhu, B.; Zhai, H.; Liao, X.; Zhu, Y.; Xu, W.; Cheng, Q.; Jayyosi, C.; Li, Z.; Zhu, J.; et al. PVDF/Palygorskite Nanowire Composite Electrolyte for 4 V Rechargeable Lithium Batteries with High Energy Density. *Nano Lett.* **2018**, *18*, 6113–6120. [CrossRef]
11. Guo, Q.; Han, Y.; Wang, H.; Xiong, S.; Li, Y.; Liu, S.; Xie, K. New Class of LAGP-Based Solid Polymer Composite Electrolyte for Efficient and Safe Solid-State Lithium Batteries. *ACS Appl. Mater. Interfaces* **2017**, *9*, 41837–41844. [CrossRef]
12. Hu, H.; Cheng, H.; Liu, Z.; Li, G.; Zhu, Q.; Yu, Y. In Situ Polymerized PAN-Assisted S/C Nanosphere with Enhanced High-Power Performance as Cathode for Lithium/Sulfur Batteries. *Nano Lett.* **2015**, *15*, 5116–5123. [CrossRef]
13. Ahmad, S.; Saxena, T.K.; Ahmad, S.; Agnihotry, S.A. The effect of nanosized TiO$_2$ addition on poly(methylmethacrylate) based polymer electrolytes. *J. Power Sources* **2006**, *159*, 205–209. [CrossRef]
14. Li, L.; Deng, Y.; Chen, G. Status and prospect of garnet/polymer solid composite electrolytes for all-solid-state lithium batteries. *J. Energy Chem.* **2020**, *50*, 154–177. [CrossRef]
15. Costa, C.M.; Lizundia, E.; Lanceros-Méndez, S. Polymers for advanced lithium-ion batteries: State of the art and future needs on polymers for the different battery components. *Prog. Energy Combust. Sci.* **2020**, *79*, 100846. [CrossRef]
16. Wieczorek, W.; Zalewska, A.; Raducha, D.; Florjańczyk, Z.; Stevens, J.R. Composite Polyether Electrolytes with Lewis Acid Type Additives. *J. Phys. Chem. B* **1998**, *102*, 352–360. [CrossRef]
17. Fan, L.; Nan, C.-W.; Zhao, S. Effect of modified SiO$_2$ on the properties of PEO-based polymer electrolytes. *Solid State Ion.* **2003**, *164*, 81–86. [CrossRef]
18. Ward, I.M.; Hubbard, H.V.S.A. Polymer gel electrolytes: Conduction mechanism and battery applications. In *Ionic Interactions in Natural and Synthetic Macromolecules*; Wiley Online Library: Hoboken, NJ, USA, 2012; pp. 817–840.
19. Xu, K. Nonaqueous Liquid Electrolytes for Lithium-Based Rechargeable Batteries. *Chem. Rev.* **2004**, *104*, 4303–4418. [CrossRef] [PubMed]
20. Choi, S.W.; Jo, S.M.; Lee, W.S.; Kim, Y.-R. An Electrospun Poly(vinylidene fluoride) Nanofibrous Membrane and Its Battery Applications. *Adv. Mater.* **2003**, *15*, 2027–2032. [CrossRef]
21. Zhang, J.; Zeng, Y.; Li, Q.; Tang, Z.; Sun, D.; Huang, D.; Zhao, L.; Tang, Y.; Wang, H. Polymer-in-salt electrolyte enables ultrahigh ionic conductivity for advanced solid-state lithium metal batteries. *Energy Storage Mater.* **2023**, *54*, 440–449. [CrossRef]

22. Li, L.; Shan, Y.; Yang, X. New insights for constructing solid polymer electrolytes with ideal lithium-ion transfer channels by using inorganic filler. *Mater. Today Commun.* **2021**, *26*, 101910. [CrossRef]
23. Zeng, J.-P.; Liu, J.-F.; Huang, H.-D.; Shi, S.-C.; Kang, B.-H.; Dai, C.; Zhang, L.-W.; Yan, Z.-C.; Stadler, F.J.; He, Y.-B.; et al. A high polarity poly(vinylidene fluoride-co-trifluoroethylene) random copolymer with an all-trans conformation for solid-state LiNi$_{0.8}$Co$_{0.1}$Mn$_{0.1}$O$_2$/lithium metal batteries. *J. Mater. Chem. A* **2022**, *10*, 18061–18069. [CrossRef]
24. Liu, W.; Yi, C.; Li, L.; Liu, S.; Gui, Q.; Ba, D.; Li, Y.; Peng, D.; Liu, J. Designing Polymer-in-Salt Electrolyte and Fully Infiltrated 3D Electrode for Integrated Solid-State Lithium Batteries. *Angew. Chem. Int. Ed.* **2021**, *60*, 12931–12940. [CrossRef]
25. Ma, Y.; Zhou, Z.; Li, C.; Wang, L.; Wang, Y.; Cheng, X.; Zuo, P.; Du, C.; Huo, H.; Gao, Y.; et al. Enabling reliable lithium metal batteries by a bifunctional anionic electrolyte additive. *Energy Storage Mater.* **2018**, *11*, 197–204. [CrossRef]
26. Lu, D.; Shao, Y.; Lozano, T.; Bennett, W.D.; Graff, G.L.; Polzin, B.; Zhang, J.; Engelhard, M.H.; Saenz, N.T.; Henderson, W.A.; et al. Failure Mechanism for Fast-Charged Lithium Metal Batteries with Liquid Electrolytes. *Adv. Energy Mater.* **2015**, *5*, 1400993. [CrossRef]
27. Peled, E.; Golodnitsky, D.; Ardel, G. Advanced Model for Solid Electrolyte Interphase Electrodes in Liquid and Polymer Electrolytes. *J. Electrochem. Soc.* **1997**, *144*, L208. [CrossRef]
28. Pang, Q.; Liang, X.; Kochetkov, I.R.; Hartmann, P.; Nazar, L.F. Stabilizing Lithium Plating by a Biphasic Surface Layer Formed In Situ. *Angew. Chem. Int. Ed.* **2018**, *57*, 9795–9798. [CrossRef] [PubMed]
29. Zhang, X.-Q.; Cheng, X.-B.; Chen, X.; Yan, C.; Zhang, Q. Fluoroethylene Carbonate Additives to Render Uniform Li Deposits in Lithium Metal Batteries. *Adv. Funct. Mater.* **2017**, *27*, 1605989. [CrossRef]
30. Shen, X.; Ji, H.; Liu, J.; Zhou, J.; Yan, C.; Qian, T. Super lithiophilic SEI derived from quinones electrolyte to guide Li uniform deposition. *Energy Storage Mater.* **2020**, *24*, 426–431. [CrossRef]
31. Song, Z.; Qian, Y.; Liu, X.; Zhang, T.; Zhu, Y.; Yu, H.; Otani, M.; Zhou, H. A quinone-based oligomeric lithium salt for superior Li-organic batteries. *Energy Sci. Eng.* **2014**, *7*, 4077–4086. [CrossRef]

Disclaimer/Publisher's Note: The statements, opinions and data contained in all publications are solely those of the individual author(s) and contributor(s) and not of MDPI and/or the editor(s). MDPI and/or the editor(s) disclaim responsibility for any injury to people or property resulting from any ideas, methods, instructions or products referred to in the content.

Article

Three-Dimensional Nanoporous CNT@Mn$_3$O$_4$ Hybrid Anode: High Pseudocapacitive Contribution and Superior Lithium Storage

Wei Zou [1], Hua Fang [1,2,3,*], Tengbo Ma [4,*], Yanhui Zhao [5], Lixia Wang [1], Xiaodong Jia [1] and Linsen Zhang [1,2,3]

[1] College of Material and Chemical Engineering, Zhengzhou University of Light Industry, Zhengzhou 450002, China; 410418512@qq.com (W.Z.); 2014050@zzuli.edu.cn (L.W.); xdjia2017@zzuli.edu.cn (X.J.); hnlinsenzhang@163.com (L.Z.)
[2] Ceramic Materials Research Center, Zhengzhou University of Light Industry, Zhengzhou 450002, China
[3] Zhengzhou Key Laboratory of Green Batteries, Zhengzhou 450002, China
[4] Institute of Nuclear, Biological, and Chemical Defiance, PLA Army, Beijing 102205, China
[5] Luohe Letone Hydraulic Technology Co., Ltd., Luohe 462300, China; zyh@letone.cn
* Correspondence: fh@zzuli.edu.cn (H.F.); mtb1988@163.com (T.M.)

Abstract: A composite electrode of carbon nanotube CNT@Mn$_3$O$_4$ nanocable was successfully synthesized via direct electrophoretic deposition onto a copper foil, followed by calcination. By uniformly depositing Mn$_3$O$_4$ nanoparticles on CNTs, a nanocable structure of CNT@Mn$_3$O$_4$ can be formed, where the CNT acts as a "highway" for electrons and ions to facilitate fast transportation. Moreover, capacitive energy storage processes play a crucial role in lithium (Li) storage, especially during high scan rates. The significant contribution of capacitance is highly advantageous for the rapid transfer of Li$^+$ ions, which ultimately results in an improved reversible capacity and prolonged cycle stability of the battery. A high specific capacity of 1367 mAh g^{-1} was maintained over 300 charge–discharge cycles at a current density of 1 A g^{-1}, indicating excellent capacity retention and an extended cycle life. Furthermore, the synthesis process was facile and cost-effective, obviating the need for complex procedures such as mixing and pasting. Additionally, no binder was required, thereby enhancing battery quality efficiency.

Keywords: Mn$_3$O$_4$; carbon nanotubes; electrophoretic deposition; lithium-ion batteries; anode material

1. Introduction

As the world confronts environmental challenges stemming from fossil fuel combustion, it is imperative that governments and scientists expeditiously explore and cultivate green renewable energy sources. Solar and wind energy are two examples of such sources; however, their intermittent nature and instability pose a threat to the security of power systems when used in large quantities. Hence, it is crucial we expand highly efficient energy conversion and energy storage systems for power produced by sustainable sources like solar, wind, and tidal energy to ensure a sustainable and secure energy future [1].

Rechargeable batteries, particularly lithium-ion batteries (LIBs), have made significant strides in the energy storage sector and renewable power technology in recent years. LIBs possess high energy density, a prolonged cycle life, and eco-friendliness, rendering them a promising option for clean energy storage systems in portable electronics, electric vehicles (EVs), and wind/solar power devices [2,3]. The speedy progress of portable consumer electronics and electric vehicles has created a critical requirement for next-generation LIBs with even higher energy density. Currently, carbonaceous materials, particularly graphite, are widely utilized as low-cost anode materials in commercial lithium-ion batteries due to their low working potential and lengthy longevity [1]. However, the graphite material currently available on the market has already reached its theoretical capacity of

372 mAh g^{-1} and is approaching its limit [2]. Hence, the quest for high-energy-density anode materials in LIBs is becoming increasingly pressing.

Transition-metal oxide compounds, such as MxOy (where M represents Ni, Mn, Co, Fe, and other elements), possess high theoretical energy density and have been identified as promising anode materials for lithium-ion batteries [3–5]. Mn_3O_4 is considered a highly attractive anode material option for LIBs due to its abundant and eco-friendly nature, as well as its significant theoretical capacity of 936 mAh g^{-1} [6]. Since the first report on Mn_3O_4 as a host for electrochemical Li insertion by Goodenough's team, Mn_3O_4 has drawn considerable focus as a qualified candidate for anode in LIBs. Nonetheless, Mn_3O_4 experiences significant expansion in volume during lithiation and delithiation processes, resulting in suboptimal cycle stability [7]. Furthermore, the low electron conductivity of Mn_3O_4 (10^{-7} to 10^{-8} S cm^{-1}) impedes its application as an anode for LIBs, resulting in an inadequate rate capability [8,9].

To address the challenges mentioned above, researchers have adopted two widely used strategies, namely the creation and production of nanocomposite materials containing a conductive matrix. The structural stability of nanoscale active materials is bolstered by the reduced mechanical stress within particles and smaller variations in volume, as has been demonstrated [5]. Additionally, the nanoscale electrode–electrolyte interface provides an increased contact area, thereby enhancing kinetic properties and reducing the Li$^+$ ion transfer pathway [3]. On the other hand, the use of conductive matrix materials including porous carbon, CNT, graphene, or conducting polymer can significantly improve electron conductivity and cushion large volume fluctuation, leading to improved rate performance and cycle stability [4]. Anchoring ultrafine Mn_3O_4 nanoparticles on a highly conductive carbon matrix can be an effective way to achieve an outstanding specific capacity, exceptional rate capability, and extended cycle life. Several studies have aimed to develop nanocomposites incorporating Mn_3O_4 nanoparticles, including the ultrasound-assisted in situ growth of Mn_3O_4 nanoplates anchored on rGO [6]. After 40 cycles of charging and discharging at 0.1 A g^{-1}, the specific capacity demonstrated was 1400 mAh g^{-1}. Several studies have reported on the development of Mn_3O_4-based nanocomposites with an improved specific capacity. For instance, Varghese et al. designed a Mn_3O_4-rGO nanocomposite via hydrothermal synthesis, which exhibited a reversible capacity of 474 mAh g^{-1} when subjected to 100 mA g^{-1} cycling [10]. A mesoporous Mn_3O_4/C nanocomposite was directly synthesized via thermolysis by Peng et al., demonstrating a high capacity of 1032 mAh g^{-1} when cycled at 200 mA g^{-1} [11]. Gangaraju reported a graphene–carbon nanotube (CNT)-Mn_3O_4 composite developed using a microwave technique; cycling the material at 1 A g^{-1} for 300 cycles resulted in a high capacity of 1337 mAh g^{-1} [12]. However, some carbon-based nanocomposites display weak interaction between Mn_3O_4 nanoparticles and the conductive carbon matrix, resulting in capacity degradation during long charge–discharge tests.

Since Taberna's initial report on the use of nanostructured current collectors in high-power electrochemical energy storage, scientists have been exploring the application of three-dimensional nanoporous electrodes to tackle issues such as pulverization and capacity degradation [13]. In this study, a three-dimensional (3D) nanoporous hybrid electrode was synthesized by the researchers using a simple electrochemical method. The 3D nanoporous electrode is composed of a CNT@Mn_3O_4 core–sheath nanocable structure, where Mn_3O_4 nanoparticles are uniformly anchored onto an interconnected matrix of carbon nanotubes. The 3D nanoporous electrode exhibits a remarkable performance as an LIB anode, being characterized by a high specific capacity, exceptional rate capability, and extended cycle stability. After undergoing 300 cycles of charge and discharge at a rate of 1 A g^{-1}, the electrode maintained an impressive specific capacity of 1367 mAh g^{-1}. Electrochemical kinetic analysis indicated that the significant contribution of pseudocapacitance had a positive impact on its superior lithium storage performance.

2. Experimental Details

2.1. Synthesis of CNT@Mn$_3$O$_4$ Hybrid Electrode

The reagents utilized in the experiments were of exceptional purity, adhering to specifications for analytical grade. The carbon nanotubes (CNTs) were obtained from Chengdu Organic Chemical Company in Chengdu, China. They had an outer diameter of approximately 50 nm and a length ranging from 0.5 to 2 µm. After undergoing refluxing in concentrated nitric acid for 6 h, the CNTs were ready for use.

A facile electrochemical synthesis strategy, which combines electrophoretic deposition with electrochemical deposition, was used to deposit a three-dimensional nanoporous CNT@Mn$_3$O$_4$ hybrid electrode on a commercial copper foil. The deposition suspension was prepared by ultrasonically dispersing 13 mg of pretreated CNTs and 120 mg Mn(NO$_3$)$_2$ (50% aqueous solution, AR) in 100 mL of absolute ethanol. A rectangular copper foil measuring 30 mm by 40 mm was utilized as the cathode, while a rectangular platinum foil of the same size served as the anode. The copper cathode and platinum anode were positioned parallel to each other with a 10 mm gap between them in the mixture. The deposition process was conducted for a duration of 300 s under a constant voltage of 50 volts. In this deposition process, the electrophoretic deposition of CNT and cathodic deposition of Mn^{2+} ions occurred synchronously, resulting in the one-step formation of a hybrid electrode. A diagram of the specific synthesis steps can be found in Figure S1 of the Supplementary Information. After deposition, the as-obtained hybrid electrode was initially rinsed with ethanol to eliminate residual Mn(NO$_3$)$_2$. Subsequently, the CNT@Mn$_3$O$_4$ hybrid electrode was obtained via calcination in a N$_2$ environment at 300 °C for 3 h. Photos of the as-prepared CNT@Mn$_3$O$_4$ sample deposited on copper foil can be found in Figure S2 of the Supplementary Information.

2.2. Material Characterization

The structure was examined using a field-emission scanning electron microscope (SEM, JSM-7001F) and transmission electron microscopy (TEM, JEOL 2010F). The crystallographic arrangement of CNT@Mn$_3$O$_4$ was identified using X-ray diffraction (XRD, Rigaku D/Max-2400). The nitrogen adsorption–desorption test at 77 K (BELSORP-Mini II) was employed to characterize the pore size distribution and specific surface area [14].

2.3. Electrochemical Characterization

The CNT@Mn$_3$O$_4$ was utilized as the active electrode and machined into a circular shape with a 14 mm diameter by means of a cutting machine. Subsequently, a CR2016-type coin cell was fabricated in a vacuum glove box (MB-10-G, MBRAUN) featuring a H$_2$O and O$_2$ content below 1 ppm. The electrolyte utilized in the experiment was a 1 M LiPF$_6$ solution dissolved in a 1:1 (v/v) blend of ethylene carbonate (EC) and dimethyl carbonate (DMC). The cell was composed of a Celgard 2300 membrane separator, with lithium foil serving as both the reference and counter electrodes.

The coin cell was subjected to galvanostatic charge–discharge (GCD) tests using a Neware battery test system. The experiments were conducted across a range of current densities spanning from 0.1 to 10 A g^{-1} and under a voltage range of 0.01 to 3 V (vs. Li/Li$^+$) in order to obtain comprehensive data. Furthermore, cyclic voltammetry (CV) experiments were performed using an electrochemical workstation (CHI660E) with a fixed voltage range of 0.01 to 3 V (vs. Li/Li$^+$). Electrochemical impedance spectroscopy (EIS) tests were conducted using a voltage amplitude of 10 mV and a frequency range spanning 10^5 to 0.01 Hz. The loading of active material in the electrode film we prepared was determined to be 0.358 mg.

3. Results and Discussion

3.1. Microstructural Studies

The morphology was examined via SEM and TEM analysis, as depicted in Figure 1. As shown in Figure 1a, the CNTs were coated with Mn$_3$O$_4$ particles, resulting in a

CNT@Mn$_3$O$_4$ nanocable with a rough surface and a consistent diameter of approximately 110 nm. The TEM images presented in Figure 1b,c demonstrate the anchoring of Mn$_3$O$_4$ nanocrystals onto the sidewall of the CNTs. As depicted in Figure 1d, the observed lattice spacings of 2.78 Å and 3.36 Å correspond to the interplanar spacing of the (103) plane of Mn$_3$O$_4$ (PDF card number 24-734) [15] and the (002) plane of CNTs (PDF card number 26-1079) [5], respectively. The Mn$_3$O$_4$ coating sheath layer had a thickness of approximately 13 nm. This nanocable structure facilitates the transport of electrons.

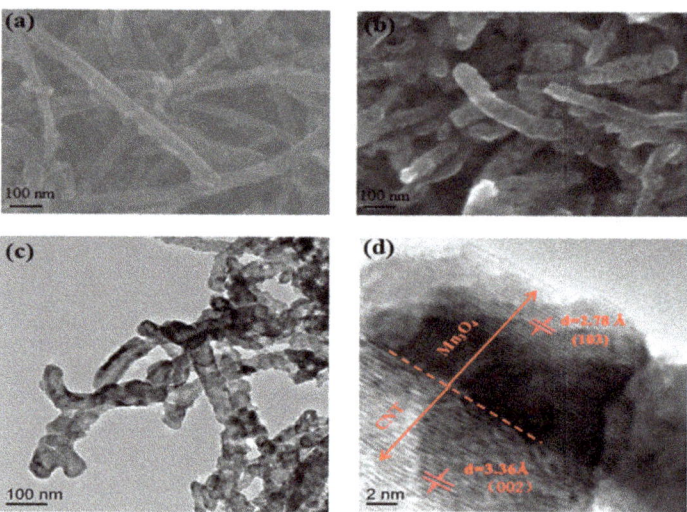

Figure 1. (a) SEM images of CNT; (b) SEM images of CNT@Mn$_3$O$_4$; (c,d) TEM images of CNT@Mn$_3$O$_4$ nanocable.

The XRD spectrum of the nanocable in Figure 2 reveals that Mn$_3$O$_4$ (PDF24-734) was the predominant crystalline phase, as evidenced by most of the observed diffraction peaks. Additionally, the diffraction peak at 2θ = 26.4° can be attributed to the (002) crystallographic plane of the hexagonal carbon structure in CNTs (PDF26-1079). The relatively broad peaks observed at the (211), (103), and (101) planes of Mn$_3$O$_4$ suggest that the sheath was composed of nanoparticles. Applying Scherrer's formula, $D = n \cdot \lambda / (\beta \cdot \cos \theta)$, we calculated a mean crystal size of ~20 nm for Mn$_3$O$_4$ [16].

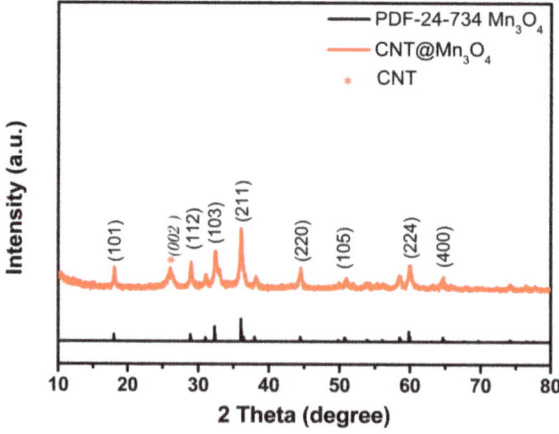

Figure 2. XRD patterns of CNT@Mn$_3$O$_4$ nanocable.

The N_2 sorption isotherm depicted in Figure 3a exhibits a type IV isotherm with a hysteresis loop, indicating the presence of mesopores [15,17]. The CNT@Mn_3O_4 material demonstrated a remarkable specific surface area of 75.477 m^2 g^{-1}, a total pore volume of 0.1805 cm^3 g^{-1}, and an average pore size of 11.711 nm. The pore size distribution curve exhibited two distinct peaks: one at approximately 2 nm indicating the presence of micropores and another broader peak above 10 nm corresponding to mesopores. The high surface area and the hierarchical pore distribution facilitate rapid migration of lithium ions, shorten solid diffusion length, and buffer volume expansion during ion insertion into the active material. Consequently, this material can provide improved power performance as an LIB anode.

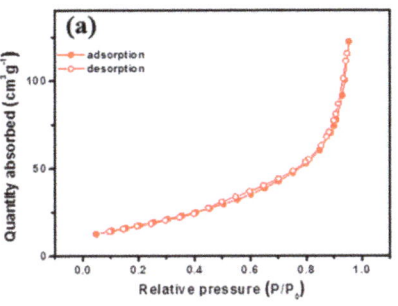
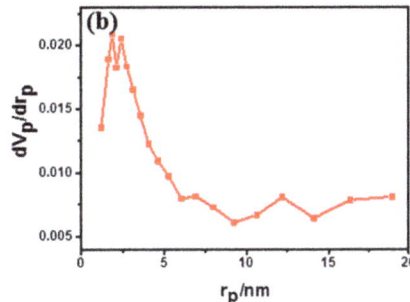

Figure 3. (a) N_2 sorption isotherm diagram; (b) pore size distribution.

3.2. Electrochemical Studies

The electrochemical behavior was initially evaluated via cyclic voltammetry (CV). Figure 4 illustrates the CV profile of the CNT@Mn_3O_4 electrode, depicting the current variation as a function of applied potential within the voltage range of 0.01 to 3.0 V vs. Li/Li$^+$ at a scan rate of 10 mV s^{-1}. A broad cathodic peak at ~1.3 V was observed in the first stage due to the reduction of Mn^{3+} to Mn^{2+} [18,19]. The cathodic peaks observed at 0.02 V during the first scan were attributed to the process of inserting Li$^+$ into the CNT framework. The reduction characteristic observed at 0.4 V corresponds to the reduction of Mn(III) to Mn(0). Regarding the anodic branch, the broad peaks observed at 1.3 V are associated with Mn(II) oxidation to Mn(III), while the other peak corresponds to metallic Mn oxidation to Mn(II). In subsequent cycles, a solid–solid interface formation caused the broad reduction peak at ~1.3 V to shift towards 1.4 V. The voltage of the 0.4 V peaks underwent a downward shift to 0.3 V, which is a characteristic feature of manganese oxide electrodes and indicates structural rearrangements associated with lithium insertion during the initial cycle [20,21]. The absence of any noticeable variation between the second and subsequent cycles, coupled with the excellent cycling stability observed, is indicative of the material's high reversibility for lithium storage [10,22]. This underscores its electrochemical reaction mechanism.

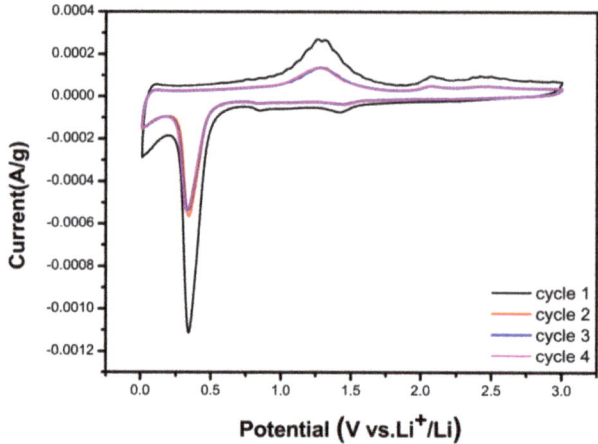

Figure 4. CV curves of the CNT@Mn$_3$O$_4$ nanocable at 10 mV s^{-1}.

Initial discharge cycle:

Mn$_3$O$_4$ + Li$^+$ + e$^-$ → LiMn$_3$O$_4$ (1.5–0.5V vs. Li/Li$^+$);
Mn$_3$O$_4$ + Li$^+$ + e$^-$ → Li$_2$O + 3MnO (1.5–0.5V vs. Li/Li$^+$);
MnO + 2Li$^+$ + 2e$^-$ → Li$_2$O + Mn (0.5–0.0 V vs. Li/Li$^+$).

Initial charge cycle:

Li$_2$O + Mn → MnO + 2Li$^+$ + 2e$^-$ (0.5–3.0 V vs. Li/Li$^+$).

Another criterion for assessing the performance of electrode materials is their rate capability under various applied currents. As depicted in Figure 5a, the component exhibited a high capacity of 869 mAh g^{-1} at a current density of 0.1 A g^{-1} and maintained a relatively impressive capacity of 309 mAh g^{-1} even when the current density was increased to 10 A g^{-1}, which is highly comparable to previously reported values [23–25]. Upon returning to a current-specific mass of 0.1 A g^{-1}, the ingredient demonstrated exceptional reversibility with a capacity retention of 900 mAh g^{-1}, indicating its superior rate performance and cycle stability.

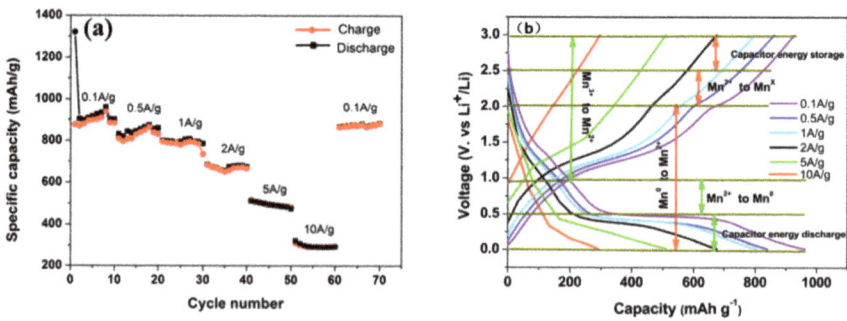

Figure 5. (a) Rate performance of the CNT@Mn$_3$O$_4$ nanocable; (b) charge–discharge curves of the CNT@Mn$_3$O$_4$ nanocable under varying current loads.

Figure 5b shows the galvanostatic charge–discharge (GCD) plots at different electric current densities. A slight variation in the polarization voltage of the electrode material was observed with an increase in current density, indicating that the electrode material has high electron and ion conductivity. The high conductivity is attributed to the unique nanocable structure, whereby CNTs provide rapid electron transport and the 3D nanoporous morphology facilitates efficient ion transportation [26,27]. Simultaneously, it served as a structural support to mitigate the volume expansion and exhibited capacitive behavior in the form of linear platforms with negative slopes ranging from 0 to 0.5 V, which may be attributed to lithium storage at the interface [14].

The stable voltage plateau observed at 0.5 V is attributed to the Faradaic involvement of the Li$^+$ insertion mechanism. As current density increases, this platform becomes slightly shorter and eventually disappears at current densities greater than 5 A g^{-1}. The sluggish reaction kinetics of the Mn_3O_4 sheath are primarily responsible for this phenomenon [28]. Encouragingly, no significant capacity reduction was observed below 0.5 V due to the capacitive nature of interfacial lithium storage. As is well known, capacitive behavior is characterized by a high rate capability, which can be attributed to the interfacial lithium storage mechanism described above.

Simultaneously, the discharge platform of capacitor performance prevails, aligning with the high current charge and discharge characteristics of capacitor energy storage. This capacitance contribution greatly facilitates rapid transmission of Li$^+$, resulting in exceptional reversible capacity and long-cycle stability [9]. This observation offers further insight into the outstanding high rate performance of these materials.

The cycling performance of the CNT@Mn_3O_4 electrode was assessed by carrying out galvanostatic charge–discharge (GCD) experiments at a current density of 1 A g^{-1}. As depicted in Figure 6a, the coulombic efficiency exhibited a sharp rise to approximately 98% and remained stable throughout subsequent cycles, which can be attributed to the enhanced reversibility of CNT@Mn_3O_4. From cycles 2 to 60, the specific capacity of the battery decreased from 768 to 511 mAh g^{-1}, but then consistently increased and reached 1389 mAh g^{-1} in the later cycles. It is not uncommon for previous studies to report an increase in capacity [29,30]. The primary potential factors contributing to this phenomenon were as follows. (1) Electrochemical grinding is caused by repeated volume expansion and the lithiation–delithiation process can result in contraction of the electroactive material. This leads to an increase in contact area between the active site and the electrolyte. (2) Lithium storage at the two-phase interface between metal and Li_2O can be achieved [13]. (3) The reversible formation and dissolution of the solid electrolyte interphase (SEI) layer enables lithium storage [31]. (4) The CNT substrate served as a scaffold to mitigate the volume fluctuations that occur during charge and discharge cycles, ensuring capacity stability [32–34]. (5) The robust architecture of the hybrid nanomembranes [35]. The TEM images of the CNT@Mn_3O_4 materials after different cycles in Figure 6b–d provide evidence to support these conclusions. In particular, Figure 6b,c demonstrate that Mn_3O_4 nanoparticles are uniformly distributed on the CNT scaffold, resulting in a highly porous three-dimensional structure. As the lithiation–delithiation cycle progressed, as shown in Figure 6d, the active material underwent increased porosity and loosening, thereby enhancing the number of active sites associated with electrochemical reactions and resulting in an optimal lithium storage capacity for the composite.

The CV splines of CNT@Mn_3O_4 were tested at a scan rate of 10 mV s^{-1} after different cycles. At the 60th cycle, the capacity reduced its minimum and then began to increase. After 60 iterations, the format of the CV document remained consistent with the output preceding the loop. After 200 and 300 cycles, the reduction peak at 0.02 V and oxidation peak at 2.1 V were significantly intensified, which can be attributed to the reaction progression and increased electrochemical grinding effect on the lithium storage interface, resulting in an increase in capacity. For more comprehensive information, please refer to the relevant figures and explanations provided in Figure 6. Furthermore, it should be noted that the length of the discharge–charge platform depicted in Figure 7b increases proportionally with the number of cycles, thus providing additional evidence to support this conclusion.

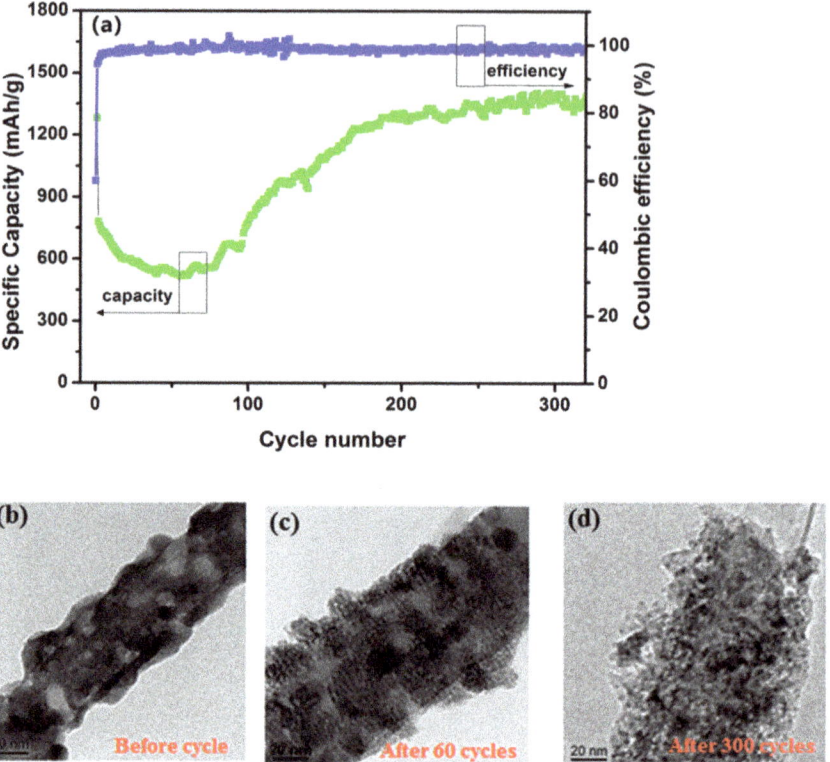

Figure 6. (**a**) The cycling stability of the CNT@Mn$_3$O$_4$ nanocable as anodes for LIBs at a current density of 1 A g^{-1}; (**b**) TEM images of the CNT@Mn$_3$O$_4$ nanocable before cycling; (**c**) after 60 cycles; (**d**) after 300 cycles.

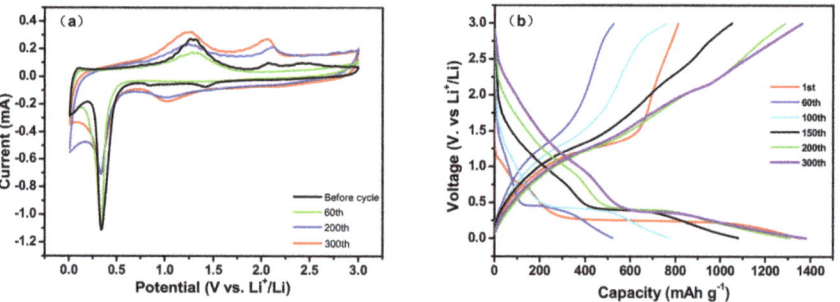

Figure 7. (**a**) CV splines of the CNT@Mn$_3$O$_4$ nanocable at 10 mV s^{-1} after different cycles; (**b**) charge–discharge curves of the CNT@Mn$_3$O$_4$ nanocable at a current density of 1 A g^{-1}.

The electrochemical performance of the CNT@Mn$_3$O$_4$ nanocable material for lithium-ion storage was evaluated using a GCD approach. Figure 7b illustrates the discharge–charge profiles of the CNT@Mn$_3$O$_4$ nanocable material at a current density of 1 A g^{-1} within the voltage range of 0.01 to 3 V (vs. Li/Li$^+$) over multiple cycles. During the initial cycle, the CNT@Mn$_3$O$_4$ nanocable material exhibited a discharge capacity of 814 mAh g^{-1} and a charge capacity of 1307 mAh g^{-1}. The primary irreversible capacity loss occurred during the formation of a solid electrolyte interphase (SEI) layer, which arose from the decomposition of the electrolyte. In the sixtieth cycle, the coulombic efficiency of the specimen increased to 98%; however, its capacity reached a nadir due to inadequate wettability at the interface between electrolyte and active material. As charging and discharging proceeded, capacity gradually recovered. After three hundred revolutions, a charging capacity of 1386 mAh g^{-1} and a discharging capacity of 1367 mAh g^{-1} were achieved due to the secondary activation and electrochemical reaction, resulting in an impressive coulombic efficiency of approximately 98.6% [30,36]. The performances exhibited superior characteristics compared to those previously reported [4,31,33,37]. The charge and discharge voltage plateaus were observed at approximately 0.4 V, 1.3 V, and 2.1 V, respectively. These findings are consistent with the results obtained from the CV curves depicted in Figure 7a.

To further investigate the factors contributing to the superior efficacy of CNT@Mn$_3$O$_4$ as an electrode material for LIBs, cyclic voltammetry (CV) measurements were conducted with scan rates ranging from 0.2 to 1.2 mV s^{-1} to analyze the kinetics of the electrode. The position of the redox peak in the CV curve remained consistent across different scanning rates, indicating a stable electrochemical performance. The high electrochemical reversibility indicates the potential of electrode materials to store energy through both Faraday and non-Faraday process [28,38], which can be qualitatively examined using CV measurements obtained at various scan rates [39]:

$$i = av^b$$

where a and b represent Ka. The slope of the log (i) to log (v) curve can be utilized for calculating the b value, which characterizes the charge storage dynamics during the charge–discharge process. A b value of 0.5 indicates a Faraday process controlled by diffusion [9], whereas a b value of 1 signifies that the electrode is governed by capacitance behavior [40,41]. As depicted in Figure 8b, the b value of the cathodic peak at 0.4 V for the CNT @ Mn$_3$O$_4$ electrode was determined to be 0.65, while that of the anodic peak was calculated as 0.8. This equation suggests that the electrochemical reaction is primarily influenced by the interplay between the diffusion mechanism and the capacitive response. Furthermore, the current output at a specific potential can be separated into two components. One is diffusion-controlled partial current ($k_1 v^{1/2}$), as illustrated by the following equation [42]:

$$i(V) = k_1 v^{1/2} + k_2 v$$

Both k_1 and k_2 are variables that can be calculated based on the slope of the $i(V)/v^{1/2}$ and $v^{1/2}$ curves, respectively. The parameter k_2 serves as a discriminant for capacitive current ($k_2 v$) from the total current. Figure 8c illustrates that the CNT@Mn$_3$O$_4$ electrode contributes to a capacitance of 73.4% at a scanning speed of 1 mVs^{-1}. Figure 8d summarizes the capacitance contributions at different scan rates, which were found to be 46.0%, 58.0%, 67.3%, 70.6%, 73%, and 77% at scan rates of 0.2, 0.4, 0.6, 0.8, 1, and 1.2 mVs^{-1}, respectively. The graph depicted in Figure 8e illustrates that the CNT@Mn$_3$O$_4$ electrode's capacitance contribution reached a remarkable 89% after undergoing 300 cycles at a scan rate of 1 mV s^{-1}, indicating the predominant role played by the energy storage process in its capacity contribution [43,44]. There are two primary factors contributing to this observation. Firstly, the CNT@Mn$_3$O$_4$ electrode material exhibits a remarkably high specific surface area of 75.477 m^2 g^{-1} based on BET testing, which enhances surface adsorption and provides Faraday capacitance. Secondly, during cycling, Mn$_3$O$_4$ generates nanoscale LiO$_2$ and M, which create numerous interfaces. These interfaces provide capacitive interfacial lithium storage and significantly contribute to the overall capacitance of the electrode [14]. The analysis of electrochemical kinetic test data suggests that capacitive energy storage processes play a crucial role in lithium (li) storage, particularly at high scan rates. The significant contribution of capacitance is highly advantageous for the quick transfer of Li$^+$ ions, ultimately resulting in improved reversible capacity and prolonged cycle stability of the battery [14].

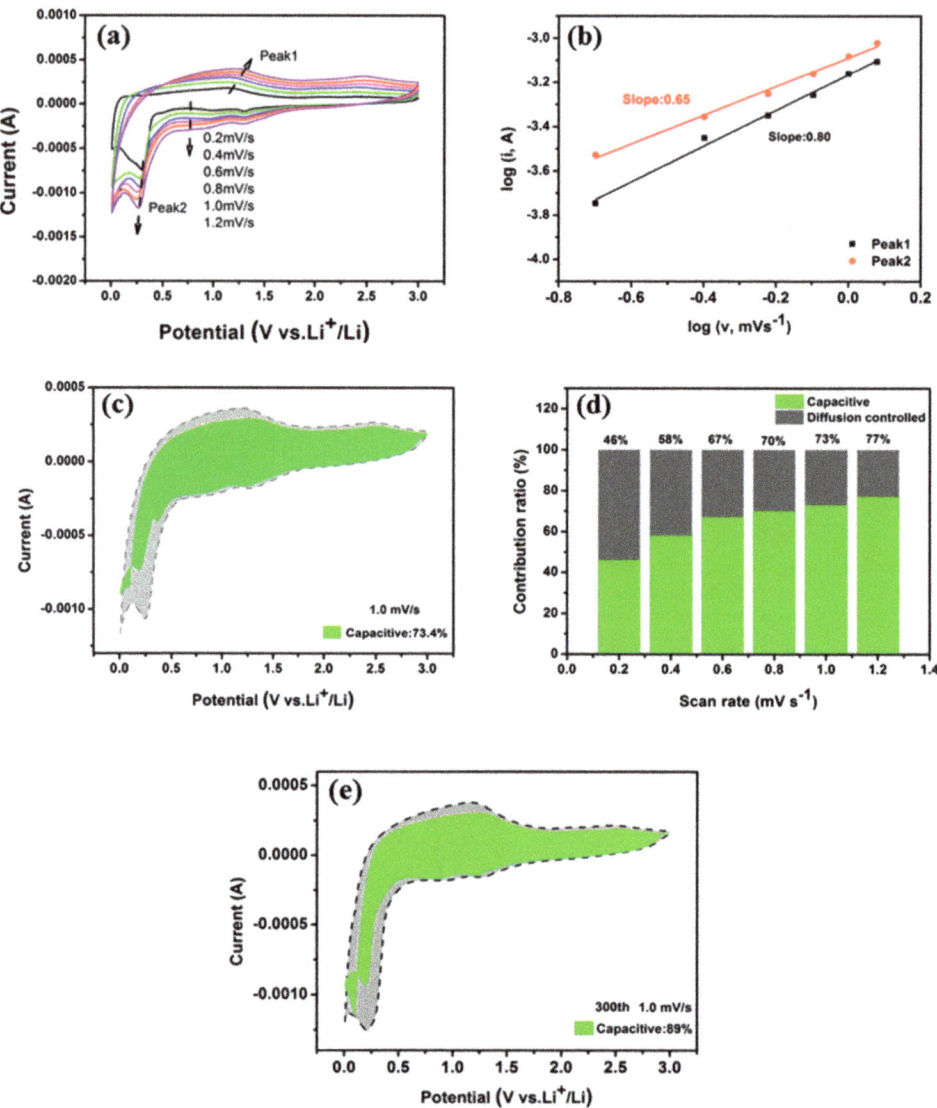

Figure 8. (a) CV curves with scan rates from 0.2 to 1.2 mV s^{-1}; (b) log i vs. log v plots at oxidation and reduction state; (c) capacitive contribution at 1 mV s^{-1}; (d) the capacitive capacity contribution at various scan rates; (e) capacitive contribution at 1 mV s^{-1} after 300 cycles.

The behavior of capacitive electrodes was analyzed using EIS. The Nyquist plots of CNT@Mn$_3$O$_4$ exhibited a partial semicircle and a straight sloping line, as depicted in Figure 9. The size of the semicircle in the diagram indicates the Faradic charge transfer resistance (Rct), which reflects the electrode's surface area and conductivity [31]. Alternatively, the diffusion of ions was correlated with the straight sloping line. In the high-frequency region, the intercept associated with electrolyte resistance (Rele) in the Nyquist plot increased over cycles, indicating enhanced lithium-ion penetration into the active component surface area during cycling. Moreover, the slope of the sloping line became steeper with cycling, indicating that CNT@Mn$_3$O$_4$ exhibited enhanced electrolyte ion diffusion during the redox reaction [31].

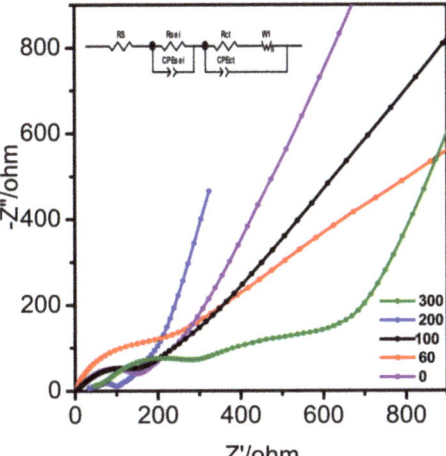

Figure 9. EIS of the CNT@Mn$_3$O$_4$ nanocable after different cycles.

The graded porous network structure of CNT@Mn$_3$O$_4$ facilitates the effective infiltration of the electrolyte into the pores of the electrode material, thereby mitigating polarization and augmenting its specific capacity. A lower Rct value was observed as a result of the graded porous network structure. However, after 300 cycles, there was a substantial increase in Rct, which was attributed to volume fluctuations in the electrode material, dissolution and aggregation of Mn$_3$O$_4$ nanoparticles, and transformation of the spinel Mn$_3$O$_4$ [45,46].

To gain a more comprehensive understanding of the electrochemical mechanism, the electrode's EIS was examined both before and after cycling. As depicted in Figure 9 by the purple line, the EIS prior to cycling displays an intersection point on the real axis at high frequency that represents electrolyte resistance—primarily caused by electrolyte and other battery components—which is ohmic in nature. The decay in the low-frequency range is indicative of the charge transfer resistance and the permanent phase element at the interface between the electrode and electrolyte. CPEct is associated with the electric double-layer capacitor, while W1 characterizes diffusion-related phenomena in the system, including the salts and lithium-ion diffusion into the active material [32,47]. The linear region at the low-frequency range depicts this phenomenon. After the 200 cycles, the EIS exhibited three distinct segments, each featuring an equivalent series resistance (R_s) in the high-frequency region. This parameter is closely related to the ohmic portion of electrode impedance and encompasses contributions from electron transport.

Several factors influence the behavior of an electrode, including its conductivity, the ionic conductivity of the electrolyte solution, and any electrical contact resistance associated with battery hardware, current collectors, or electrode materials. The arc in the mid-to high-frequency range represents the resistance (R_{SEI}) and capacitance (C_{SEI}) of the SEI layer, while the arc in the medium-frequency range corresponds to charge transfer resistance (R_{ct}) and a constant phase element at the electrode–electrolyte interface. The constant phase element (CPE) is associated with double-layer capacitance, whereas linear decay observed in the low-frequency region can be attributed to W, which represents lithium-ion diffusion through an electrode [32].

Based on the blue lines in Figure 9, the semicircle radius of the electrode was significantly reduced after cyclic activation, indicating a smaller Rs due to the presence of CNT@Mn$_3$O$_4$ nanostructures during cycling. The accumulation of CNTs hindered the aggregation of Mn$_3$O$_4$, resulting in increased interlayer spacing and a higher number of electroactive sites. This facilitates charge transport and electrolyte infiltration, as evidenced by the steeper slope and shorter line lengths in the low-frequency region after 200 cycles.

The lithium diffusion rate was accelerated, while the change in the diffusion path was minimized. These observations are consistent with those from TEM analysis shown in Figure 6, which indicate that the electrode became more porous and independent after cycling. Moreover, the nanostructures retained their original morphology, providing further evidence of their exceptional lithium storage performance.

4. Conclusions

In this study, the researchers employed the electrophoretic deposition technique to fabricate a CNT@Mn$_3$O$_4$ nanocable electrode material. This approach enabled direct deposition of active material onto commercially available copper foil, which can function as a current collector for LIBs. Moreover, the absence of a binder contributed to the enhanced quality and efficiency of the battery. The CNT@Mn$_3$O$_4$ material exhibited an impressive stable reversible capacity of 1335 mAh g^{-1} at a current density of 1 A g^{-1}, demonstrating exceptional cycling stability and rate performance. The unique 3D nanoporous architecture of this material enables effective absorption of large volume fluctuations and enhances the speed of electron transfer and transport of lithium ions during charging and discharging processes. Incorporating a porous carbon coating can effectively protect the active material from pulverization, while also enhancing cycling stability and enabling faster charging/discharging rates through increased electrical conductivity. Therefore, the CNT@Mn$_3$O$_4$ composite material exhibits great potential as a superior anode material for high-performance lithium-ion batteries. This study offers a novel perspective on synthesizing high-performance electrode materials using simple and cost-effective methods. The results of this study are expected to advance the development of rechargeable lithium-ion batteries with high power and a long cycle life for energy storage applications.

Supplementary Materials: The following supporting information can be downloaded at: https://www.mdpi.com/article/10.3390/batteries9070389/s1. Figure S1: a simplified schematic diagram of the synthesized CNT@Mn$_3$O$_4$ sample; Figure S2: photos of the as-prepared CNT@Mn$_3$O$_4$ sample deposited on copper foil; Figure S3: physical properties and electrochemical LIB cycling data of Mn$_3$O$_4$-based nanocomposites [5,9,18,19,48]; Figure S4: (a) Raman spectra and (b,c) xps of CNT@Mn$_3$O$_4$ composite materials [10,49,50]; Figure S5: (a) CV curves of the CNT at 10 mV s^{-1}; (b) rate performance of the CNT.

Author Contributions: Conceptualization, H.F. and W.Z.; methodology, H.F. and W.Z.; software, validation, formal analysis, and investigation, W.Z. and Y.Z.; resources, W.Z. and H.F.; data curation, W.Z. and H.F.; writing—original draft preparation, W.Z. and H.F.; writing—review and editing, H.F.; visualization, X.J. and L.W.; supervision, L.Z., T.M. and H.F.; project administration, H.F., T.M. and L.Z.; funding acquisition, H.F. and L.Z. All authors have read and agreed to the published version of the manuscript.

Funding: The Henan Provincial Department of Education sponsored this study through the Key Scientific Research Project in Colleges and Universities of Henan Province, China (grant number 20B530006). Additionally, the Henan Provincial Department of Science and Technology provided funding through the Science and Technology Project of Henan Province, China (grant number 222102240122).

Data Availability Statement: Not applicable.

Conflicts of Interest: The authors declare no conflict of interest.

References

1. Zheng, S.; Dong, F.; Yue, P.; Han, P.; Yang, J. Research Progress on Nanostructured Metal Oxides as Anode Materials for Li-ion Battery. *J. Inorg. Mater.* **2020**, *35*, 134.
2. Marques, O.J.B.J.; Walter, M.D.; Timofeeva, E.V.; Segre, C.U. Effect of Initial Structure on Performance of High-Entropy Oxide Anodes for Li-Ion Batteries. *Batteries* **2023**, *9*, 115. [CrossRef]
3. Cao, K.; Jin, T.; Yang, L.; Jiao, L. Recent progress in conversion reaction metal oxide anodes for Li-ion batteries. *Mater. Chem. Front.* **2017**, *1*, 2213–2242. [CrossRef]

4. Rahman, M.M.; Marwani, H.M.; Algethami, F.K.; Asiri, A.M. Comparative performance of hydrazine sensors developed with Mn_3O_4/carbon-nanotubes, Mn_3O_4/graphene-oxides and Mn_3O_4/carbon-black nanocomposites. *Mater. Express* **2017**, *7*, 169–179. [CrossRef]
5. Xu, L.; Chen, X.; Zeng, L.; Liu, R.; Zheng, C.; Qian, Q.; Chen, Q. Synthesis of hierarchical Mn_3O_4 microsphere composed of ultrathin nanosheets and its excellent long-term cycling performance for lithium-ion batteries. *J. Mater. Sci. Mater. Electron.* **2019**, *30*, 3055–3060. [CrossRef]
6. Huang, H.; Zhao, Z.; Hu, W.; Liu, C.; Wang, X.; Zhao, Z.; Ye, W. Microwave-assisted hydrothermal synthesis of Mn3O4/reduced graphene oxide composites for efficiently catalytic reduction of 4-nitrophenol in wastewater. *J. Taiwan Inst. Chem. Eng.* **2018**, *84*, 101–109. [CrossRef]
7. Zhang, X.; Li, S.; Wang, S.; Wang, Z.; Wen, Z.; Ji, S.; Sun, J. An amorphous hierarchical MnO_2/acetylene black composite with boosted rate performance as an anode for lithium-ion batteries. *Dalton Trans.* **2021**, *50*, 10749–10757. [CrossRef] [PubMed]
8. Wu, L.L.; Zhao, D.L.; Cheng, X.W.; Ding, Z.W.; Hu, T.; Meng, S. Nanorod Mn_3O_4 anchored on graphene nanosheet as anode of lithium ion batteries with enhanced reversible capacity and cyclic performance. *J. Alloys Compd.* **2017**, *728*, 383–390. [CrossRef]
9. Wang, B.; Li, F.; Wang, X.; Wang, G.; Wang, H.; Bai, J. Mn_3O_4 nanotubes encapsulated by porous graphene sheets with enhanced electrochemical properties for lithium/sodium-ion batteries. *Chem. Eng. J.* **2019**, *364*, 57–69. [CrossRef]
10. Varghese, S.P.; Babu, B.; Prasannachandran, R.; Antony, R.; Shaijumon, M.M. Enhanced electrochemical properties of Mn_3O_4/graphene nanocomposite as efficient anode material for lithium ion batteries. *J. Alloys Compd.* **2019**, *780*, 588–596. [CrossRef]
11. Peng, H.-J.; Hao, G.-X.; Chu, Z.-H.; Lin, J.; Lin, X.-M.; Cai, Y.-P. Mesoporous Mn_3O_4/C Microspheres Fabricated from MOF Template as Advanced Lithium-Ion Battery Anode. *Cryst. Growth Des.* **2017**, *17*, 5881–5886. [CrossRef]
12. Gangaraju, D.; Sridhar, V.; Lee, I.; Park, H. Graphene—Carbon nanotube—Mn_3O_4 mesoporous nano-alloys as high capacity anodes for lithium-ion batteries. *J. Alloys Compd.* **2017**, *699*, 106–111. [CrossRef]
13. Deng, Y.; Wan, L.; Xie, Y.; Qin, X.; Chen, G. Recent advances in Mn-based oxides as anode materials for lithium ion batteries. *RSC Adv.* **2014**, *4*, 23914–23935. [CrossRef]
14. Fang, W.; Zou, W.; Yan, J.; Xing, Y.; Zhang, S. Facile Fabrication of Fe_2O_3 Nanoparticles Anchored on Carbon Nanotubes as High-Performance Anode for Lithium-Ion Batteries. *ChemElectroChem* **2018**, *5*, 2458–2463. [CrossRef]
15. Kuila, B.K.; Zaeem, S.M.; Daripa, S.; Kaushik, K.; Gupta, S.K.; Das, S. Mesoporous Mn_3O_4 coated reduced graphene oxide for high-performance supercapacitor applications. *Mater. Res. Express* **2018**, *6*, 015037–015046. [CrossRef]
16. Rosaiah, P.; Zhu, J.; Shaik, D.P.; Hussain, O.M.; Qiu, Y.; Zhao, L. Reduced graphene oxide/Mn_3O_4 nanocomposite electrodes with enhanced electrochemical performance for energy storage applications. *J. Electroanal. Chem.* **2017**, *794*, 78–85.
17. Zhou, Y.; Guo, L.; Shi, W.; Zou, X.; Xiang, B.; Xing, S. Rapid Production of Mn_3O_4/rGO as an Efficient Electrode Material for Supercapacitor by Flame Plasma. *Materials* **2018**, *11*, 881. [CrossRef]
18. Li, X.; Yue, W.; Li, W.; Zhao, J.; Zhang, T.; Gao, Y.; Gao, N.; Feng, D.; Wu, B.; Wang, B. Rational design of 3D net-like carbon based Mn_3O_4 anode materials with enhanced lithium storage performance. *New J. Chem.* **2022**, *46*, 13220–13227. [CrossRef]
19. Cao, K. Mn_3O_4 nanoparticles anchored on carbon nanotubes as anode material with enhanced lithium storage. *J. Alloys Compd. Interdiscip. J. Mater. Sci. Solid-State Chem. Phys.* **2021**, *854*, 157176–157179. [CrossRef]
20. Seong, C.-Y.; Park, S.-K.; Bae, Y.; Yoo, S.; Piao, Y. An acid-treated reduced graphene oxide/Mn_3O_4 nanorod nanocomposite as an enhanced anode material for lithium ion batteries. *RSC Adv.* **2017**, *7*, 37502–37507. [CrossRef]
21. Yao, J.; Yue, J.L.; Guo, Q.; Xia, Q.; Hui, X. Highly Porous Mn_3O_4 Micro/Nanocuboids with In Situ Coated Carbon as Advanced Anode Material for Lithium-Ion Batteries. *Small* **2018**, *14*, 1704296.
22. Shah, H.U.; Wang, F.; Javed, M.S.; Shaheen, N.; Saleem, M.; Li, Y. Hydrothermal synthesis of reduced graphene oxide-Mn_3O_4 nanocomposite as an efficient electrode materials for supercapacitors. *Ceram. Int.* **2017**, *44*, 3580–3584. [CrossRef]
23. Alfaruqi, M.H.; Gim, J.; Kim, S.; Song, J.; Duong, P.T.; Jo, J.; Baboo, J.P.; Xiu, Z.; Mathew, V.; Kim, J. One-Step Pyro-Synthesis of a Nanostructured Mn_3O_4/C Electrode with Long Cycle Stability for Rechargeable Lithium-Ion Batteries. *Chemistry* **2016**, *22*, 2039–2045. [CrossRef] [PubMed]
24. Jing, M.; Hou, H.; Yang, Y.; Zhang, Y.; Yang, X.; Chen, Q.; Ji, X. Electrochemically Alternating Voltage Induced Mn_3O_4/Graphite Powder Composite with Enhanced Electrochemical Performances for Lithium-ion Batteries. *Electrochim. Acta* **2015**, *155*, 157–163. [CrossRef]
25. Yang, Z.; Lu, D.; Zhao, R.; Gao, A.; Chen, H. Synthesis of a novel structured Mn_3O_4@C composite and its performance as anode for lithium ion battery. *Mater. Lett.* **2017**, *198*, 97–100. [CrossRef]
26. Guo, W.; Wang, Y.; Li, Q.; Wang, D.; Zhang, F.; Yang, Y.; Yu, Y. SnO_2@C@VO_2 Composite Hollow Nanospheres as an Anode Material for Lithium-Ion Batteries. *ACS Appl. Mater. Interfaces* **2018**, *10*, 14993–14998. [CrossRef]
27. Thauer, E.; Shi, X.; Zhang, S.; Chen, X.; Deeg, L.; Wenelska, K.; Mijowska, E.; Lund, H.; Kaiser, M.J. Mn_3O_4 encapsulated in hollow carbon spheres coated by graphene layer for enhanced magnetization and lithium-ion batteries performance. *Energy* **2021**, *217*, 119399–119420. [CrossRef]
28. Oloore, L.E.; Gondal, M.A.; Popoola, A.; Popoola, I.K. Pseudocapacitive contributions to enhanced electrochemical energy storage in hybrid perovskite-nickel oxide nanoparticles composites electrodes. *Electrochim. Acta* **2020**, *361*, 137082–137091. [CrossRef]
29. Peng, L.; Hao, Q.; Xia, X.; Wu, L.; Xin, W. Hollow Amorphous $MnSnO_3$ Nanohybrid with Nitrogen-Doped Graphene for High-Performance Lithium Storage. *Electrochim. Acta* **2016**, *214*, 1–10.

30. Wang, Y.; Rao, S.; Mao, P.; Zhang, F.; Xiao, P.; Peng, L.; Zhu, Q. Controlled synthesis of Fe_3O_4@C@manganese oxides (MnO_2, Mn_3O_4 and MnO) hierarchical hollow nanospheres and their superior lithium storage properties. *Electrochim. Acta* **2020**, *337*, 135739–135777. [CrossRef]
31. Yan, D.-J.; Zhu, X.-D.; Mao, Y.-C.; Qiu, S.-Y.; Gu, L.-L.; Feng, Y.-J.; Sun, K.-N. Hierarchically organized $CNT@TiO_2@Mn_3O_4$ nanostructures for enhanced lithium storage performance. *J. Mater. Chem. A* **2017**, *5*, 17048–17055. [CrossRef]
32. Gao, D.D.; Luo, S.S.; Zhang, Y.H.; Liu, J.Y.; Wu, H.M.; Wang, S.Q.; He, P.X. Mn_3O_4/carbon nanotubes nanocomposites as improved anode materials for lithium-ion batteries. *J. Solid State Electrochem.* **2018**, *22*, 3409–3417. [CrossRef]
33. Liu, P.; Xia, X.; Lei, W.; Hao, Q. Rational synthesis of highly uniform hollow core–shell Mn_3O_4/$CuO@TiO_2$ submicroboxes for enhanced lithium storage performance. *Chem. Eng. J.* **2017**, *316*, 214–224. [CrossRef]
34. Zhu, S.; Li, J.; Deng, X.; He, C.; Liu, E.; He, F.; Shi, C.; Zhao, N. Ultrathin-Nanosheet-Induced Synthesis of 3D Transition Metal Oxides Networks for Lithium Ion Battery Anodes. *Adv. Funct. Mater.* **2017**, *27*, 1605017. [CrossRef]
35. Sun, X.; Si, W.; Liu, X.; Deng, J.; Schmidt, O.G. Multifunctional ni/nio hybrid nanomembranes as anode materials for high-rate li-ion batteries. *Nano Energy* **2014**, *9*, 168–175. [CrossRef]
36. Zhao, Y.; Ma, C.; Li, Y. One-step microwave preparation of a Mn_3O_4 nanoparticles/exfoliated graphite composite as superior anode materials for Li-ion batteries. *Chem. Phys. Lett.* **2017**, *673*, 19–23. [CrossRef]
37. Wang, Z.-H.; Yuan, L.-X.; Shao, Q.-G.; Huang, F.; Huang, Y.-H. Mn_3O_4 nanocrystals anchored on multi-walled carbon nanotubes as high-performance anode materials for lithium-ion batteries. *Mater. Lett.* **2012**, *80*, 110–113. [CrossRef]
38. Gan, Z.; Yin, J.; Xu, X.; Cheng, Y.; Yu, T. Nanostructure and Advanced Energy Storage: Elaborate Material Designs Lead to High-Rate Pseudocapacitive Ion Storage. *ACS Nano* **2022**, *16*, 5131–5152. [CrossRef]
39. Xiong, P.; Ma, R.; Sakai, N.; Sasaki, T. Genuine Unilamellar Metal Oxide Nanosheets Confined in a Superlattice-like Structure for Superior Energy Storage. *ACS Nano* **2018**, *12*, 1768–1777. [CrossRef]
40. Yoon, E.S.; Choi, B.G.; Jeon, H.J. Highly ordered nanoscale phosphomolybdate-grafted polyaniline/metal hybrid layered structures prepared via secondary sputtering phenomenon as high-performance pseudocapacitor electrodes. *Phys. Scr.* **2021**, *96*, 125882. [CrossRef]
41. Meng, X.; Guan, Z.; Zhao, J.; Cai, Z.; Li, S.; Bian, L.; Song, Y.; Guo, D.; Liu, X. Lithium-pre-intercalated T-Nb2O5/graphene composite promoting pseudocapacitive performance for ultralong lifespan capacitors. *Chem. Eng. J.* **2022**, *43*, 438–446. [CrossRef]
42. Chu, Y.; Guo, L.; Xi, B.; Feng, Z.; Wu, F.; Lin, Y.; Liu, J.; Sun, D.; Feng, J.; Qian, Y.; et al. Embedding $MnO@Mn_3O_4$ Nanoparticles in an N-Doped-Carbon Framework Derived from Mn-Organic Clusters for Efficient Lithium Storage. *Adv. Mater.* **2018**, *30*, 1704244. [CrossRef] [PubMed]
43. Avvaru, V.S.; Fernandez, I.J.; Feng, W.; Hinder, S.J.; Rodríguez, M.C.; Etacheri, V. Extremely pseudocapacitive interface engineered CoO@3D-NRGO hybrid anodes for high energy/power density and ultralong life lithium-ion batteries. *Carbon* **2021**, *171*, 869–881. [CrossRef]
44. Vincent, M.; Avvaru, V.S.; Rodriguez, M.C.; Haranczyk, M.; Etacheri, V. High-rate and ultralong-life Mg-Li hybrid batteries based on highly pseudocapacitive dual-phase TiO_2 nanosheet cathodes. *J. Power Sources* **2021**, *15*, 506. [CrossRef]
45. Wang, X. Facile synthesis of Mn_3O_4 hollow polyhedron wrapped by multiwalled carbon nanotubes as a high-efficiency microwave absorber. *Ceram. Int.* **2020**, *46*, 6550–6559. [CrossRef]
46. Choudhury, B.J.; Roy, K.; Moholkar, V.S. Improvement of Supercapacitor Performance through Enhanced Interfacial Interactions Induced by Sonication. *Ind. Eng. Chem. Res.* **2021**, *60*, 7611–7623. [CrossRef]
47. Bhagwan, J.; Sahoo, A.; Yadav, K.L.; Sharma, Y. Porous, one dimensional and High Aspect Ratio Mn_3O_4 Nanofibers: Fabrication and Optimization for Enhanced Supercapacitive Properties. *Electrochim. Acta* **2015**, *174*, 992–1001. [CrossRef]
48. Wang, J.G.; Jin, D.; Zhou, R.; Li, X.; Liu, X.R.; Shen, C.; Xie, K.; Li, B.; Kang, F.; Wei, B. Highly flexible graphene/Mn_3O_4 nanocomposite membrane as advanced anodes for li-ion batteries. *Acs Nano* **2016**, *10*, 6227–6234. [CrossRef]
49. Dong, S.; Chang, C.; Liang, Z.; Zhang, Z.; An, L. Electroless deposition of carbon nanotubes doped with nickel and their electrical contact properties. *J. Xian Univ.* **2020**, *47*, 88–94.
50. Zhang, J.; Chu, R.X.; Chen, Y.L.; Zeng, Y.; Zhang, Y.; Guo, H. Porous carbon encapsulated Mn_3O_4 for stable lithium storage and its ex-situ XPS study. *Electrochim. Acta* **2019**, *319*, 518–526. [CrossRef]

Disclaimer/Publisher's Note: The statements, opinions and data contained in all publications are solely those of the individual author(s) and contributor(s) and not of MDPI and/or the editor(s). MDPI and/or the editor(s) disclaim responsibility for any injury to people or property resulting from any ideas, methods, instructions or products referred to in the content.

MDPI
St. Alban-Anlage 66
4052 Basel
Switzerland
www.mdpi.com

Batteries Editorial Office
E-mail: batteries@mdpi.com
www.mdpi.com/journal/batteries

Disclaimer/Publisher's Note: The statements, opinions and data contained in all publications are solely those of the individual author(s) and contributor(s) and not of MDPI and/or the editor(s). MDPI and/or the editor(s) disclaim responsibility for any injury to people or property resulting from any ideas, methods, instructions or products referred to in the content.

www.ingramcontent.com/pod-product-compliance
Lightning Source LLC
LaVergne TN
LVHW070634100526
838202LV00012B/807